ENERGY SCIENCE, ENGINEERING AND TECHNOLOGY

ENERGY AND ENVIRONMENT NOWADAYS

ENERGY SCIENCE, ENGINEERING AND TECHNOLOGY

Additional books in this series can be found on Nova's website under the Series tab.

Additional e-books in this series can be found on Nova's website under the e-book tab.

ENERGY SCIENCE, ENGINEERING AND TECHNOLOGY

ENERGY AND ENVIRONMENT NOWADAYS

LUIS G. TORRES
AND
ERICK R. BANDALA
EDITORS

nova publishers
New York

For permission to use material from this book please contact us:
Telephone 631-231-7269; Fax 631-231-8175
Web Site: http://www.novapublishers.com

Library of Congress Cataloging-in-Publication Data

ISBN 978-1-63117-398-1

Published by Nova Science Publishers, Inc. † New York

CONTENTS

LIST OF REVIEWERS

We are infinitely thankful to the following professionals who helped in the review process of the chapters that comprehend this book:

Erick R. Bandala
Nathalie Cabirol
Karla Campos Díaz
Sandra Carpinteyro Urbán
Luis C. Fernández Linares
Mario Giraldi
Hugo Méndez Acosta
Marcia Morales Ibarría
Francisco Rodríguez Valadez
Luis G Torres Bustillos
Luis Torres Torres

INTRODUCTION

Energy and Environment form up a binomium that can hardly be separated. Our world is facing several problems nowadays such as accelerated demographic growing, lack of sufficient food and health for all the inhabitants, environmental disasters including global warming, energetic crisis, and a long series of etceteras. From this group, energy and environment problems are quite considerable and are linked in many ways. For example, the use of fossil-derived energetics has become in a serious problem due to the production of greenhouse gases and the subsequent global warming effect. On the other hand, production of goods for human life has led to the generation of liquid, solid and gaseous residues which contaminate all kind of matrixes (water, air, soil pollution). Many of these residues can be employed in order to produce sustainable bio-energies. Examples of these are the bio-methane from wastewaters and sludges, bio-ethanol from agro industrial residues and bio-hydrogen from organic municipal residues.

The particular situation in Mexico regarding energy issues was presented in the National Energy Strategy in February 2012 (British Chamber of Commerce, 2012). Among the many interesting data, the hydrocarbon reserves were considered as 13.810 billion barrel of oil equivalent (Bboe), from which 30.612 Billion of oil barrels (Bbo) corresponds to oil and 61,641 Billion of cubic feet (Bcf) to gas. Mexico is among the 10 largest oil producing countries in the world and ranks among the top 20 in terms of gas production. Regarding the renewable energy sector in Mexico for 2012, there is a total estimated potential of 71,000 MW related to wind and solar energy. Regarding the small hydropower generation, from the 3,250 MW potential, only 109 MW are in use. Mexico installed capacity for geothermal energy is about 1,000 MW (the third largest in the world). The country has an average daily sun exposure of 5 kWh/m^2 available to install PV systems for rural electrification and pumping and solar water heating systems for residential and commercial use. Waste to energy includes methane in landfills, manure gasification in farms, among others (British Chamber of Commerce, 2012).

In the Environmental Sector, with the environmental legislation and awareness constantly increasing in the country, Mexico`s environmental market has been rapidly and steadily growing in recent years. According to the Mexican National Institute of Statistics, Geography and Information (INEGI) and the National Institute of Ecology (INE), Mexico is the second most important environmental market in Latin America after Brazil and it is estimated to be worth approximately $8 billion USD in total. More than 2,800 environmental service companies are operating in Mexico. In regards to climate change, it is priority for the

Mexican government, which has made it a cornerstone of its 2007-2012 National Development Plan (the basis for all the current administration's policies). Mexico is one of the first developing countries to commit to a voluntary carbon reduction target to combat climate change. The goal is to reduce 51 million tonnes of CO_2, by 2012. However, according to SEMARNAT, by 2010 the country had already reduced 21 million, which represents more than 40% of the total. (British Chamber of Commerce, 2012).In 2008, Mexico contributed 1.3% of global greenhouse gas (GHG) emissions excluding land use, land-use change and forestry, the 13th-highest level in the world (OECD, 2013)

This book present an overview of the research carried out in Mexico in respect to Energy and environment in the last years. Researchers form different national and local universities offer a wide range of subjects and approaches. The different chapters the possibility to treat residual sludges solving an environmental problem and producing biogas, composed basically by methane; the potential of bio-refineries in the production of natural-coagulant, as well as oil for biodiesel production, energy and heat under the scheme of Industrial Ecology; the use of microalgae biomass for producing lipids and biofuels and an affordable rural ecologic home for marginal communities. Other interesting information included is related with problems associated with the use of fossil fuels have prompted a search for alternative fuel sources and the production of the three main biofuels currently used: biodiesel, biogas, and bio-ethanol; humic acids production from wastewater sludge; the work life cycle analysis and GHG emissions assessment in biofuels production; methane production from tequila wastewater; the direct generation of electricity by solar irradiation using PV panels and the mathematical modeling and simulation of hydrogen production by dark fermentation technology.

We hope that under- and graduated students, teachers, researchers, technicians and general public will find interesting data regarding the problem of energy-environment nowadays in these pages. Maybe these chapters may suggest new solutions for national or local problems, inspire new research lines or future research projects. This is our modest contribution to this area of knowledge which needs multidisciplinary team works and proper national policies urgently. If this book contributes in any way to the knowledge of the problem and suggest solutions, we will be pleased and satisfied.

Luis G Torres and Erick R. Bandala
Coeditors

REFERENCES

British Chamber of Commerce (2012) Mexico Energy, Natural Resources and Environment Report.Consulted in june, 2013 at http://www.britchamexico.com/ckeditorArchivos/files/March%202012%20Mexico%20Energy%20Natural%20Resources%20and%20Environment%20Report.pdf

OECD (2013) OECD Environmental performance Reviews: Mexico 2013. Consulted in June, 2013 at htpp:/dx.doi.org/10.1787/9789264180109-en.

In: Energy and Environment Nowadays
Editors: Luis G. Torres and Erick R. Bandala
ISBN: 978-1-63117-398-1

Chapter 1

BIOGAS PRODUCTION FROM WASTEWATER SLUDGE

Nathalie Cabirol†[1], Marcelo Rojas-Oropeza[1] and Bernd Weber[2]

[1] Facultad de Ciencias, Departamento de Ecología y Recursos Naturales,
Universidad Nacional Autónoma de México (UNAM),
Ciudad Universitaria, Mexico, D.F., Mexico
[2] Facultad de Ingeniería, Universidad Autónoma
del Estado de México (UAEM),
Toluca, Mexico

INTRODUCTION

Biogas is an energy resource produced by the anaerobic digestion of various types of organic matter, such as sewage sludge. The mud can have different origins, depending on the type of treatment used for the wastewater from which it comes. Sewage sludge is a highly important waste. In fact, the produced sludge is a concentrated suspension of the contaminants (substrate) present in the wastewater and, in consequence, it contains a large amount of organic and inorganic compounds and microorganisms (pathogens and parasites). In this case, organic compounds have the capacity of putrefaction. In addition, the presence of pathogens and parasites makes it indispensable to stabilize the sludge. This consists of reducing the volatile fraction of the sludge and destroying microorganisms that present a risk to public and environmental health. Stabilized sludge must present physical, chemical, and biological characteristics suitable for its possible use in agriculture. The most feasible solution is to stabilize the sludge through anaerobic digestion, which allows generating a biosolid that complies with both the agronomic needs and the recommendations of the World Health Organization (WHO). In addition, the produced biogas is a high energy product that can be used in electrification, as transportation fuel, or for heating purposes.

† e-mail: natcabirol@yahoo.com.

HISTORY

Although wastewater sewage systems have been known for some millennia, they became popular again one and a half centuries ago, when European cities began to install their sewage systems. Basel, in Switzerland, today considered one of the most developed and richest cities of the world, started following this trend in 1896 (Lofrano and Brown, 2010; Leonard et al., 1910). The implemented technologies were able to provide cleaner inner cities but caused contamination due to the discharge of untreated wastewater into water bodies, which rapidly called for the design of waste water treatment plants. This scheme is still common and can be described as an “end of pipe” technology. The main objective is to protect the environment and to secure the delivery of clean drinking water. The first applied methods consisted of dispersing the untreated wastewater over special reserved fields or placing it into primary settling tanks (Cooper, 2007). The natural occurrence of organic matter degradation on these irrigation fields by aerobic and anaerobic microorganism was observed. The same natural degradation has been adapted technically in subsequent biological stages of the treatment process, which is still with some modifications the state of art in wastewater treatment plants.

MUNICIPAL WASTEWATER TREATMENT

In the first stage of a modern municipal wastewater treatment plant, suspended inorganic and organic matter with a critical diameter of 100 μm can be separated from its liquid phase in the primary settling tanks. Organic matter that is dissolved, or in colloid and supra-colloid phase is treated during the subsequent biological treatment, where aerobic or anaerobic microorganisms build up biomass in parallel to the release of mineralization products, such as carbon dioxide and methane, if methanogens are present. This biomass is characterized by an increased diameter and can be settled again, giving rise to a second stream of organic matter-containing sludge. Further sludge management is a great challenge for sustainable plant operation due to high water content, contamination of heavy metals, pathogens, and emergent pollutants. However, the energy content of organic matter should be used to minimize the energy demand of operating plants or even running them autonomously from electricity demand (Haberkern et al., 2006).

POTENTIAL OF BIOGAS PRODUCTION IN MUNICIPAL WASTEWATER TREATMENT

Currently 52.1% of the world’s population, that is $6.97 \cdot 10^9$ people live in urban agglomerations (ESA, 2011). All of the cities should be connected to state of art wastewater treatment plants, to be able to purify 150 to 200 L of wastewater generated daily per capita. Total amount of wastewater to be treated is $654 \cdot 10^6$ m^3/d in the world. Additionally, discharges of industrial wastewater have to be considered. The potential for energy recovery is given by the content of organic matter in the wastewater parameterized as biological oxygen demand (BOD), chemical oxygen demand (COD) or, nowadays, more frequently by the total organic carbon content (TOC). Typical loads contributed by one person are in the range of 50 to 75 g_{BOD}/d, 100 to 150 g_{COD}/d, or 30 to 80 g_{TOC}/d (Tchobanoglous, 2003). If, for wastewater, a typical composition is assumed, the theoretical potential to produce biogas is about 0.6 m^3/kg_{TOC}. Useful for this estimation is the equation provided by Symons and Buswell (1933), which describes the composition of biogas as depending on the concentration of the elements C, H, N, S in the substrate. This gives a total theoretical biogas yield of $110 \cdot 10^6\ m^3_{Biogas}/d$ (30 L_{Biogas}/d for one person) from wastewater generated by the world's population living in cities.

Figure 1. Berlin Wassmannsburg (Deutschland).

Figure 2. Digester of wastewater sludge (courtesy of Puebla municipality, Mexico).

However, the current situation is far away from this benchmark. Besides the technical limitations discussed so far, the fact is that a lot of treatment plants do not have an anaerobic digester for the treatment of sludge. In addition, many developing countries have not implemented yet wastewater treatment systems for their cities. Studies about coverage in developed countries show that about 95% of wastewater is treated in advanced treatments systems (OCDE, 2011). The source reports that 1.2% of wastewater is treated in primary, 30.2% in secondary, and 36.8% in tertiary treatment systems in the United States of America. In Germany, nearly 1200 wastewater treatment plants are equipped with an anaerobic sludge digester and produce $1.9 \cdot 10^6\ m^3_{Biogas}/d$ ($19\ L_{Biogas}/d$ per person) revealing that the determined performance is close to the theoretical potential (Figure 1) (Haberkern, 2007). Implementation of wastewater treatment systems is in progress in emerging countries, such as Brazil and Mexico, where the coverage of treatment for urban areas is only 47.0% and 44.8%, respectively (Figure 2) (SEMARNAT, 2011; IANAS, 2012). Even lower is the percentage (37%) of population of low income countries connected to a sewerage system (WHO, 2011).

Biogas Production by Methanogenic Archaea

Anaerobic digestion of organic waste removes organic pollutants and reduces the organic waste volume (although only marginally for wet digestion with a high moisture content) in a system that provides optimal living conditions for microorganisms and produces biogas containing mainly high calorific methane (Narihiro et al., 2007; McKeown et al., 2012). The following four stages comprise the digestion process through the sequential and coordinated activity of various microbial trophic groups: hydrolysis, acidogenesis, acetogenesis, and methanogenesis (McKeown et al., 2012; Traversi et al., 2012). A high diversity of microorganisms is involved in each step. During the first stage, a group of microorganisms secretes extracellular enzymes that hydrolyze complex polymers into monomers in order to convert particulate materials into dissolved materials (Whitman et al., 2006). This group of microorganisms removes the small amounts of O_2 present and creates anaerobic conditions. Subsequently, the acidogenic phase includes the action of a large and diverse group of fermentative bacteria that usually belong to the Clostridia class and the Bacteroidaceae family. These bacteria hydrolyze and ferment organic materials and produce organic acids, CO_2 and H_2. Next, in the third phase, acetogenic bacteria convert these monomers to acetic acids. The final step in biogas production is performed by acetoclastic methanogens (*Methanosarcina* primarily in high-acetate environments [$> 10^3$ M], and *Methanosaeta*, which grow only by the acetoclastic reaction) and hydrogenotrophic methanogens (McKeown et al., 2012).

Therefore, among the microorganisms involved in digestion, methanogens are the major microbiological group responsible for methane production. Methanogenic archaea are a phylogenetically diverse group, with an energy metabolism derived from CO_2, H_2, formate, methanol, methylamines, or acetate (Table 1). They are classified in 5 well-established orders, Methanobacteriales, Methanococcales, Methanomicrobiales, Methanosarcinales, and Methanopyrales, but only some genera are represented in anaerobic digesters: *Methanobacterium*, *Methanobrevibacter*, *Methanothermobacter*, *Methanomicrobium*, *Methanocolleus*, *Methanofollis*, *Methanospirillum*, *Methanocorpusculum*, *Methanosarcina*, and *Methanosaeta* (Liu and Whitman, 2008; Thauer et al., 2008).

Table 1. Methanogenic taxa. (Liu and Whitman, 2008)

Order	Family	Genus	Substrate
Methanobacteriales	Methanobacteriaceae	*Methanobacterium*	CO_2/H_2, formate
		Methanobrevibacter	CO_2/H_2, formate
		Methanosphaera	CO_2/H_2 + methanol
	Methanothermaceae	*Methanothermobacter*	CO_2/H_2, formate
		Methanothermus	CO_2/H_2
Methanococcales	Methanococcaceae	*ethanococcus*	CO_2/H_2, formate
		Methanothermococcus	CO_2/H_2, formate
	Methanocaldococcaceae	*Methanocaldococcus*	CO_2/H_2
		Methanotorris	CO_2/H_2
Methanomicrobiales	Methanomicrobiaceae	*Methanomicrobium*	CO_2/H_2, formate
		Methanoculleus	CO_2/H_2, formate
		Methanofollis	CO_2/H_2, formate
		Methanogenium	CO_2/H_2, formate
		Methanolacinia	CO_2/H_2
		Methanoplanus	CO_2/H_2, formate
	Methanospirillaceae	*Methanospirillum*	CO_2/H_2, formate
	Methanocorpusculaceae	*Methanocorpusculum*	CO_2/H_2, formate
		Methanocalculus	CO_2/H_2, formate
Methanosarcinales	Methanosarcinaceae	*Methanosarcina*	CO_2/H_2, $MeNH_2$, Acetate
		Methanococcoides	$MeNH_2$
		Methanohalobium	$MeNH_2$
		MethanohalophilusMethanolobus	$MeNH_2$
		Methanomethylovorans	$MeNH_2$
		Methanimicrococcus	Acetate
		Methanosalsum	CO_2/H_2+ $MeNH_2$
			$MeNH_2$
		Methanosaeta	
	Methanosaetaceae		Acetate
Methanopyrales	Methanopyraceae	*Methanopyrus*	CO_2/H_2

$MeNH_2$ methylamine.

Only two genera are known to use acetate as substrate. Therefore, they are called acetoclastic methanogens: *Methanosarcina* and *Methanosaeta*. The first is relatively versatile, preferably uses methanol and methylamines as substrates, rather than acetate; many species of this genus use H_2 and present faster growth rates. On the other hand, the genus *Methanosaeta* is a specialist that only uses acetate as substrate with slower growth rates (Thauer et al., 2008).

The methanogenic archaea that use CO_2/H_2 and formate as substrates, are known as hydrogenotrophic methanogens. A variety of this type of methanogens belongs to the order Methanomicrobiales and Methanobacteriales. A low level of H_2 indicates an efficient hydrogenotrophic methanogenesis and is usually associated with a stable activity ($< 10^{-4}$ Pa) (Thauer et al., 2008). The hydrogenotrophic methanogens that use formate as an electron donor are among the microorganisms that have faster growth rates within the microbial community of anaerobic digesters (Thauer et al., 2008).

LIMITATIONS OF BIOGAS PRODUCTION

The diversity, structure, and activity of microbial communities depend on various physical, chemical, and biological factors like the origin of the sludge (for example municipal or industrial or treatment type), hydraulic and cell retention time, temperature, redox, substrate homogeneity, presence of minerals (such as iron-oxide, sulphate), microbial competition, microbial syntrophy, and microbial depredation, among other conditions. Each treatment and each reactor plant uses microbial consortia that are unique and their study is of high importance for future improvements in environmental biotechnology (Schink, 1997; Lozada et al., 2004; Kato et al., 2012).

The microbial diversity of anaerobic systems has mainly been studied from a taxonomic and functional point of view. Knowledge about the assembly of methanogenic microorganisms along with the methanotrophics is still incipient. It is, at this stage, where the production of methane within biogas is defined. However, the efficiency of methane production is determined according to biotic and abiotic factors (Caldwell et al., 2008; Hook et al., 2010; Segarra et al., 2013). The participation of methanotrophic archaea, carrying out "*reverse methanogenesis*" could change the concentration of methane in biogas. This activity can have a negative consequence on the energetic potential of the biogas, by decreasing the methane content.

Therefore, methane is both produced and consumed in anaerobic environments, as in sediment from coastal wetlands and the continental margin, *via* microbial processes. The association between methanogenesis and anaerobic oxidation of methane (AOM) plays a pivotal role in regulating earth's climate. Consequently, AOM moderates the input of methane, an important greenhouse gas, to the atmosphere by consuming methane produced in various marine, terrestrial, and subsurface environments (Caldwell et al., 2008; Segarra et al., 2013). However, this consumption of methane can have a negative effect in decreasing the CH_4 concentration when it comes to biogas production within reactor systems.

Microbiological Control of Biogas Production

Nowadays, there are more tools for a better control of biogas production. Traversi et al., (2012) demonstrated that it is possible to achieve a positive and significant correlation between the biogas production rate and methanogens abundance. They suggest that real-time qPCR could be used to measure methanogens concentration during anaerobic digestion and, thereby, biogas production capacity can be determined. These authors determined that a higher mean methanogens concentration (log methanogen concentration, *mrc*A copies/μL, equivalent to 7.5) was observed for production rates above 0.6m^3 biogas/kg VS, in a pilot mesophilic reactor of municipal solid waste and wastewater sludge with a hydraulic retention time of 20 days.

Conventional Anaerobic Sludge Digestion in Mesophilic Conditions

The sludge produced in wastewater treatment plants is classified into primary sludge, secondary sludge, and tertiary sludge. The organic matter of dried primary sludge reaches 60% to 75%. The fermentation of still visible components like excrements, fruits, vegetables and paper starts rapidly. Thickening of this sludge establishes solid concentrations of about

5% to 10% in the sludge. Secondary sludge with a higher content of organic matter (55% to 80%) also starts fermentation processes rapidly, but thickening is not very effective, achieving, in the best cases, only 3% of solid concentration (Boehnke, 1993). The original reason for anaerobic digestion in wastewater treatment plants was to obtain sludge with more favorable characteristics for further processing, this is called "stabilizing" sludge. Production of biogas, when implementing such systems, was considered an added value. Well digested sludge obtained with residence times in the digester of over 30 days reduces organic matter concentration of solids down to approximately 30%. Sludge digested for shorter periods can have organic dry matter content close to that of a secondary sludge. The stabilized sludge containing 4% to 12% of solids, when leaving the thickener, can be *dewatered* by filters, presses or centrifuges. The main mineralization product of digesting sludge is biogas bubbling out of the liquid phase and captured on the top of digester. From an economic point of view, digesters are designed for residence times of approximately 30 days, because longer residence times only increase slightly biogas yield. Another important parameter is the organic loading rate of the digester, which is defined as the kilogram of organic dry matter fed daily into one cubic meter of digester per day. For smaller digesters, the organic loading rate is near to 2 kg TS /(m^3 d) and digesters with volumes more than 1500 m^3 can be loaded with up to 3.5 kg TS /(m^3 d) (Boehnke, 1993). The main criteria that define this value are the concentration of organic acids in the digester. Experience from digester operation has shown that, for stable operation, this intermediate step in the degradation pathway should not be over 1000 mg/L. Higher concentration of organic acids in the digester can reduce the pH to below 6.8, which stops methanogenic activity. However, due to pH-buffering of digesters, unstable operating conditions can be detected earlier with the concentration of CO_2 in the biogas or by the FOS/TAC rate, which correlates the concentration of organic acids with total alkalinity (Lili, 2011). The concentration of methane in biogas reaches near 70% when produced only by methanogens, whereas CO_2 is produced by acidogenic and acetogenic bacteria. As a consequence, a perturbation in one of the subsequent process steps may influence a change in the concentration of CO_2. Biogas yield is not an adequate parameter to observe stability, because a reduction of organic matter feed can also reduce production.

THERMOPHILIC ANAEROBIC SLUDGE DIGESTION

In artificial systems of sewage sludge treatment, an alternative is thermophilic anaerobic digestion that takes place in the range of 40 to 75 °C (Angelidaki et al., 2003). On an industrial scale, under this parameter, anaerobic digestion is more efficient: smaller volume digesters, higher ranges of load, increased production of methane (80% CH_4 in the biogas) and efficient removal of pathogenic and parasitic microorganisms (Kim et al., 2002; Angelidaki et al., 2003; Cabirol et al., 2003; Nielsen et al., 2004; De la Rubia et al., 2005). However, the startup of a thermophilic reactor is extremely unstable and prolonged, when the thermophilic inoculum is not available. It is possible to use a mesophilic inoculum, but the period of acclimation at thermophilic temperatures can be long (a year or more) (Van Lier, 1996; De la Rubia et al., 2005).

Sekiguchi et al., (1998) compared the microbial composition of mesophilic (35°C) and thermophilic granular (55°C) sludges from a laboratory reactor fed with sucrose, propionate, and acetate. They found that, in mesophilic conditions, *Methanosaeta concilii* predominate in the consortium, whereas in the thermophilic reactor predominate species of *Methanosaeta thermophila* and *Methanobacterium thermoformicicum.* Cabirol et al., (2003) noted that it is feasible to obtain a thermophilic inoculum from a mesophilic one after 2 months with a direct increase of temperature; they also reported that the activity of hydrogenotrophic methanogens tended to predominate under these conditions. Angenent et al., (2002) investigated the effect of temperature on the diversity of methanogens in a reactor of farm waste treatment during three months of sludge acclimatation (fed with ammonia and AGV´s). Based on the study of 16S rDNA, the abundance of *Methanosarcina* (acetoclastic methanogen) increased from 1.2 to 3.8% and *Methanosaeta concilii* (acetoclastic methanogen) remained below 2.2%. On the other hand, levels of Methanomicrobiales, hydrogenotrophic methanogens, increased from 2.3 to 7%.

Therefore, in different countries the thermophilic anaerobic process is considered as being effective to achieve a good stabilization and disinfection of sewage sludge, reducing the risk to health as it is applied to land cultivation and restoration. This process has been applied in the laboratory, at pilot scale and large scale, obtaining a high removal of volatile solids (60-85%), with a high level of disinfection. In such cases, the process is operated using different designs, such as 1) conventional thermophilic anaerobic digestion, followed by an extended (in several stages) thermophilic anaerobic process, with continuous, mixed flow full and short HRT; 2) treatment in two stages (using a thermophilic and a mesophilic anaerobic digestion) (Table 2).

Technical Improvements

Specific biogas yields from primary and secondary sludge is normally related to the concentration of organic dry matter (oDM) in the sludge. Under reasonable retention times (< 30 days) of sludge in the digester, the common yield of biogas is about 400 L/kg oDM. While a longer retention time increases the biogas yield only slightly, the cost for construction of bigger digesters grows at different orders of magnitude. Hence, technical improvements focus on different sludge pretreatments to achieve a better performance of the anaerobic digestion with higher biogas yields. Technologies like thermal and mechanical disintegration, as well as ultrasonic destruction, are available (Koeppke, 2007). Specifically mechanical disintegration produces destruction of cell walls characterized by low biodegradability. The first benefit is that the area of cell walls vulnerable for biological attack is increased and inner cellular juice is dissolved. In addition, some lysates are released to a liquid phase acting as an agent for faster hydrolysis (Dohányos et al., 1997). When treating solids in anaerobic digesters, in most cases, the hydrolysis step is the velocity-limiting factor. In consequence, by using one of the described disintegration processes, the retention time of sludge in the digester can be reduced. At the same time, an increase in biogas yield of over 20% is observed sometimes.

Table 2. Thermophilic anaerobic digestion of wastewater sludge

Design and type of process	Volume	TRH	Temperature	Food and/orinoculum	Results	Ref.
Annacis Island Plant, (Vancouver) Conventional thermophilic anaerobic, followed by a thermophilic process extended (in 3 reactors, full continuous, mixed flow) process		21.5-d (for the process)	54.6°C	Food: mixture of primary and secondary sludge thickened, applied continuously	Gas prod 1.00m^3/kgSVrem E (sv) 60% % *AGV*452 mg/L CO_2 37.4% *Salmonella*: Not detected Fecal coliforms:<limit established for class A sludge	Krügel et al., 1998
Lions Gate plant (Vancouver) Anaerobic thermophilic, 2 digesters in series, with continuous flow and complete mixing process	2 digesters of 3100 m^{3}each	44-d for the entire process	55.8°C	Inoculum: sludge digested under mesofilia Food: Municipal primary sludge.	Gas prod 0.65 m^3/kgSV des E (SV) 80% % *AGV*567 mg/L CO_2 35.4% *Salmonella*: Not detected. Fecal coliforms:<limit established for class A sludge	Krügel et al., 1998
TITP San Pedro Plant (California) (*Terminal Island Treatment Plant*) thermophilic process, in 2 digesters with continuous flow egg-shaped	2 digesters of 4500 m^{3}each	12-d	one to 55°C another 65 ° c	Inoculum: sludge digested under mesophile Food:Raw sludge (70% secondary + primary 30%) coming from the treatment of municipal wastewater with 10% sea water and 40-60% of industrial Organic loading:3 kgSV/m^{3}d	Prod of CH_4: 0.2 m^3.kg^{-1}SV.d^{-1} to 55 ° C Prod of CH_4: 0.16 m^3.kg^{-1}SV.d^{-1} to 65 ° C SV removal: 28% at 55 °c and 22% at 65 ° C Effluent at 55 ° C: fecal coliforms 100 MPN/gST (a few days > 1000) *Salmonella*< 0.3 MPN/4gST Helminths< 1 HH viable/4gST	Iranpour et al., 2002 Ahring et al., 2002
Pilot scale of full mix	1 m^3	11.6-d	55.5°C	Inoculum:Municipal sludge produced at a pilot plant of anaerobic thermophilic. ST 150 g/l STV 55 % Food: municipal secondary sludge ST 84.4 g/l STV 65.1 % Organic load:(operating conditions) 4.8 kgSTV/m^{3}d	*AGVT*: 159 mgHAc/L Gas production: 0.9-1.3 m^3/m^3 d Prod. specify gas: 0.19 m^3/kg *AGVT* CH_4 65.4%	Cecchi et al., 1992

Table 2. (Continued)

Design and type of process	Volume	TRH	Temperature	Food and/orinoculum	Results	Ref.
Laboratory scale two reactors with continuous flow	Reactor-1 2.15 L Reactor 2 2,075 L	9.41 d	55°C	Inoculum: mesophilic sludge SV 4.8 % Food: 25% of wet weight, Organic load6.8 kgSV/m^3.d (R-2 received 30g more than the R-1)	Greater efficiency of stabilization in R-2 to R-1	Lepisto and Rintala, 1995
Laboratory scale, 1 egg-shaped reactor	5-L	16-d	55°C	Inoculum: mesophilic sludge Food: municipal secondary sludge	Helminths 1.12 H. viable/4gST < limit established for sludge class B (very close to the class A: 1Hv/4gST) Fecal coliforms:<limit established for class A sludge	Rojas Oropeza et al., 2001
Design of 10 plants on a large scale in Germany Only 5 plants have high efficienciy	Therm Meso	Therm Meso		Inoculum: municipal sludge	Greater efficiency of the two-stage	
			Mesophilic 35 a 67°C Thermophilic 55 a 60°C	ST kg/m3 — m3/d	Gas Prod m3/d — CO_2 Prod % — TS Degra %	Oles et al., 1997
Altenmarkt	1x60 1x300	3.5 17		49 — 17	233 — 32 — 54	
Auenheim	1x126 1x600	3 15		55-65 — 40	450 — - — 60	
Erkelenz	2x110 1x1600	2.6 19		35-50 — 84	550 — 30 — 60	
Geseke	1x120 1x1650	4 55		10 — 30	260 — 34 — 47	
Osterode	2x130 1x900	4 14		25-35 — 63	755 — 34 — 55	
Egg-Shaped Reactors (1 thermophilic and 4 mesophilic) Köln-Stammheimplant	Thermo Meso 1x11,000 4x11,000	Thermo Meso 5.5 21	Thermophilic 55 a 60°C Mesophilic 35 a 37°C	Inoculum: municipal sludge ST kg/m3 60 V m3/d 2.01	Gas Prod m3/d — CO_2 Prod % — TS Degra % 32.106 — 34 — 60	Oles et al., 1997
Laboratory scale, 2 egg-shaped reactors	5-L	Reactor T1 14-d	55°C	Inoculum: mesophilic sludge Food: T1 municipal secondary sludge	Removal of the T1: SV 63% Fecal coliforms:<limit class A	Cabirol et al., 2002

Despite all the biogas that can be produced form wastewater treatment sludge, one thing has to be mentioned: treating wastewater is an energy-intensive process. This is the reason why almost all wastewater treatment plants show a negative balance on electric energy even though anaerobic digesters are part of the treatment schemes. As a result, for example, energy consumption of wastewater treatment plants in the United States of America amounts to only 3% of the total electric energy demand (EPA, 2006). Various studies revealed that bigger scale treatment plants are more energy-efficient than the smaller scale plants. It seems that additional equipment like tertiary treatments should have the effect of higher specific energy demand. However, factors like well-designed plant, higher automatation grade, energy-efficient equipment, optimized aeration, and integration of anaerobic digesters compensate this additional feature reporting a specific electricity demand of 32 kWh/person, whereas small scale plants have a specific demand of 75kWh/person (Haberkern et al., 2006). The main consumers are the aeration tanks in the aerobic stage and other equipment providing circulation in basins. Thus, investments into energy efficient aeration systems show short pay-back times. Frequently, aeration is driven to excess; in this case, the energy demand is increased not only during this step. In addition, as a consequence of an increase in sludge age, the biogas yield in the anaerobic stage is affected (Kapp, 1991). Typical values for sludge aged over 20 days; this can be reduced to 14 days when special care of the process is considered. The process' stability must be controlled more carefully by the personnel, and additional equipment to control floating sludge in the aeration tank is another requirement.

Primary settling plays an important role in energy savings in the context of organic matter that can be removed at this stage, because oxygen rate in the following aerobic stage is reduced accordingly. A second advantage is that the potential of biogas is increased because partial mineralization by aerobic microorganisms does not occur. Accordingly, new strategies are followed by the Scandinavian way of wastewater treatment to enhance removal of suspended particles by upgrading primary settlers into advanced primary treatment (Karlsson 1996). An advanced primary treatment under favorable conditions can achieve removals of up 80% of BOD and 95% of phosphate.

Improvements in the electrification of biogas can contribute also to reach autonomous running with respect to meeting the electricity demands of wastewater treatment plants. Today, motor technology offers electrification efficiency of over 40% and residual energy in motor escapes can be additionally electrified by organic ranking cycles (ORC Process) giving supplementary electrification efficiency of 10%. Additionally, the same performance is shown by fuel cell technologies operating for a couple of years at some sites (Kishinevsky and Zelingher, 2003).

It is known that dewatering of anaerobic sludge transfers considerable flows of ammonium to the liquid phase that, in most cases, is recycled to wastewater and has to be nitrified/denitrified twice (Descoins, 2012). Newer designs of plants take care of these streams with an extra treatment that leads only to nitrite in the oxidation phase. The silver bullet produces fertilizers by magnesium ammonium precipitation (MAP) from the anaerobic digester juice (Kern, 2008). Even today the crystallization mechanisms are not totally understood. Specifically, it is not known why crystallization occurs in some digesters and not in others. Enterprises, like Ostara, offer applications to produce this kind of fertilizers from digested wastewater sludge.

The principal challenge for future biofuels is based on logistic management, because the handicap of biomass collection is its very low specific density of energy per area in comparison to the exploitation of fossil fuels. Anaerobic sludge digesters, however, may be able to be distributed across the country. From a geographic point of view, they have the potential to convert, via co-digestion, other organic residues generated near to the plant. This idea is not new, in Germany, during World War II, in some cities waste recollection was achieved with vehicles driven with digester gas. To enhance biogas production, co-digestion with foliage was practiced (Biogas Handbuch). However, in some countries, legislation may be an obstacle, because sludge residue from co-digestion has to be managed still according to legislation for wastewater treatment sludge. From a technical point of view, co-substrates have to be similar to sludge. Examples for co-substrates are foliage, rumen content from slaughterhouses, residues from canteens, and collected grease from traps. Co-digestion of the latter may make sludge dewatering more difficult (Koeppke, 2007).

However, it must be considered that the net energy gain must take into account the energy needed for pretreatment. The production of biogas with high methane energy efficiency should be weighed against the energy used at each stage of the sludge treatment process to calculate the net rate of methane production: for example, the sludge transfer to the system of treatment, possible pre-treatment (physical or chemical), the possibility of caloric energy increase of psychro- to meso and, lastly, to thermophilic anaerobic digestion, and, finally, the costs of biogas transfer to its final use (Kwon et al., 2013; Kuglarz et al., 2013).

REFERENCES

Ahring, B., Mladenovska, Z., Iranpour, R., Westermann, P. 2002. State of the art and future perspectives of thermophilic anaerobic digestion. *Water Science and Technology.* 45 (10): 293-298.

Angelidaki, I., Ellegaard, L., Ahring, B. K. 2003. Applications of the anaerobic digestion process. *Advances Biology, Engineering and Biotechnology* 82: 1-10.

Angenent, L. T., Sung, S., Raskin, L. 2002. Methanogenic population dynamics during start-up of a full-scale anaerobic sequencing batch reactor treating swine waste. *Water Research* 36: 4648-4654.

Böhnke, B. 1993. Anaerobtechnik. Handbuch der anaeroben Behandlung von Abwasser und Schlamm. Hrsg. B. Böhnke, Berlin [u.a.]: Springer.

Cabirol, N., Fernández, F. J., Mendoza, L., Noyola, A. 2003. Acclimation of mesophilic anaerobic sludge to thermophilic conditions: PCR detection methodology. *Water Science Technology* 48 (6): 81-86.

Cabirol, N., Rojas-Oropeza, M., Noyola, A. 2002. Removal of helminth eggs and fecal coliforms by anaerobic thermophilic sludge digestion. *Water Science and Technology,* 45:269-274.

Caldwell, S.L., Laidler, J.R., Brewer, E.A., Eberly, J.O., Sandborgh, S.C., Colwell, F.S. 2008. Anaerobic oxidation of methane: mechanisms, bioenergetics, and the ecology of associated microorganisms. *Environmental Science Technology.* 42(18):6791-6799.

Cecchi, F., Pavan, P., Musacco, A., Mata-Alvarez, J., Sans, C. 1992. Comparison between thermophilic and mesophilic anaerobic digestion of sewage sludge coming from urban wastewater treatment plants. *Water Science and Technology,* 26 (9-11): 2409-2412.

Cooper PF. 2007. Historical aspects of wastewater treatment. In: Lens P, Zeeman G, Lettinga G, editors. Decentralised sanitation and reuse: concepts, systems and implementation. *IWA Publishing.*

De la Rubia, M. A., Romero, L. I., Sales, D., Pérez, M. 2005. Temperature conversion (mesophilic to thermophilic) of municipal sludge digestion. *AICHE J.* 51 (9): 2581-2593.

Descoins, N., Deleris, S., Lestienne, R., Trouvé, E., Maréchal, F. 2012. Energy efficiency in waste water treatments plants: Optimization of activated sludge process coupled with anaerobic digestion. In: *Energy* 41 (1), S. 153–164.

Dohányos, M., Zábranská, J., Jenícek, P. 1997. Innovative technology for the improvement of the anaerobic methane fermentation. *Water Science and Technology* 36: 333-340

EPA, 2006. Wastewater Management Fact Sheet-heat Drying. EPA 832-F-06-024.http://water.epa.gov/scitech/wastetech/upload/2006_10_16_mtb_heat-drying.pdf accessed 05.10.12.

ESA, 2011. United Nations, Department of Economic and Social Affairs. World Urbanization Prospects the 2011 Revision. http://esa.un.org/unup/Documentation/faq.htm, accessed 17.05.2013

Haberkern, B. 2007. Neue Techniken der Faulgasverwertung. *Fachgespraech BMU/UBA* 29.01.2007, Bonn, Germany

Haberkern, B., Maier W., Schneider, U. 2006. Steigerung der Energieeffizienz auf Klaeranlagen. Report 20526307, Umweltbundesamt

Hook, S.E., Wright, A.D., McBride, B.W. 2010. Methanogens: methane producers of the rumen and mitigation strategies. *Archaea.*1-11.

IANAS. 2012. Diagnóstico del agua en las Américas. Programa del agua de IANAS *Interamerican Network of Academies of Sciences* and Foro Consultivo Científico y Tecnológico AC. ISBN 978-607-9217-04-4.

Iranpour, R., Oh, S., Cox, H., Shao, Y., Moghaddam, O., Kearney, R., Deshusses, M., Stenstrom, M., Ahring, B. 2002. Changing mesophilic wastewater sludge digestion into thermophilic operation at terminal island treatment plant. *Water Environment Research.* 74 (5): 494-507.

Kapp H., 1991. Beitrag der Abwasserbehandlung zur CO_2-Belastung der Umwelt; StuttgarterBerichtezurSiedlungswasserwirtschaft Bd. 114

Karlsson, I. 1996. Environmental and energy efficiency of different sewage treatment processes. *Water Science and Technology.* 34, S: 203–211.

Kato, S., Hashimoto, K., Watanabe, K. 2012. Methanogenesis facilitated by electric syntrophy via (semi)conductive iron-oxide minerals. *Environmental Microbiology* 14:1646-1654.

Kern, J., Heinzmann, B., Markus, B., Kaufmann, A.C., Soethe, N., Engels C. 2008. Recycling and assessment of struvite phosphorus from upgraded sewage sludge. *Agricultural EngInt: The CIGR Journal.* 10: CE 12 01.

Kim, M., Ahn, Y. H., Speece, R. E. 2002. Comparative process stability and efficiency of anaerobic digestion: mesophilic vs. thermophilic. *Water Research.* 36: 4369-4385.

Kishinevsky, Y., Zelingher, S. 2003. Coming clean with fuel cells. In: IEEE power / *energy magazine.* (6), S: 20–25.

Koeppke, K.E. 2007. Vorschlaege zur Konkretisierung der Anforderung zur Energieeffizienz in Anhang 1. *Fachgespraech BMU/UBA,* Bonn, Germany

Krugel, S., Nemeth, L., Peddie, C. 1998. Extending thermophilic anaerobic digestion for producing class A biosolids at the greater Vancouver Regional District Annacis Island Wastewater Treatment Plant. *Water Science and Technology.* 38 (8-9): 409-416.

Kuglarz, M., Karakashev, D., Angelidaki, I. 2013. Microwave and thermal pretreatment as methods for increasing the biogas potential of secondary sludge from municipal wastewater treatment plants. *Bioresource Technology.* 134:290-297.

Kwon, E.E., Yi, H., Kwon, H.H. 2013. Urban energy mining from sewage sludge. *Chemosphere.* 90:1508-1513.

Leonard, B., Kinnicutt, C., Winslow, E.A., Pratt W. 1910. Chapter VIII: Purification of Sewage by Broad Irrigation and Sewage Farming, *Sewage Disposal*, New York*: John Wiley & Sons,* 1910, pp. 193-221.

Lepisto Satu, S., Rintala Jukka, A. 1995. Thermophilic anaerobic digestion of the organic fraction of municipal solid waste: start-up with digested material from a mesophilic process. *Environmental Technology.* 16: 157-164.

Lili, M., Biro, G., Sulyok, E., Petis, M., Borbely, J., Tames J. 2011. Novel Approach on the Basis of FOS/TAC Method. In: Analele Universităţii din Oradea, *Fascicula Protecţia Mediului* 17: 713-717.

Liu, Y.C., Whitman, W.B. 2008. Metabolic, phylogenetic, and ecological diversity of the methanogenic archaea. In: Incredible Anaerobes. *Physiology to Genomics to Fuels* 1125:171-189.

Lofrano, G., Brown, J. 2010. Wastewater management through the ages: A history of mankind. *Science of the Total Environment.* 408: 5254-5264.

Lozada, M., Itria, R.F., Figuerola, E.L.M., Tullio, L.A., Erijman, L. 2004. Dinámica, selección y estabilidad de poblaciones bacterianas en lodos activados. 5° *Jornadas de Desarrollo e Innovación,* Mexico.

McKeown, R.M., Hughes, D., Collins, G., Mahony, T., O'Flaherty, V. 2012. Low-temperature anaerobic digestion for wastewater treatment. *Current Opinions in Biotechnology.* 23:444-451.

Narihiro, T., Sekiguchi, Y. 2007. Microbial communities in anaerobic digestion processes for waste and wastewater treatment: a microbial update. *Biotechnology.* 18: 273-278.

Nielsen, H.B., Mladenovska, Z., Westermann, P. y Ahring, B.K. 2004. Comparison of two-stage thermophilic (68ºC/55ºC) anaerobic digestion with one-stage thermophilic (55ºC) digestion of cattle manure. *Biotechnology and Bioengineering.* 86 (3): 291-300.

OECD, 2011. Environment Statistics, http://www.oecd-ilibrary.org/environment/data/oecd-environment-statistics_19/02/2013.

Oles, J., Dichtl, N., Niehoff, H.-H. 1997. Full scale experience of two stage thermophilic/mesophilic sludge digestion. *Water Science and Technology.* 36(6 7): 449-456.

Rojas-Oropeza, M., Cabirol, N., Ortega, S., Castro Ortiz, L.P., Noyola, A. 2001. Removal of fecal indicator organisms and parasites (fecal coliforms and helminth eggs) from municipal biologic sludge by anaerobic mesophilic and thermophilic digestion. *Water Science and Technology.* 44:97-101.

Schink, B. 1997. Energetics of syntrophic cooperation in methanogenic degradation. Microbiology and Molecular biology. *Reviews.* 61:262-280.

Segarra, K.E.A., Comeford, C., Slaughter, J., Joye, S.B. 2013. Impact of electron acceptor availability on the anaerobic oxidation of methane in coastal freshwater and brackish wetland sediments. *Geochimica et Cosmochimica Acta.* 115:15-30.

Sekiguchi, Y., Kamagata, Y.,Syutsubo, K.,Ohashi, A., Harada, H., Nakamura, K. 1998. Phylogenetic diversity of mesophilic and thermophilic granular sludges determined by 16S rRNA gene analysis. *Microbiology.* 144: 2655-2665.

SEMARNAT. 2011. Inventario nacional de plantas municipales de potabilización y de tratamiento de aguas residuales en operación. *Secretaria de Medio Ambiente y Recursos Naturales,* Mexico.

Symons, G. E.; Buswell, A. M. 1933. The methane fermentation of carbohydrates. *Journal of American Chemistry Society. 55* (5), S: 2028–2036.

Tchobanoglous, G. 2003. Wastewater engineering - Treatment and Reuse. New York, *Metcalf & Eddy, Inc.*

Thauer, R.K., Kaster, A.K., Seedorf, H., Buckel, W., Hedderich, R. 2008. Methanogenicarchaea: ecologically relevant differences in energy conservation. *Nature Reviews Microbiology.* 6: 579-591.

Traversi, D., Villa, S., Lorenzi, E., Degan, R., Gilli, G. 2012. Application of a real-time qPCR method to measure the methanogen concentration during anaerobic digestion as an indicator of biogas production capacity. *Journal of Environmental Management.* 111:173-177.

Van Lier, J.B. 1996. Limitations of thermophilic anaerobic wastewater treatment and the consequences for process design. *Antonie van Leeuwenhoek.* 69:1-14.

Wellinger, Artur 1991. Biogas-Handbuch. Grundlagen - Planung – *Betrieblandwirts chaftlicher Anlagen.* 2. Eds. Aarau: Wirz.

Whitman, W.B., Bowen, T.L., Boone, D.R. 2006. The Methanogenic bacteria. *Prokaryotes* 3:165-207.

WHO. 2011. WHO/UNICEF Joint Monitoring Programme (JMP) for Water Supply and Sanitation http://www.wssinfo.org/data-estimates/introduction/

In: Energy and Environment Nowadays
Editors: Luis G. Torres and Erick R. Bandala ISBN: 978-1-63117-398-1

Chapter 2

BIOREFINERY: USING MICROALGAL BIOMASS FOR PRODUCING LIPIDS, BIOFUELS AND OTHER CHEMICALS

Alma Toledo-Cervantes[1] and Marcia Morales[2]
[1]Doctorado en Ciencias Biológicas y de la Salud,
Universidad Autónoma Metropolitana UAM- Iztapalapa, Mexico
[2]Departamento de Procesos y Tecnología,
Universidad Autónoma Metropolitana UAM-Cuajimalpa, Mexico

1. INTRODUCTION

The strong dependence on oil for production of fuels and chemicals and the undeniable diminishing of petroleum sources causes a global concern on environmental, social, politic and economic terms. In order to simultaneously reduce the dependence on oil and mitigate the climate change, the use of microalgae biomass to produce biofuels such as biodiesel has been proposed. It is well know that the process is technically feasible and the potential for CO_2 sequestration is high. At the present, the economic feasibility of algal-to-biofuel seems to be dependent on government subsidies and the future price of oil (Gallagher, 2011). However, based on the fact that all current commercial algae production is motivated by the sale of high-value products for pharmaceutical and nutritional uses, and the available evidence suggests that multiple product extraction of the microalgal biomass is possible; it is currently agreed that co-product sales will be critical in achieving economically feasible algal biofuels production (Wijffels et al., 2010) in a biorefinery concept for production of bioenergy, biofuels and biochemicals (Cherubini, 2010).

Biorefining integrates multiple unit processes for biomass conversion that are analogous to petroleum refinery processes. Biomass is the organic matter produced from transformation of sun energy to chemical energy including wood, agricultural crops, food, even wastes, sewages or microalgal biomass.

Crude oil is refined to produce useable products such as: diesel oil, fuel oil, LPG (Liquefied Petroleum Gas), kerosene, naphtha, and gasoline among others; and from biomass it can be obtained: 1) gaseous biofuels (syngas, biohydrogen, biomethane), 2) solid biofuels (biomass, charcoal) and 3) liquid biofuels (bioethanol, biodiesel, FT-fuels, bio-oil). Furthermore, the products from biomass are: 1) chemicals (fine chemicals, building blocks, bulk chemicals), 2) organic acids (succinic, lactic and other sugar derivatives), 3) polymers and resins (starch-based plastics, phenol resins, furan resins), 4) biomaterials (wood panels, pulp, paper, cellulose) and 5) food, animal feed and fertilizers (Cherubini, 2010).

However, in order to maximize the potential of microalgae biofuels, it is necessary to define the whole chain of process development in an integrated way starting from an adequate supply of CO_2, good harvesting methods, de-watering, and downstream processing (Mata et al., 2010). For this, it is necessary to develop a deep knowledge of the potential added value products that can be obtained, the processing technology and the sequence of separation to maintain the integrity of the compounds and maximize their recovery.

2. Microalgae as Feedstock

Biomass is the only C-rich organic source available on Earth, it appears to be an attractive feedstock (the term "feedstock" refers to raw materials used in the biorefinery) to produce fuel since it is a potentially renewable source that could be sustainably developed, and is environmentally friendly.

The sustainable biomass production is a crucial issue in order to develop a biofuel production process, especially concerning a possible fertile land competition with food and feed industries (Cherubini, 2010).

Microalgae as feedstock is gaining interest due to their fast growth potential coupled with their relatively high lipid, carbohydrate and nutrients content. All of these properties make them an excellent source for biofuels such as biodiesel, bioethanol, biohydrogen and biomethane; as well as a number of other valuable pharmaceutical and nutraceutical products (Singh et al., 2010). Microalgae are a large and diverse group of photosynthetic eukaryotes with a simple cellular structure, ranging from unicellular to multicellular forms. The elemental composition of microalgae biomass is a mixture of C, H and O in a proportion of $CO_{0.48}H_{1.83}N_{0.11}P_{0.01}$ (calculated by Chisti in 2007, from data presented by Grobbelar in 2004).

The main advantages of microalgae versus terrestrial plants as feedstock are (modified from Gouveia, 2011):

1) A higher CO_2 sequestration capacity so accordingly higher biomass yields can be reached (1.8 tons of CO_2 is needed to produce 1 ton of microalgal biomass), which contributes to reduce the greenhouse effect (Wijffels and Barbosa, 2010).
2) Photon conversion efficiency approximately between 3–8% against 0.5% for terrestrial plants.
3) The water usage is lower than crops. To produce 1 L of biofuel from crops, approximately 10,000 L of water is necessary (de Fraiture et al., 2008) in contrast with the ~0.75 L of water needed per kg of microalgal biomass produced (Kliphuis et al., 2010).

4) Microalgae cultivation does not compete with arable land for food production; they are able to grow in liquid medium, wastewater or seawater in marginal areas unsuitable for agricultural purposes.
5) Their production is not seasonal and the systems can be easily adapted to various levels of operational and technological skills.
6) They produce value-added metabolites (e.g., proteins, polysaccharides, pigments, biopolymers, animal feed).
7) Microalgae growth does not require the use of herbicides or pesticides.

Furthermore, the above-mentioned advantages the biochemical composition of microalgae can be modified by different growth conditions; and the algae-based biofuels do not generate toxins and are biodegradable (Gendy and El-Temtamy, 2013).

Particularly, the fast growth required of carbon-neutral renewable alternatives makes microalgae one of the most promising sources of biofuels, which could meet the global demand for transport fuels (Chisti, 2007).

3. Metabolites from Microalgae

There are a great variety of metabolites produced by the microalgae and many have commercial importance. The first use of microalgae by humans dates back 2000 years, in China, where *Nostoc* sp. was used as food. However, algal biotechnology did not begin to develop until the middle of last century (Spolaore et al., 2006). Today, there are numerous applications for microalgae biomass: 1) as a supplement to increase the nutritional value of food for humans and animals, 2) to play a crucial role in aquaculture, 3) it can be incorporated in cosmetics and 4) they are industrially produced to obtain high value products, such as: drugs, polyunsaturated fatty acids and pigments. Also, microalgae serve as feedstock to produce renewable biofuels and they are obtained from conversion of microalgal biomass basic products like starch and lipids.

Lipids

Many lipids are accumulated by microalgae under several specific culture conditions. According to an application point of view, microalgal lipids are classified into two types depending on their carbon numbers such as fatty acids containing 14–20 carbons used for biodiesel production, and polyunsaturated fatty acids (with over 20 carbons) used as health and food supplements (Yen et al., 2012).

Typical microalgal fatty acid composition is a mixture of unsaturated fatty acids, such as palmitoleic (16:1), oleic (18:1), linoleic (18:2) and linolenic (18:3) acid, and certain saturated fatty acids, like stearic (18:0) and palmitic (16:0).

Microalgae are a potential source of various food supplements and biomaterials used in the pharmaceutical industry. Some of these are omega-3 fatty acids, eicosapentanoic acid (EPA), and decosahexaenoic acid (DHA). EPA has been used in clinical purposes such as the treatment of heart and inflammatory diseases, asthma, arthritis, migraine headache and psoriasis (Singh et al., 2005). Studies have shown that marine microalgae has significantly more decosahexaenoic acid (DHA) content compared to fresh water microalgae (Patil et al., 2007).

Carbohydrates

They are one of the most important sources of energy in microalgae. In general, carbohydrates are mainly composed of starch, glucose, cellulose, hemicelluloses, and several kinds of polysaccharides with important biological functions in algae cells, mainly as storage, protection and structural molecules.

Also, starch and glucose are used for biofuel production, especially for bioethanol and biohydrogen. Some microalgae are able to accumulate a high carbohydrate content, more than 50% of its dry weight (Ho et al., 2012).

Proteins

The high protein content in some species of microalgae is one of the reasons it can be considered as a source of protein for food and feed. However, many of the data presented in the literature are estimations based on the total nitrogen content (using the factor 6.25), however, 10% of the total nitrogen corresponds to non-protein content. So it is necessary to evaluate the nutritional value of microalgal biomass, before suggesting its use in animal feed or human food (Richmond, 2004).

Pigments

Pigments are available to improve the efficiency of light energy utilization and protect microalgae against solar radiation and related effects. The function as antioxidants inside the microalgae leads their potential role as antioxidants in foods and humans.

Carotenoids are essential for many functions in the human body such as vision, growth, and bone development. They are also potent antioxidants and help balance the effect of hormones. The concentration of carotenoids in microalgae varies from 0.1 to 0.2% of dry weight.

Chlorophyll can be found under optimum conditions in about 4% of microalgal dry weight (Harun et al., 2010a). It has chelating properties and is used for therapeutic treatment of liver and ulcers; it also repairs cells, increases hemoglobin in the blood and increases cell growth. In the food industry, chlorophyll is used as a natural pigment ingredient in processed foods (Singh and Gu, 2010).

Cyanobacteria, Rhodophyta and Cryptomonads microalgae contain phycobiliproteins. Phycobiliproteins are formed by a protein backbone covalently linked to phycobilins, which are used for fluorescence applications, as highly sensitive fluorescence markers in clinical diagnosis and for labeling antibodies used in multicolor immunofluorescence.

4. Biodiesel from Microalgal Lipids

At the moment there is growing interest in the use of microalgae for the production of biodiesel, although this is not the only producible fuel: biogas can also be produced, as well as bioethanol, biohydrogen or hydrocarbons. The quality and composition of the microalgal biomass represents the best option for the biofuel to be produced. A biomass rich in lipids will be suitable for the production of bio-oil and biodiesel, while a biomass rich in sugars will be better suited for the production of bioethanol. The anaerobic fermentation of sugars, proteins, and organic acids will produce biogas or bioethanol.

Microalgae biodiesel is produced primarily through the trans-esterification reaction of previously extracted triglycerides and alcohol usually in the presence of a metal catalyst. The reaction products are a glycerol molecule and one to three fatty acid esters (almost always either ethyl or methyl alcohol are used yielding a methyl or ethyl ester).

Several species of microalgae are very rich in lipids (up to 70-80% dry weight, with a good average standard of 30-40%). After extraction of the microalgae oil, the residue will normally represent more than 50% of the initial biomass, which could be transformed in a biorefinery concept.

5. Biorefinery Definition and Strategy on Microalgal Biomass

The International Energy Agency's Bioenergy Task 42 on Biorefineries (IEA, 2007) defines biorefining as the sustainable processing of biomass into a spectrum of bio-based products (food, feed, chemicals) and bioenergy (biofuels, power and/or heat). The aim of a biorefinery is to deliver multiple products by depolymerizing and deoxygenating the feedstock components, thereby maximizing the value derived for the biomass feedstock (Cherubini, 2010). It was pointed out by Valderrama et al., (2002) that microalgae have emerged as one of the most promising biorefinery feedstocks because of their widespread availability and high oil yields. The biorefinery approach is a combination of technologies looking for zero waste process by conversion of biomass into several final products.

It must be economically profitable, socially acceptable, and environmentally friendly. Therefore, the design objectives that must be achieved in a microalgae biorefinery (Sánchez-Tuirán et al., 2013) are:

1. Use as much of the microalgae biomass as possible in the process.
2. Perform mass and energy integration analysis to reduce consumption of fresh raw materials, reduce waste effluents and energy consumption.
3. Perform an economic analysis after mass and heat integration to check the profitability of the process.

6. Biorefinery Strategy

A strategy to establish a microalgae-based biorefinery should consider:

a) *To isolate microalgae strains rich in the target products.* To obtain the best microalgae strains that combine high lipid content or other products, high growth rate and wide-ranging environmental tolerances for biofuels production, researchers have focused on species selection and cultivation techniques. Microalgae with a high tolerance to wide ranges of temperature can become predominant in a culture environment. Nevertheless, microalgal biomass feedstock will experience seasonal changes in both quantity (biomass and lipid productivity) and quality (unsaturation degree); this may result in a change of biorefinery capacity towards products that can be derived from biomass all year round and leave niche products to specialized producers (Sivakumar et al., 2012).

   Genetic modification of microalgae has received attention in order to obtain efficient processes in biofuel production and bioremediation. Flynn et al., (2013) have explored routes to enhance production through modifications to a range of generic microalgal physiological characteristics. However, the consequent risks to ecosystems by a harmful algal bloom should be considered.

b) *To establish the cultivation conditions and operation strategies.* Since the accumulation of different components (such as lipids, carbohydrates, proteins and pigments) in microalgae requires different cultivation conditions and operation strategies (Chen et al., 2011). The study of the culture conditions for improving target-product production as well as designing efficient and economically feasible microalgae cultivation systems (open systems or photobioreactors) are critical to improve the productivity of microalgal biomass or target products. The synthesis and accumulation of a large amount of high-value metabolites occurs under stress due to chemical and physical environmental changes, either acting individually or in combination. The main chemical stresses are nutrient starvation (nitrogen, phosphorous, sulphur, silicon). Under low-nutrient conditions algae rapidly start to convert energy from the sun into chemical energy stored as lipids (Gouveia and Oliveira, 2009). The effect of the depletion of the N source is associated with the decreases in the intracellular chlorophyll and chloroplast numbers, where a large amount of phospholipids and glycolipids lie within the chloroplast. Also the depletion of the N source affects the intracellular consumption of the nitrogen pool to support the synthesis of cell material for further cell division. Thus, the concentration and duration of N are important parameters to optimize the biomass and lipid productivity of microalgae. Salinity, pH and the physical stresses such as temperature and light intensity also affect the lipid production. Operation strategies include: i) *a two-stage culture* of microalgae. The first stage is for cell growth and the second stage is to increase the metabolite production with a desired co-product rather than increasing the overall biomass (Suali and Sarbatly, 2012). An additional CO_2 supply is necessary in the second stage to increase the carbon-base metabolites; ii) *Hybrid systems,* first microalgae are growing continuously in photobioreactors under nutrient sufficient conditions and then a portion is transferred to nutrient-limited open ponds to induce the accumulation of reserve products after ponds are harvested, cleaned and then re-inoculated (Huntley and Redalje, 2007).

c) *To define the conversion processes of whole microalgae/defatted microalgal biomass to biofuels.* To design a biorefinery, it is important to understand the potential of microalgae, based on the specific species, since its biochemical profiles could aim at a specific process, either biochemical, thermochemical or chemical separation. Each process has certain feedstock requirements that are shown in figure 1. More specific information is given in the next section.

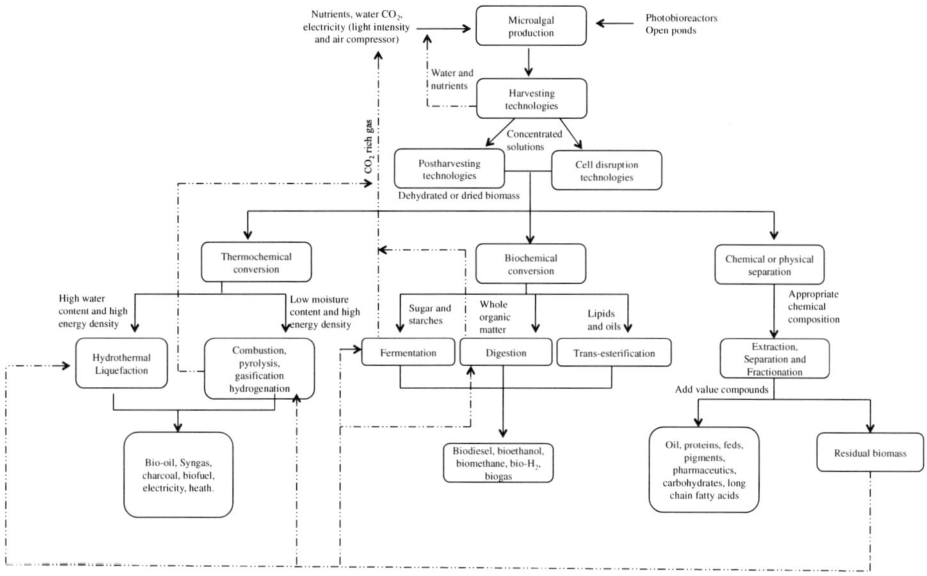

Figure 1. Microalgae biorefinery for production of biofuels and added-value compounds by thermochemical and biochemical processing.

d) *To define the biomass harvesting, post-harvesting technology.* Microalgae cultures have high water contents and removal is the first challenge (Gouveia, 2011). Dewatering is complicated due to the small size of algal cells (3−30 μm) (Bhave et al., 2012). Many studies have focused on harvesting technologies such as centrifugation, filtration, flotation, coagulation, flocculation and scraping by using an attached culture system (Ahmad et al., 2011; Divakaran and Pillai, 2002; Zheng et al., 2012; Johnson and Wen, 2010). Table 1 describes the main characteristics and equipment required in the harvesting processes. The biomass downstream processing to concentrate/purify the metabolite can be substantially more expensive than the culturing stage (Molina et al., 2003).

   The post-harvest processing depends strongly on the desired product and the selected biomass transformation technology. Dehydration or drying of the biomass is necessary for certain thermochemical and biochemical processes to produce biofuels (Molina et al., 2003). Drying methods (Table 1) for microalgae include sun drying (Prakash et al., 1997), spray drying (Richmond, 2004; Murugesan and Orsat, 2012), drum drying, and freeze-drying (Molina et al., 1996; Lee et al., 2012). Freezing has the added advantage of making the cell walls more porous by the formation of ice crystals within the cells and enhancing the extraction of intracellular products (Wiltshire et al., 2000), however it is limited to a small scale.

e) *To define the sequence extraction of co-products and processing the whole biomass or the pre-extracted product to final products.* Cell disruption is an integral part of the downstream operation required to produce biofuels from microalgae. An important approach to consider is the nature and type of microalgal strain. The shape of cells, cell wall structure and added-value metabolites composition vary from one strain to another, even when the same strain is cultivated at different conditions (Li et al., 2011).

The disruption methods use mechanical and non-mechanical techniques (Figure 2). Although, several methods have already been used for cell disruption, the most efficient method for microalgae has not yet been settled (Lee et al., 2010).

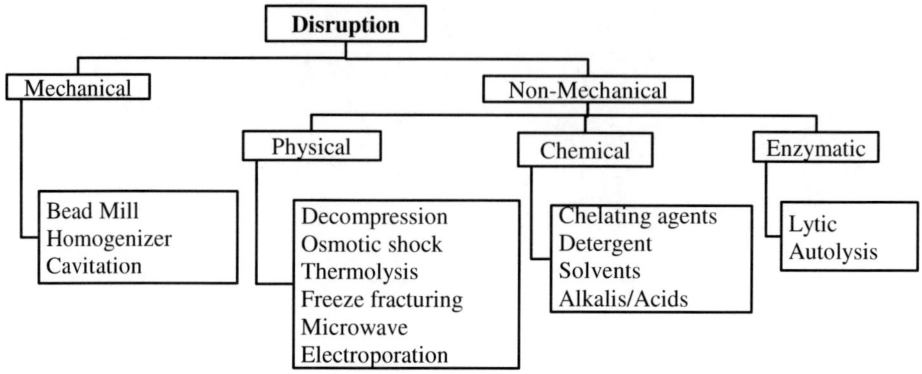

Adapted from Lee et al., 2012.

Figure 2. Classification of cell disruption methods.

For the production of microalgal lipids or pigments for pharma/nutraceutical uses, the high required disruption energy can be justified by the high product values and process energy is an important but not critical factor. However, for biofuel production, the energy input for biomass disruption is a critical consideration since a net negative energy output is not sustainable (Lee et al., 2012).

Extractions of interesting products are done by physical and chemical methods in the form of solvent extractions, or combinations. The extraction methodology should be fast, easily scalable, effective and should not damage the extracted product.

The most common solvents include hexane and alcohol (Soxhlet extractor). The supercritical fluid extraction and subcritical water extraction techniques are other processes that have received attention as an alternative, since these extraction techniques provide higher selectivity, shorter extraction times and do not use toxic organic solvents (Herrero et al., 2006). Physical techniques include expeller presses, electromechanical methods and ultrasonic extraction. The selection of the proper solvent system is still essential to the extraction process because it may weaken the cell wall structure facilitating cell disruption and therefore extraction (Araujo et al., 2013). More detailed information is presented on the biorefinery processing of metabolites separation/extraction in next section.

f) *To maximize the process integration of stream and recycling of materials, reducing wastes.* A microalgal biorefinery must be designed to maximize multiple product recovery from the biomass as high-value phytochemicals, combined with process chain integration to include carbon capture and storage, and bioremediation of

wastewater, therefore it could enhance the viability and future sustainability of biofuel-directed microalgae production (Brennan and Owede, 2010).
Coupled to the microalgae production of biofuel other purposes can be considered like (Wang et al., 2008):

1) The reduction of greenhouse gases by removing CO_2 from industrial gas emissions.
2) Removal of NH_4^+, NO_3^{1-} and PO_4^{3-} from wastewater, when they are used as a nutrient source.
3) After extracting the oil, residual biomass can be processed to produce bioethanol, biomethane, livestock feed, organic fertilizer, or simply burned to co-generate energy.

g) *To perform a Life cycle analysis.* Life cycle analysis is a method to quantify the environmental impact of products that appears to be a relevant tool to evaluate new technologies for bioenergy production. This tool identifies the technological bottlenecks and therefore supports the ecodesign of an efficient and sustainable production chain (Gouveia, 2011).
An evaluation of energy and emission balance for biofuel production and its uses it is essential to the identification of energy utilization and emission reductions. This can only be done using a systematic approach to investigate the entire production cycle such as Life cycle assessment that gives an overall picture of the superior quality of energy dynamics reliability and environmental impacts and hence, can help in the decision-making process for implementation of the potential microalgae biofuel production.

7. Processes and Technologies for Biorefineries

There are two different motivations recognized for the development of biorefineries: 1) Energy or biofuel production and, 2) the production of a portfolio of bio-based products.

In the energy or biofuel-driven biorefinery the biomass is primarily used for the production of transportation biofuels, power and heat. Process residues are sold or even further upgraded to value-added bio-base products to provide further economic and environmental benefits.

Processes involved in biomass transformation in the biofuel-driven biorefineries are classified in Thermochemical and Biochemical processing and they are explained below.

7.1. Thermochemical Processing

The technology, named as thermochemical, processes the biomass at elevated temperatures and pressures in order to break the chemical bonds of the cell wall structure to create compounds similar to petroleum that can be refined to produce specific fuels; therefore there is no need for modification of engines.

Table 1. Harvesting, post-harvesting and cell disruption technologies for microalgal biomass

	Technical Principles	Advantage/Disadvantage	Equipment required	Observations	References
HARVESTING TECHNOLOGIES: The goal of this stage is dewatering the biomass to obtain solutions between 20 and 100% more concentrated than the starting solution.					
Centrifugation	Phase separation applying centripetal force.	Rapid, easy, efficient / Very high energy input.	Centrifuge.	It can be affected by culture features including cell concentration, pH and ionic strength. Recovery of the biomass depends on the settling characteristics of the cells, the residence time of the biomass in the centrifuge, and the settling depth.	Molina et al., 2003; Rawat et al., 2013
Sedimentation	Forces of gravity and relative density; Stokes' law.	Low cost, potential for water recycling / Slow process (0.1-2.6 cm/h), product deterioration.	Lamella separators and sedimentation tanks.	Among the factors that affect sedimentation are gravity, relative density, and the most important are diameter and radius of the cell, the first affects the drag of cells and the second one the sedimentation velocity.	Brennan and Owende, 2010; Greenwell et al., 2010; Rawat et al., 2013
Flocculation	Negative charges of microalgae cell walls, which prevent self-aggregation, can be neutralized by addition of chemicals known as flocculants.	Low cost, low cell damage / Biomass contamination, no water reuse, inefficient in saline environments, produces large quantities of biomass sludge.	Lamella separators and sedimentation tanks	Frequently used to increase the efficiency of gravity sedimentation.	Danquah et al., 2009a; Sukenik and Shelef, 1984
Electro-coagulation	An electric field drives charged microalgae to the top.	Environmental compatibility, versatility, energy efficiency, safety, selectivity, and cost effectiveness.	Settling tank, or gravity clarifier.	This mechanism is used to remove microalgae in the water treatment industry.	Mollah et al. 2004; Gouveia, 2011; Chen et al., 2011

	Technical Principles	Advantage/Disadvantage	Equipment required	Observations	References
Flotation	Flotation can capture particles with a diameter of <500 μm by collision between a bubble and a particle and the subsequent adhesion of the bubble and the particle.	Low cost, easy application at large scale / Water reuse and product extraction may be negatively affected.	In dispersed air flotation bubbles are mechanically formed by a combination of a high-speed mechanical agitator and an air injection system. Separation tanks.	The bubble generation is produced by a decompression of pressurized fluid. Based on the bubble, sizes can be divided into dissolved air flotation (fine bubbles <10 mm), dispersed air flotation and electrolytic flotation (bubbles size from 700–1500 μm).	Chen et al., 1998; Molina et al., 2003; Yoon and Luttrell, 1989; Uduman et al., 2010
Filtration	Liquid-solid separation through a porous surface, which retains the solids and allows the passage of fluid.	Low cost, water reuse, recommended as a harvesting tool to filamentous or large colonial microalgae / Slow, membrane fouling and clogging, limited volume, cell damage.	Filter presses, vacuum filters.	Often applied at a laboratory scale, but in large-scale applications presents problems, such as formation of compressible filter cakes and the high maintenance costs. There are different kinds of filtration, such as dead end filtration, microfiltration, ultra filtration, pressure filtration, vacuum filtration and tangential flow filtration (TFF).	Molina et al., 2003; Gouveia, 2011; Rawat et al., 2013; Harun et al., 2010a; Danquah et al., 2009b
Microalgal biofilms	Biofilm formation occurs due to the concentration of cations, proteins, and organic molecules on submerged surfaces. Microalgae colonize a surface then they secrete extracellular polymeric substance composed of polysaccharides, proteins, nucleic acids, and phospholipids..	Biomass attachment is influenced by pH, organic film, culture age, culture density, cell viability and the presence of bacterial films.	Material supports.	Microalgal biofilm system can better integrate production, harvesting, and dewatering operations, potentially leading to a more efficient process with reduced downstream processing costs.	Christenson and Sims, 2011; Olguin, 2012; Qureshi et al., 2005

Table 1. Harvesting, post-harvesting and cell disruption technologies for microalgal biomass (Continued)

	Technical Principles	Advantage/Disadvantage	Equipment required	Observations	References
POSTHARVESTING TECHNOLOGIES: Dehydration or drying of the biomass is necessary for certain thermochemical and biochemical processes to produce biofuels. It is the conversion of a solid, semi-solid or liquid feedstock into a solid product by evaporation of the liquid into a vapor phase via application of heat.					
Sun drying	The sun heats up the microalgal cake and also the surrounding air. This causes water evaporation from the solids.	Cheap / It is difficult to maintain the quality, slow drying rate, biomass degradation and bacterial contamination.	Pavement and partial mechanization.	Sun drying is a traditional method for reducing the moisture content by spreading the microalgae in the sun.	Prakash et al., 1997; Mata et al., 2010
Spray drying	A fine spray of solution droplets is brought into continuous contact with hot air in a large (1-10m diameter) chamber. Large droplet surface area and small droplet size ensure high evaporation rates.	Continuous operation; no further size reduction of the dry biomass is required and an aseptic rapid drying is achieved / Sensitivity of the biomass to oxidative processes imposes, for some applications, the addition of antioxidants. Thermo-labile components can be damaged or inactivated during the spray drying process.	Spray drier.	Spray-drying is the choice method for higher value products where the entire biomass is desired for rapid and continuous drying of biomass production.	Richmond, 2004; Murugesan and Orsat, 2012; Orset et al., 1999; Molina et al., 2003
Drum drying	High temperatures causes an instant vaporization of the protoplasmic water and the sudden increase in pressure cause a rupture in the cell walls and releases the desired products. The thermal shock can be achieved with a spray drier, direct cooking, or steam injection.	Rapid / Compared to spray drying, drum drying is a more intense heat treatment which results in more denatured proteins.	Drum drier.	In this process the algal biomass (usually algal slurry coming from a centrifugation step) is subjected to a thermal shock in a short time. It is used to dry heavy pastes.	Richmond, 2004; Rodriguez et al., 1996
Freeze-drying, or lyophilization.	The biomass first is frozen, then the frozen samples are subjected to low pressure of about 1 kPa and at temperatures of about -40°C at which the ice crystals sublimate	To maintain the biochemical composition of the biomass, and where a breakage of the cells ease the downstream process / Too expensive for large scale applications	Liofilizer.	It is the most gentle of the drying algal biomass methods.	Lee et al., 2012; Richmond, 2004; Wiltshire et al., 2000

	Technical Principles	Advantage/Disadvantage	Equipment required	Observations	References
CELL DISRUPTION TECHNOLOGIES: the aim is to increase the high-value metabolites release from the microalgae cells using mechanical and non-mechanical techniques.					
Bead mills	Bead mills are based on the rapid stirring of a thickened suspension of cells in the presence of beads. Cells are disrupted in the contact zones of the beads by compaction or shearing action and by energy transfer from beads to cells.	Suitable for industrial scale applications / The disruption efficiency depends on the size and composition of the beads.	Shaking vessels (disrupts cells by shaking the entire vessel). Agitated beads (rotating agitator in a fixed vessel equipped with cooling jackets).	The combination of agitation, collision and grinding of the beads produce an effective disruption process. The optimal bead size for microalgae cells is 0.5mm.	Middelberg, 1995; Gouveia, 2011; Lee et al., 2012; Kula et al., 1987; Richmond, 2004
Homogenization	The cell walls are broken by a combination of cavitation and mechanical forces such as turbulence or shear present in the homogenizer.	Successful to large scale / Degradation of the product, formation of cellular debris particles, potential risk of the destruction of large molecules.	Manton-Gaulin homogenizer (incorporates an adjustable valve with restricted orifice through which cells are forced at pressures up to 550 atm).	It is a common method for liberating intracellular products in fermentations using high pressure homogenizers.	Kelly and Muske, 2004; Clarke et al., 2010
Sonication	Compression/decompression cycles of the sonic waves generate both transient and stable cavitation (from an ultrasonic or hydrodynamic source). Microbubbles implosion releases a large amount of mechanical energy as shock waves.	The power required to create cavitation in the large volumes of culture medium may convert this method as energetically unfeasible for industrial applications.	Sonicator	The application of ultrasound, also known only as sonication, it is usually utilized as a cell disruption method in protein extraction from microalgae. It works by generating sonic waves, with frequencies around 25 kHz or a high flow velocity passing through a constriction.	Lee et al., 2012; Halim et al., 2012; Richmond, 2001
Microwave	The intracellular heating causes water evaporation and the vapor disrupts the cells from the inside.	Most simple and most effective for microalgae lipid extraction in comparison with other extraction methods.	Microwave ovens.		Middelberg, 1995; Kaufmann et al., 2001; Lee et al., 2010

Table 1. Harvesting, post-harvesting and cell disruption technologies for microalgal biomass (Continued)

	Technical Principles	Advantage/Disadvantage	Equipment required	Observations	References
Thermolysis	Thermolysis is a chemical decomposition caused by heat.	Lower disruption effectiveness than microwaves, the heat can only be diffused from the surroundings through the cell membrane to the interior of the cells, the extraction dependent on the type of strain and its growth phase.	Thermolysis reactor, heater, thermocouples.	It is used to disrupt the outer membrane and release periplasmic metabolites.	Middelberg, 1995; Lee et al., 2012; Campanella et al., 2012
Frozen in liquid nitrogen	The frozen-cells are fractured under the mortar because of their breakable nature. In addition, ice crystals at low temperatures may act as an abrasive.	Feasible to small scale only.	Mortar and pestle.	Microalgae can be frozen in liquid nitrogen and then pulverized.	Gouveia, 2011
Use of chemicals	Solvents increase the cell wall permeability allowing to the soluble metabolites to be released.	Chemicals can contaminate the products or destroy the activity of valuable cellular components; the efficiency depends on the cell wall composition.	Solid-Liquid extraction equipment like Soxhlet extractor.	This methodology uses different chemicals such as solvents, detergent or EDTA, acids or alkalis to increase the cell wall permeability.	Lee et al., 2012; Middelberg, 1995 Halim et al., 2012
Electroporation	It is a technique in which high electric field strengths are used to perforate the cell wall. Depending on the strength and length of the electric field this can cause reversible or irreversible electroporation.	Low cost of energy consumption; effective separation without harming the intracellular content.	A substance is placed between two electrodes, between which the pulsed electric field is applied.	Electroporation includes: Pulsed Electric Field (PEF), and Supersonic Flow Fluid Processing. Many companies are utilizing it in their systems for algae extraction (e.g., Origin Energy, OpenCEL, Diverse Technologies and Open algae).	Sheng et al., 2012; Vanthoor-Koopmans et al., 2013; de Boer et al., 2012

	Technical Principles	Advantage/Disadvantage	Equipment required	Observations	References
Supersonic Flow Fluid Processing	Supersonic steam is introduced into a special annular ''conditioning'' chamber that is wrapped around the core of the unit and injected through nanopore channels into the biomass, whereby a thermo fluid interaction with the cells is obtained for an optimal mass transfer. With increased steam flow, the steam exit velocity becomes supersonic and starts to form a controllable shock wave. This shockwave continues to grow and forms a low density, low temperature and low pressure, supersonic velocity zone across the bore diameter increasing energy transfer and cell disruption.	Using this technique a homogeneous disruption is obtained; Temperature does not exceed 35°C, therefore does not cause denaturation of valuable components.	The company PDX has patented this technique as a disinfection technique in food industry.	Not much information can be found on using this novel technique specifically to disrupt cells. However, as it causes electroporation it can be a promising technique for the biorefinery of microalgae.	Vanthoor-Koopmans et al., 2013

Table 2. Thermochemical processing

	Pyrolysis	Gasification	Hydrothermal Liquefaction
Feedstock conditions	Biomass should be free from water content	Moisture less than 40%, the biomass particle must carry less nitrogen and alkali components to reduce the impurities of products	Tolerable moisture content of biomass up to 65%
Operation temperature	300-750 °C	>700 °C	200-500 °C
Operation pressure	Atmospheric	Atmospheric	2-25 MPa
Pretreatment of feedstock	Drying microalgae biomass	Microalgae biomass does not need pretreatment	The feedstock of biomass is pre-treated with catalyst in surge bin to make microalgae slurry
Main product yield	Bio oil: 28.6-57.9% (fast pyrolysis) Pyrolysis gas: 13-25% (slow pyrolysis)	H_2: 5-56% CO: 9-52%	Bio oil
Co-products	Bio char: 10-25% (slow pyrolysis)	CH_4 and other HC products, CO_2, tar: up to 20%, ash	Gaseous product: methane (6–20%) with 21 MJ/kg heating value
Recyclable waste	Fluidizing gas can be recycled to pyrolysis reactor	The hydrocarbon can be synthesized for MeOH production (Yield: 64% of biomass)	Aqueous co product can be used as feed for microalgae culture

The thermochemical processes can be divided as pyrolysis, hydrothermal liquefaction, gasification and hydrogenation (Figure 2). The pyrolysis and hydrothermal liquefaction processes produce bio-oil fuel as the main product, whereas gasification produces syngas and hydrogenation (Table 2) and is a process to improve the biofuel or feedstock properties (Sualiand Sarbatly, 2012).

Bio-oil production is more effective from microalgae than lignocellulosic biomass, due to microalgae containing higher amounts of cellular lipid, polysaccharide and protein which are more easily processed and result in higher quality bio-oil production than bio-oil from wood (Demirbas, 2006; Huang et al., 2010). This technology can use the whole microalgae cells or biomass after lipid extraction for biodiesel production. However, the major problem is the high water content, which can lead to increased cost for the biomass drying or the generation of undesirable compounds by unwanted reactions.

Pyrolysis

It is carried out in the absence of oxygen followed by rapid quenching to produce char, gas and oils. During the pyrolysis process, approximately 10–25% of the biomass is converted into char and between 10–30% into a non-condensable gas (Suali and Sarbatly, 2012). Slow pyrolysis of biomass (350–700°C) is associated with a high charcoal content. Fast pyrolysis involved temperatures around 500°C at a heating rate over 100°C per minute (Sharar et al., 2012) and is associated with the production of liquid fuels, and/or gas at higher temperatures (Encinar et al., 1998).

Pyrolysis is extremely fast, which confers one major advantage over other conversion methods, with reaction times in the order of seconds to minutes. Although no economic evaluation of the process exists, studies only demonstrated the potential to recover energy and nutrients from microalgae residues after lipid extraction.

Gasification

This is achieved by reacting the material at high temperatures (>700°C), without combustion, with a controlled sub-stoichiometric amount of oxygen and/or steam. The resulting gas mixture is CO, CO_2 and H_2 called syngas (from synthesis gas or synthetic gas) (Klasson, 2012). Syngas can be used directly as a stationary biofuel or can be a chemical intermediate for the production of liquid fuel via Fischer–Tropsch synthesis (FTS) (FT-fuels, dimethyl ether, ethanol, isobutene) or chemicals (alcohols, organic acids, ammonia, methanol), which makes the process more flexible (Gouveia, 2011).

The gasification process is applicable for biomass with moisture content of less than 15%. However, moisture contents of up to 40% in microalgal biomass were reported to be tolerable for the gasification process (Ramzan et al., 2011) therefore it is possible to integrate the microalgae biomass feedstock into an existing thermochemical infrastructure (Clark and Deswarte, 2008).

Hydrogenation

Hydrogenation of a feedstock is the process of adding or reacting H_2 into the double bonds of hydrocarbons. The aim of this reaction is to obtain better feedstocks for other processes or products that have lower molecular weights. It can also be applied directly to convert biomass into bio-oil. In the hydrogenation process of biomass or unsaturated hydrocarbons, they usually react with hydrogen in a catalytic fixed-bed reactor, resulting in the production of steam, and later the hydrogenated feeds are separated into two components: light gasses, such as untreated hydrogen, methane and propane, and a liquid fraction (Suali and Sarbatly, 2012). Chin and Engel (1981) proved that microalgal biomass, specifically *Chlorella pyrenoidosa*, is capable of achieving 50% oil yield in a batch autoclave with a hydrogen pressure of 0.98–147 bar.

Hydrothermal Liquefaction

Due to the high fraction (>80%) of water in microalgae biomass, thermochemical processes like pyrolysis and gasification could not be economically interesting (Amin, 2009). Thermochemical processes for wet biomass, such as hydrothermal treatments appear to be more suitable for microalgae feedstock (López et al., 2013). It consists of bringing biomass into contact with liquid water at moderate thermal conversion temperatures such as 250-300°C and medium pressures of 2-25MPa with residence time up to 60 minutes to produce an oily liquid with a certain water content (Demirbas, 2010).

However, oils produced from hydrothermal liquefaction contained high oxygen, which decrease the caloric value and storage stability of oils. In order to overcome these problems, the use of organic solvents instead of water has been tested (Duan et al., 2013; Harun, 2010b).

A schematic representation of the hydrothermal liquefaction pathways in a biorefinery concept is shown in Figure 2, where either the whole algae or the residual fraction after extraction of value-added products (e.g., lipids, saccharides or protein) can be used. Table 3 shows main results of studies done with different microalgae submitted to hydrothermal liquefaction. de Boer et al., 2012 presented an analysis focused on energy consumption associated with thermochemical processing. They demonstrated that thermochemical conversion of the whole biomass including hydrothermal liquefaction, supercritical and catalyzed subcritical gasification are feasible.

Table 3. Hydrothermal liquefaction of microalgae cultures

Microalgae (% moisture content)	Lipids (%)	T (°C) and P (MPa)	Catalyst	Yield (%)	Hvv (MJ/Kg)	Reference
Botryococcus braunii (90-92 %)	>49	200-340° 2-10 MPa	Na_2CO_3 (0-5%)	64	42-50	Dote et al., 1994; Sawayama et al., 1995
Dunaliella tertiolecta (dried-78.4%)	2.9-20.5	126-340° 2-10MPa	Na_2CO_3 (0-5%) Ethyleneglycol acidified with H_2SO_4	25.8--97	28.42-37.8	Minowa et al., 1995; Sawayama et al., 1995; Shuping et al., 2009
Spirulina sp. (dried 7.8%)	5-12	300 -380° 5MPa H_2, N_2,CO	No catalyst Fe $(CO)_5S$ KOH, Na_2CO_3. CH_3COOH and HCOOH	42-78.3	32-36.8 Methanol 39.83 Ethanol 38.32 1-4dioxane 36.76	Matsui et al., 1997; Ross et al., 2010; Biller and Ross, 2011; Yuan et al., 2011
Mycrocystis viridis (95%)	Not reported	300 and 340° (3MPa with N_2)	Na_2CO_3 (0-5%)	33	27.6-39.5	Yang et al., 2004
Chlorella vulgaris (5.2-5.9%)	25	300 -350°	KOH, Na_2CO_3 CH_3COOH and HCOOH No catalyst Pt/Al_2O_3 Ni/Al_2O_3 Co/Mo/Al_2O_3	18-39	32-42	Ross et al., 2010; Biller and Ross, 2011; Biller et al., 2011
Nannochloropsis sp. (dried 78%)	28	200-500° (70-3500 kPa) with He or H_2	No catalyst Pd/C, Pt/C, Ry/C, Ni/SiO_2-Al_2O_3	16-57	37-40.1	Brown et al., 2010; Duan, 2010
Porphyridium cruentum (5.1%)	8	350°	HCOOH	20	35.7	Biller and Ross, 2011
Nannochloropsis occulata (7.2%)	17-32	150-350°	Na_2CO_3 Pt/Al_2O_3 Ni/Al_2O_3 Co/Mo/Al_2O_3	18-60	21-42	Biller and Ross, 2011; Biller et al., 2011; Du et al., 2012
Chlorogloeopsis fritschii	7	300°	No catalyst Biomass was pretreated with microwave	18-49.5	35-44.4	Biller et al., 2013
Pseudochoricystis ellipsoidea	67	300°	No catalyst Biomass was pretreated with microwave	18-49.5	35-44.4	Biller et al., 2013
Desmodesmus sp. (90.62%)	10-14	175-450°	No catalyst	10-46	30.1-36.9	García Alba, 2011

Microalgae (% moisture content)	Lipids (%)	T (°C) and P (MPa)	Catalyst	Yield (%)	Hvv (MJ/Kg)	Reference
Mycrocystis (9.59%)	5.22	300-700°	No catalyst	54.97	31.9	Hu et al., 2013
Chlorella sorokiniana		300°	The first step involves the subcritical water extraction of valuable algal polysaccharides at 160 °C.	20-27.8	40.8-43.7	Chakraborty et al., 2012
Spirulina platensis (4.54%)	13.3	Liquefaction 350° 2MPa N_2 Pyrolisis 350-500	No catalyst	23-41	29.3- 33.62	Jena and Das, 2011
Chlorella pyrenoidosa	0.1-19	200–300°, 2.8–9.0 MPa H_2 or N_2	Heterogeneous catalyst Ammonium ZSM-5 type zeolite catalyst HZSM-5	35-70	28.4-42.7	Zhang et al., 2013; Duan et al., 2013
Scenedesmus sp.	20	300°	No catalyst	Not specified	36	Campanella et al., 2012

Wood=21 KJ/kg, microalgae 29 KJ/kg, fossil oil=42 KJ/kg, Biodiesel from microalgae 41 KJ/kg, diesel fuel 40-45 KJ/kg (Amin, 2009).

These methods require large economies of scale, have high capital costs and produce a low volume of fuel relative to the lipid content in the algal biomass. Furthermore, the product of the process is typically of low value (methane, syngas or crude bio-oil) and requires upgrading before marketing.

Despite these drawbacks, the ability to convert the whole biomass to liquids could have possibilities for using this method to convert rapid-growing cultures or naturally occurring algal blooms that are low in lipid content to fuel.

7.2. Biochemical Processing

The most common types of biochemical processes are fermentation and anaerobic digestion.

Biomass Pretreatment

For biochemical processing there is a necessary pretreatment of the microalgal biomass, either whole or residual defatted biomass, to reduce the processing time and increase the biomass conversion.

Different technologies have been proposed to pretreat microalgal biomass, which includes thermal, alkaline, acidic, ionic liquid, ultrasound, hydrodynamic, pulsed electric field, biological pretreatment among others. One of the most common is the biomass conversion by hydrolysis. Hydrolysis uses acids, alkalis or enzymes to depolymerise polysaccharides and proteins into their component sugars or derivate chemicals (Sun and Cheng, 2002).

The acid hydrolysis is faster, easier and cheaper than other types of hydrolysis, nevertheless, it may lead to decomposition of the sugars into unwanted compounds that inhibit the fermentation process (Harun et al., 2010).

Enzymatic hydrolysis is slower and more expensive than acidic hydrolysis, but it is environmentally friendly, since the process does not produce inhibitory products. However, it requires physical or chemical pretreatments of the biomass to enhance the hydrolysis efficiency (Ho et al., 2012).

Certain variations in the traditional pretreatment steps for lignocellulosic material must be considered because microalgae residues do not have lignin polymers.

Fermentation

Bioethanol is usually obtained by alcoholic fermentation of starch (cereal grains, such as corn, wheat and sweet sorghum), sugar (sugar cane and sugar beet) and previously hydrolyzed lignocellulosic feedstocks (Antolin et al. 2002). Microalgal bioethanol can be produced through two distinct processes: via dark fermentation or yeast fermentation.

The dark fermentation of microalgae consists of the anaerobic production of bioethanol by the microalgae itself through the consumption of intracellular starch (Hu et al., 2008). The yeast fermentation process of microalgal biomass is well-known industrially and to achieve higher yields, it is necessary to screen microalgal strains with high carbohydrate content or induce accumulation of intracellular starch.

The polysaccharides on cell walls of microalgae are not easily fermentable for bioethanol production by microorganisms. For fermentation, an acid pre-treatment has been proposed as the best option compared to other pre-treatment methods, namely in terms of cost-effectiveness and low energy consumption (Harun and Danquah, 2011). During the bioethanol fermentation process the pH is maintained in the range of 6–9 during the fermentation, because a pH below 6 or over 9 could slow down the bioethanol production. The fermentation process consumes less energy and the process is much simpler in comparison to the biodiesel production system. In addition, the CO_2 produced as a by-product from the fermentation process can be recycled as carbon sources for microalgae cultivation, thus reducing greenhouse gas emissions as well.

Anaerobic Digestion

Anaerobic digestion is the bacterial decomposition of organic biopolymers (i.e., carbohydrates, lipids and proteins) into monomers in the absence of oxygen over a temperature range from about 30 to 65°C. These monomers are easier to convert into a methane-rich gas via fermentation (typically 50–75% CH_4) and CO_2 is the second main component found in biogas (approximately 25–50%). Biogas can be upgraded up to >97% methane content and used as a substitute for natural gas (Romano and Zhang, 2008) for the generation of electricity.

The biogas production from the anaerobic digestion of microalgae biomass is primarily affected by its organic loadings, temperatures, pH and retention time of the reactor used.

Screening several species of microalgae demonstrated that the quantity of biogas potential is strongly dependent on the species and on biomass pretreatment (Mussgnug et al.,

2010). Sialve et al., (2009) reported that the methane content of the biogas from microalgae is 7 to 13% higher compared with the biogas from maize.

Biomethane from microalgae biomass can be used as fuel gas and to generate electricity while the spent biomass can be used to make biofertilizers or in a thermochemical approach a wide range of biofuels and chemicals. Although microalgae offers good potential for biogas production, commercial production has still not been fully implemented.

Biohydrogen Production

Whole microalgae biomass after oil or starch removal can be converted into biohydrogen (bio-H_2) by dark-fermentation, which is one of the major bioprocesses using anaerobic organisms for bio-H_2 production. Microorganism such *Enterobacter* and *Clostridium* are well known as good producers of bio-H_2 and are also capable of consuming several carbon sources. Bio-H_2 has 2.75 times more energy density than any other existing biofuels (Rashid et al., 2012).

Yang et al., (2011) demonstrated the potential of *Scenedesmus* biomass derived from oil extraction processes as feedstock for biohydrogen production using a heat-treated aerobic digested sludge as inoculum; the biohydrogen production was sensitive to initial pH, and the optimal initial pH range was found as 6.0–6.5. An excellent biohydrogen producing performance (bio-H_2 production rate of $2.82 mLh^{-1}$ and bio-H_2yield of $30.03 mLgVS^{-1}$) was obtained when the culture was operated at the initial pH 6.0–6.5. Their findings provide fundamental knowledge for the design of a high efficiency biohydrogen bioreactor, and for practical application of continuous biohydrogen production systems in the future.

A more in-depth discussion of biochemical transformation for bioethanol, biomethane and biohydrogen can be found in the recent review article (Lakaniemi et al., 2013).

7.3. Metabolites Extraction

From all the above mentioned, it could be seen that biorefineries might adopt and integrate a range of materials handling and preprocessing equipment, thermochemical and biochemical conversion technologies. However, there is another approach, the product-driven biorefineries, where biomass is typically fractionated into a portfolio of bio-based products with the focus being to derive the highest economic value from the biomass. A cascade approach is often used where residues are used for power and/or heat production. It is based on the fact that microalgae contain high amounts of lipids, proteins and carbohydrates, which all can be used for different markets (food, feed, chemical and the pharmaceutical sector as well).

In a feasibility analysis presented by Wijffels et al., (2010) for microalgal biodiesel they valorized all the products of the algal biomass considering their functionality and bulk markets values. They assumed that lipid fraction could not only be used for production of biodiesel but also as a feedstock for the chemical industry or for edible oils (valorized in €2/kg). In this case, they assumed that 25% of the lipids are used for functional products and 75% are used for biodiesel production; therefore the lipid fraction of the algae globally increases in value. For protein analysis, they assumed that soluble fraction (20%) could be used for food (valorized in €5/kg) and insoluble fraction (80%) for feed (valorized in €0.75/kg) production, and finally a fraction of 10% of carbohydrates (valorized in €1/kg) was considered. Considering all the above-mentioned, they estimated that the total value is higher (€1.65/kg) than the total cost for algae production (€0.40/kg) and concluded that the integrated biorefinery concept of microalgae with biodiesel is one of the products can lead to a feasible process for production of only biodiesel from microalgae.

The main bottleneck of this approach is the separation of the different fractions without damaging one or more of the product fractions. Conventional cell disruption, extraction and separation techniques, such as bead milling, homogenizers, high pressure, heating, osmotic shock and chemicals, are focused on obtaining one product while damaging the other fractions. Technologies to overcome these bottlenecks need to be developed, and they should be applicable for a variety of end products of sufficient quality at large quantities.

From the cell disruption techniques previously shown in Table 1 pulsed electric field, supersonic flow fluid processing, ultrasound and enzyme are suitable as mild cell disruption techniques that do not harm the intracellular content and are effective for cell inactivation (Vanthoor-Koopmans et al., 2013).

There are some promising technologies under research for microalgal metabolites extraction such as ionic liquids and surfactants to separate all fractions without losing any product.

Ionic liquids are made of cations and anions as they are simple molten salts. The advantage and novelty of ionic liquids are their low melting temperature, generally below 100°C. Ionic liquids can be an alternative to classical organic solvents by their ability to extract hydrophilic from hydrophobic compounds. They are easily recyclable and therefore produce minimum pollution compared to organic solvents.

Vanthoor-Koopmans et al., (2013) proposed a rupture of the cell wall in a mild way so the lipids and proteins will be available for extraction. Ionic liquids can then be applied as an additive to the aqueous phase before separating the biomass from the proteins and lipids. Adding the ionic liquid in this stage makes that both, hydrophobic and hydrophilic compounds are soluble. After this, separation of the remaining biomass can take place by micro or ultrafiltration, or centrifugation. Other separation techniques such as gas chromatography, liquid chromatography, and capillary electrophoresis have been successfully used in combination with ionic liquids. However, further research and development will be necessary to make ionic liquids an applicable tool in separating fractions selectively.

On the other hand, surfactants have been used to separate specific proteins. Recently, anionic surfactants have been used to bind specific proteins (lysozyme) in order to make them insoluble in aqueous solutions, and therefore the protein–surfactant bond precipitated without losing protein activity. Surfactant precipitation reduces the amount of surfactant required in comparison with reverse micellar extraction. Moreover, less energy is required to recover the precipitated proteins in comparison with solubilized proteins.

A microalgal biorefinery scheme was proposed by Nobre et al., (2013) and Ferreira et al., (2013) based on laboratory scale results. The microalga *Nannochloropsis* sp. was used as biomass feedstock for the production of fatty acids for biodiesel, biohydrogen and high added-value compounds such as carotenoids. In order to separate lipids from carotenoids, they assayed the supercritical fluid extraction (SFE) using ethanol as a co-solvent. It was possible to extract the lipids and recover 70% of the pigment. Anaerobic dark fermentation was carried out with the remaining biomass by *Enterobacter aerogenes* and the resulting biohydrogen production was 60.6 mL /g dry biomass. An economic evaluation and life cycle analysis were also performed considering two extraction methods, SFE and soxhlet; the biorefinery for oil, pigment and bio-H_2 production via SFE was the best energy/CO_2/economy compromise (Ferreira et al., 2013).

Another biorefinery scheme was proposed by García et al., (2013) for valorization of microalgae biomass. They specified that the order of treatments is fundamental to maximize the recovery of products. They suggest that first step is the protein recovery through enzymatic hydrolysis, after carbohydrates, through the hydrolysis and fermentation to obtain bioethanol, direct transesterification of lipids and anaerobic digestion of residual biomass. However, this study was carried out at lab scale and no economic analysis was performed.

Vanthoor-Koopmans et al., (2013) proposed a conceptual biorefinery to obtain all available and useful products produced by algae. In the flow diagram there is no concentration step previous to cell disruption, this might be possible using a continuous flow cell disruption technique that will decrease energy consumption significantly, although larger process flows will need to be processed. The extraction method was the use of ionic liquids. It must be said that in most parts of the total proposed scheme, research is still necessary to find optimal process conditions for efficient cell disruption and separation and to find the best techniques and process flows necessary to make biorefinery for microalgae successful.

8. Integration of Process

The recent need for valorization of microalgal biomass has re-opened interest in combining CO_2 removal systems with wastewater treatment and production of metabolites or biofuels; all in order to obtain an economically attractive and environmentally sustainable process.

Nevertheless, the major challenge in achieving the double purpose of microalgae is to determine a way for downstream processing that is suitable for producing biofuel and other bioproducts (Christenson and Sims, 2011). Olguín (2012) affirmed that the economic viability of the dual-purpose systems within a biorefinery depends on the efficiency of the microalgae-bacteria systems at removing nutrients.

Some algae species, such as *Chlorella*, *Scenedesmus*, *Phormidium*, *Botryococcus*, *Chlamydomonas*and *Spirulina*, have been widely applied for wastewater treatment and had proven abilities for removing nitrogen, phosphorus, and reducing chemical oxygen demand (COD).

While the initial purpose for introducing the algae pond process was to further treat the secondary effluent in order to prevent eutrophication (Olguín 2003), it was observed that the treatment removed organic nutrients from settled domestic sewage more efficiently than activated sewage process (Wang et al., 2009; Kong et al., 2010), suggesting that it is possible to employ the algal system as the secondary rather than the tertiary treatment process.

The organic substances may function directly as an essential organic nutrient through heterotrophic growth by directly incorporating organic substrate in the oxidative assimilation process for storage material production reducing COD. But it must be considered that the specie selection depends on various factors like: wastewater characteristics, the climatic conditions in the treatment plant and the original habitat of the algal strain, among others (Olguín, 2012).

Christenson and Sims, (2011) affirmed that the major challenge is to implement an integrated system, which includes the large-scale production of algae and the harvesting of microalgae in a way that allows the production of biofuels and other bioproducts of value.

Table 4. Biorefinery technological status

Country/ company	Classification	Feedstocks	Products	Capacity
Commercial scale biofuel driven biorefineries: these biorefineries are state of the art and are worldwide in commercial operation under current economic conditions.				
Crop Energies AG (Germany)	C6 sugar biorefinery for bioethanol and animal feed from sugar and starch crops	Cereals, sugar	Bioethanol, feed (DDGS "ProtiGrain"), electricity	In Europe's largest bioethanol plant in Zeitz (capacity 2008: 360,000 m^3)
Permolex (Alberta, Canada)	C6 sugar biorefinery for bioethanol, animal feed and food from starch crops	Wheat	Bioethanol, animal feed, food (bakery flour, gluten)	
Sofiproteol (France)	Bio-oil biorefinery for biodiesel, glycerine, animal feed, chemicals and polymers from oil crops	Rape, sunflower oilseed	Biodiesel, glycerine, chemicals and polymers (coatings, biolubricants, biopolymers), animal feed	4.1 Mt a year of oilseeds to: 2.5 Mt of biodiesel, 250 000 tons of glycerine and 2.3 Mt of oilseed meal used as a source of protein in animal feed
Ensyn (Ontario, Canada)	Pyrolytic liquid biorefinery for food flavorings, polymers, fuels and heat from lignocellulosic residue	Residual woody biomass from a hardwood flooring plant and sawmill	Food (flavorings), polymers (resins), heat, electricity plus (in the future) synthetic biofuels (upgraded bio-oil)	Ensyn's commercial facility in Renfrew processes 100 dry ton/d of wood residue
Zellstoff Stendal GmbH (Germany)	Lignin biorefinery for biomaterials, electricity and heat from lignocellulosic crops or residues	Softwood, small logs, sawmill chips	Biomaterials (paper products), electricity and heat	
Lenzing AG (Austria)	C5 sugars and lignin biorefinery for biomaterials, chemicals, food, electricity and heat from lignocellulosic crops or residues	Beech wood	Food (food grade acetic acid, xylose-based artificial sweetener), chemicals (furfural), biomaterials (cellulose fibers), electricity and Heat	
Highmark Renewables (Canada)	C6 sugars and biogas biorefinery for bioethanol, animal feed, fertilizer, electricity and heat from starch crops and organic residues	Wheat, manure, slaughtering waste	Bioethanol, animal feed, fertilizer, electricity and heat	BioRefinery™ at full scale, should generate 40 million L of bioethanol, 10,000 tons of biofertilizer, and over 75,000 tons of greenhouse gas emissions credits each year

Country/ company	Classification	Feedstocks	Products	Capacity
Roquette Frères (France)	C6 sugar biorefinery for starch, chemicals and animal feed from starch-rich crops	Maize, wheat, potatoes, peas and microalgae	Food, feed, paper-board, pharmacy-cosmetology and biochemistry industries	
EcoMotion (Germany)	1-platform (oil) biorefinery using oilseed crops or oil based residues for biodiesel, glycerin and feed	Vegetable oils Rapeseed 170,000 ton/y Plant oil 40,000 ton/y	Biodiesel, rapeseed cake	Bioethanol production capacity between 100,000 ton /y 1000,000 ton/y rapeseed cake for animal feed, glycerin 10,000 ton /y, plant oil 40,000 ton/y
AGRANA (Austria)	C6 sugars and starch crops for bioethanol and feed	Wheat, corn, triticale, barley and syrup made from sugar beet		190,000 ton/y protein-rich feed 100,000 ton/y
Abengoa (Hugoton, KS, USA)	Agricultural residues would be converted via enzymatic hydrolysis to sugars and fermented into cellulosic bioethanol	Agriculture residues	Cellulosic bioethanol, green power	1,200 ton/d
Bluefire Renewables (Fulton MS, USA)	The feedstock is mixed with sulfuric acid producing liquor. The lignin-cake is used to generate steam for the process. The acid is recovered and reused in the process. The sugar stream is used to produce gypsum and converted into bioethanol through fermentation by non-GMO (genetically modified organism) yeasts. The bioethanol is purified and the yeast residues are sold as animal feed	Un-merchantable lumber, logging residues or chips.	Bioethanol, animal feed, internal heat/power, gypsum	More than 18 million gal/y of denatured bioethanol
POET/DSM Advanced Biofuels, LLC (Emmetsburg, IA)	Integration of an innovative lignocellulose-to-ethanol biochemical process into an existing dry-grind corn processing infrastructure on a commercial scale	Primarily Corn Stover	Lignocellulosic ethanol and renewable Heat	20 Mgal /y of lignocellulosic ethanol
Mascoma (Kinross, MI, USA)	The pretreated hardwood pulpwood is submitted to an enzymatic hydrolysis to sugars and simultaneous fermentation to bioethanol in one step. Lignin residues would be recovered dewatered, and to generate both steam and electricity	Hardwood pulpwood	Bioethanol, lignin and bark used to produce heat and electricity	20 M gal/y of denatured bioethanol (with long- term potential for doubling to 40 M gal/ y)

Table 4. Biorefinery technological status (Continued)

Demonstration biofuel-driven biorefineries: in these biorefineries their main processes are demonstrated on a technical scale at one or more locations worldwide, but they need further technical optimization and cannot be operated under current commercial conditions				
Country/ company	**Classification**	**Feedstocks**	**Products**	**Capacity**
Inbicon IBUS (Denmark)	C6/C5 sugars and lignin biorefinery for bioethanol, animal feed, electricity and heat from lignocellulosic residues	Straw	Bioethanol, animal feed (molasses with hemicellulose sugars), lignin, electricity and heat	
Lignol (Canada)	C6/C5 sugars and lignin biorefinery for bioethanol, chemicals and biomaterials from lignocellulosic crops or residues	Softwood, hardwood, annual fibers, agriculture-residuals	Bioethanol, chemicals (furfural) and biomaterials (HP-L™ High Purity Lignin)	
CHOREN (Germany)	Syngas biorefinery for synthetic biofuels (or electricity and heat) from lignocellulosic residues	Woody biomass	Synthetic biofuels (BTL diesel, BTL naphtha) or electricity and heat	Annual capacity of the plant is 18 M L of BTL generated from 65,000 tons of wood
European Bio-Hub Rotterdam (the Netherlands)	Syngas platform biorefinery for fuels and chemicals from imported lignocellulosic biomass. C5/C6 sugar, lignin and protein platform biorefinery for feed, chemicals and fuels from biofuel process residues	Lignocellulosic biomass (densified raw biomass, intermediates, agro-residues)	Base-chemicals, biofuels, Combined Heat and Power (CHP)	
SEKAB-E technology (SWEDEN)	Wood chips are hydrolyzed and C5 and C6 are fermented for bioethanol, the lignin is used to produce bio-oil via pyrolysis. The phenols from bio oil are separated and the residues are combusted to produce heat and electricity.	Wood	Bioethanol, phenols	
Utzelnach (Austria)	Biorefinery using grass and manure for biomethane, amino acids, lactic acid, biomaterials and fertilizer, electricity and heat	Grass and manure	Biomethane, amino acids, lactic acid	
BioSNG (Austria)	Biorefinery uses the gasification of wood chips to produce synthesis gas; the biomethane is produced via methanization of the natural gas via steam reforming the syngas is used to produce hydrogen. CO_2 produced is separated and used for various industrial applications. Residues and gaseous streams are used to produce CHP	Wood chips	Biomethane, carbon dioxide and biohydrogen	

Country/ company	Classification	Feedstocks	Products	Capacity
Chemrec (Sweden)	The straw is used to produce pyrolysis oil and char afterwards, they are mixed and treated in a gasification plant to produce syngas, which is converted to FT-biofuels and methanol and process residues are used to produce CHP.	Straw	Bioliq®, bioliqSyncrude ® and methanol	
Chemrec (Sweden)	Uses woodchips for FT-biofuels, CHP and waxes with steam gasification. The wood chips are gasified with steam to produce a gas, which is used to produce raw FT biofuels via catalytic reaction. The final quality of the transportation FT biofuels is reached in the upgrading step with hydroprocessing. The process residues are combusted to produce electricity and heat	Wood chips	Waxes, FT-biofuels, electricity and heat	
Verenium (Jennings, LA, USA)	Feedstock is pretreated and subjected to proprietary enzymatic hydrolysis systems and the subsequent hexose and pentose sugars are fermented separately with proprietary ethanologens to produce cellulosic ethanol	Sugar cane bagasse, energy cane, sorghum	Cellulosic ethanol	1.4 M gal of bioethanol
Enerkem (Pontotoc, MS, USA)	Conversion of wastes through gasification to produce syngas and Catalytic synthesis for the production of biofuels and chemicals	Municipal solid waste and wood residues	Ethanol and methanol	10 M gal/ y
NEOS Bio/New Planet Bioenergy (Vero Beach, FL, USA)	Utilizing a unique combination of gasification and fermentation processes of vegetative, yard, and municipal solid waste as feedstock, which are heated to produce a synthesis gas and fed to naturally occurring bacteria that produce bioethanol	Vegetative & Yard waste, municipal solid waste	Cellulosic ethanol, renewable power	8 M gal/Y and 6MW (gross) of electricity generation
Myriant (Lake Providence, LA, USA)	Myriant's technology is based on a proprietary platform that involves modified (non-genetically modified organisms) *Escherichia coli* strains to produce bio-succinic acid	Multi-feedstock capability	Bio-succinic acid	30 million pounds per year of bio-succinic acid
Red Shield Acquisition, LLC (RSA) (Old Town, ME, USA)	The process creates clean lignocellulosic sugars; they will then become feedstock in a sugar conversion process to biofuels, biochemicals, and/or bioplastics.	Woody biomass	Lignocellulosic sugars, algal oil	28,000 tons/y of sugars; 555,000 gal/y of green oil

Table 4. Biorefinery technological status (Continued)

Country/ company	Classification	Feedstocks	Products	Capacity
Sapphire Energy, Inc. (Columbus, NM, USA)	The technology involves a process where CO_2, brackish water, and nutrients are used to cultivate and harvest oil-rich algae. Green crude oil is then extracted and refined into liquid transportation fuel. The residual solid biomass provides nutrients and water from the harvest process that is then recycled back into production ponds	Carbon dioxide, algae, sunlight	Jet fuel and diesel fuel	1 M gal/y of finished product
Pilot scale: Promising technologies are screened and validated through pilot-scale projects, which typically process at least one dry metric ton of feedstock per day.				
Algenol Biofuels, Inc (Fort Myers, FL, USA)	DIRECT TO ETHANOL® technology is based on over expressing the genes in blue-green algae for certain enzymes found widely in nature. The algae utilize CO_2 to make bioethanol that diffuses through the cell wall into the culture medium and then evaporates into the headspace of a patented photobioreactor. The ethanol-water vapor condenses on the inner surface of the photobioreactor and is collected as a liquid and then further concentrated into fuel ethanol	Carbon dioxide, algae, seawater	Biothanol	Greater than 100,000 gal/y of bioethanol
American Process, Inc. (API) (Alpena, MI, USA)	Woody biomass is concentrated, hydrolyzed, and simultaneous fermented of five- and six-carbon sugars to produce bioethanol. Sugar solution also reacts and is separated by reverse osmosis to produce potassium acetate	Mixed northern hardwood and aspen	Bioethanol, potassium acetate	894,200 gal/y of cellulosic ethanol and 696,000 gal/ y of aqueous potassium acetate
Amyris, Inc. (Emeryville, CA, USA)	It uses industrially proven yeast-based fermentation of traditional or lignocellulosic-derived sugar feedstocks. The fermentation intermediate is readily recovered as a water-immiscible long-chain liquid hydrocarbon (trans-ß-farnesene), then anaerobic digestion is used to reduce effluent and utilize residual sugars for biogas production. Biogas is then converted to hydrogen via reformation.	Sweet sorghum	Renewable diesel produced from farnesene, a renewable hydrocarbon precursor to fuels and specialty chemical products	

Country/ company	Classification	Feedstocks	Products	Capacity
Archer Daniels Midland (ADAM) (Decatur, IL, USA)	ADM developed a process to pretreat and pelletize corn stover. Hydrolysis converts cellulose and hemicellulose fractions into sugars, while lignin is utilized as an energy source. Some of the sugars are fermented to bioethanol, and the remaining sugars are hydrogenated to polyols, which are submitted to catalytic conversion to produce butyl acrylate chemical, that is used to make plastics, adhesives, paints, coatings, and a range of other materials.	Corn Stover	Ethanol Fuel, Butyl Acrylate, and Process Heat	25,800 gal of bioethanol and 21,000 lb/y of butyl acrylate
Haldor Topsoe, Inc. (Des Plaines, I, USA)	Haldor Topsoe, Inc. will demonstrate a new, economical thermochemical process for the conversion of wood waste and woody biomass to gasoline.	Wood pellets	Renewable gasoline	345,000 gal/ y (approximate)
ICM, Inc. (St. Joseph, MO, USA)	The process uses a biochemical platform pretreatment and enzymatic hydrolysis technology coupled with a robust C5/C6 co-fermenting organism to refine cellulosic biomass into fuel ethanol and co-products.	Corn Fiber, switchgrass, and energy sorghum	Cellulosic ethanol	245,000 gal/y fuel or product
Logos/Edeniq Technologies (Visalia, CA, USA)	The project utilizes a suite of Edeniq, Inc.'s proprietary technologies: the Cellunator™ (mechanical pretreatment), advanced enzymes for conversion of cellulose to sugars, and high-yielding yeasts to ferment the sugars to bioethanol.	Various cellulosic feedstocks: corn stover, switchgrass, wood chips, etc.	Cellulosic ethanol	50,000 gal/y
Renewable Energy Institute International (Toledo, OH, USA)	The production of high-quality, synthetic diesel fuels from agriculture and forest residues uses advanced thermochemical and catalytic conversion technologies.	Agriculture and forest residues	Synthetic diesel fuel	54 gal of drop-in synthetic diesel fuel from one dry ton of biomass processing 25 dry ton/d
Rentech ClearFuels (Commerce City, CO, USA)	Woodwaste and bagasse, are transformed into renewable diesel and jet fuel using a single facility design for producing a wide range of low tar syngas combined with FischerTropsch conversion and upgrading to diesel and jet fuel.	Woodwaste, Bagasse	Renewable F-T Diesel and F-T Jet Fuel	420 gal/d

Table 4. Biorefinery technological status (Continued)

Country/ company	Classification	Feedstocks	Products	Capacity
UOP, LLC (Kapolei, HI, USA)	The feeds will be converted to fuels via integrated pyrolysis and hydroconversion	Agricultural and forestry residue, wood, energy crops, and algae	Gasoline, diesel, and jet fuels	Four barrels per day
ZeaChem, Inc. (Boardman, OR, USA)	The technology uses chemical fractionation to separate the feedstock into a sugar-rich stream is biochemically converted into acetic acid. The acetic acid is processed into an acetate ester, an intermediate bio-based chemical that can be marketed and sold or converted into ethanol via hydrogenolysis	Hybrid poplar and other cellulosic feedstocks	Ethanol and intermediate chemicals	250,000 gal/y
Solazyme, Inc. (Peoria, IL, USA)	Algae grow efficiently in the dark in industrial fermentation vessels to very high cell densities. They ingest and metabolize carbon substrates provided in the growth media and convert them to triglycerides	Sucrose (from cane), municipal green waste, switchgrass	Biodiesel and Renewable Diesel from Purified Algal Oil	300 kG/Y Purified Algal Oil
Conceptual biofuel-driven refineries: these biorefineries have not been demonstrated on a technical scale so far, but it is expected that they will be further technically developed.				
Avantium Furanics (The Netherlands)	C6/C5 sugars and lignin biorefinery for synthetic biofuels, chemicals and polymers, and electricity and heat from starch crops or lignocellulosic crops or residues	Cellulose, hemi-cellulose, starch, sucrose	Synthetic biofuels, chemicals and polymers (furan dicarboxylic acid, furan diamine organic acids, solvents, flavors & fragrances), electricity and heat	
Green Biorefinery (Austria)	Biogas and organic solution biorefinery for organic acids, fertilizer, biomaterials, biomethane and electricity and heat from grasses	Mixtures of grass, clover, lucerne silage	Organic acids (lactic and amino acids), biomaterials (fiber products), fertilizer, biomethane or electricity and heat	
Country/ company	Classification	Feedstocks	Products	Capacity
M-real, Hallein AG (Austria)	C6 sugars biorefinery for bioethanol, biomaterials, electricity and heat from lignocellulosic crops or residues	Spruce wood	Bioethanol, biomaterials (paper products), electricity and heat	

Country/ company	Classification	Feedstocks	Products	Capacity
Borreggard (Norway)	The wood chips and saw mill residues are pretreated for the hydrolysis to separate the sugars and the lignin. The C5 and C6 sugars are fermented to bioethanol. The produced CO_2 can be separated for the food industry. The lignin is combusted to produce electricity and heat. For bioethanol production also C6 sugars in the sulphite liquor from pulp production can be used	Wood and liquor	Bioethanol, pulp and paper, electricity and heat	
Abengoa (Spain)	Starch is transported to the biorefinery, where the straw is pretreated for the hydrolysis to separate the sugars and the lignin. The C6 sugars from starch crops and the C5 and C6 sugars from the straw are fermented to bioethanol, which is purified using distillation. The fermentation solids, mainly proteins are dried and pelleted for animal feed. CO_2 from fermentations is separated and used for food industry or as industrial gas. The lignin is gasified and the syngas is used to produce FT-biofuels. The residues are combusted to produce electricity and heat	Starch crops and straw	Bioethanol. Feed. FT biofuels, electricity and heat	
WUR Micro-algae Biorefinery (The Netherlands)	Oil and organic solution biorefinery for base-chemicals, biodiesel, power and/or heat from micro-algae	Microalgae	Base-chemicals, biofuels, (CH) power	

Growing microalgae in municipal wastewater for biofuel production such as anaerobic digestion to biogas, transesterification of lipids to biodiesel, fermentation of carbohydrates to bioethanol and high temperature conversion to bio-oil, represents a minimum environmental impact.

Olguín (2012) recently published a review detailing information about several aspects of the treatment of municipal and animal wastewater with microalgae. Also, several studies have already focused on utilizing streams in wastewater treatment plant for dual purpose microalgae cultivation (Cabanelas et al., 2013; Chiu and Wu, 2013; Cho et al., 2013 Dickinson et al., 2013; McGinn et al., 2012; Rawat et al., 2011; Sivakumar et al., 2012; Sturm et al., 2011).

9. Current Status on Microalgal Biorefinery

There are academic, governmental and enterprise efforts to develop, demonstrate, and deploy technologies for advanced biofuels production from different biomass feedstock.

Many projects to produce chemicals, material and fuels from biomass were launched in the 1970's and 1980's after the huge increase of petroleum processes, however very few of these projects survived the mid-80's when petroleum prices decreased to just a few US dollars per barrel. New technologies are being developed that use biomass to make not only low value products such as fuels but also high value materials. The main goal of IEA Bioenergy-Task 42 *Biorefineries: Co-production of Fuels, Chemicals, Power and Materials from Biomass* is to initiate and actively promote information exchange on all aspects of the energy-driven biorefinery concept. It commenced in 2007 with 11 countries Australia, Austria, Canada, Denmark, France, Germany, Ireland, Italy, the Netherlands, Turkey and USA; and according to the IEA Bioenergy Annual Report 2012, the budget during that year for a Biorefinery was 165,000 USD from a total of 1,847,740 USD for Bioenergy projects.

Table 4 shows biorefineries classified in commercial, demonstration, pilot and conceptual scales using information from the IEA Bioenergy-Task 42 Biorefinery (Jungmeler et al., 2013: Schavan and Aigner, 2012) and the U.S. Department of Energy (DOE 2013). Most of the commercial scale biorefineries produce bioethanol from wood or sugar processing; or biodiesel from vegetable oils using biochemical processing. Microalgae biorefineries, used to produce biodiesel from microalgae lipids are classified at a conceptual level in Europe. Wageningen University and research Centre (Wageningen Ur) at the Netherlands leads the project. Sapphire Energy is working at a demonstration scale and proclaims a production of 1 million of gallon per year of Jet and diesel fuel. Two companies are working at pilot scale in the USA. Algenol Biofuels produces 100,000 gallons of ethanol per year using genetically-modified cyanobacteria to transform carbon dioxide into ethanol and Solazyme Inc. uses sucrose as carbon source to heterothrophically produce lipids; the biodiesel production is 300,000 gallons per year.

Conclusion

In the future, microalgae biomass might contribute to the production of transport fuels in a significant volume. Nevertheless, if microalgae biomass is used only to produce energy, the cost of derived fuels is not competitive with that of fossil fuels. The application of the biorefinery concept for production of additional chemicals and fuels from microalgae biomass is a key factor to positively balance the economics and offers environmental mitigation associated with carbon sequestration, waste minimization and material recycling.

However, there are several research and technological challenges associated with the development of the integrated production chain necessary for a microalgae-based biorefinery, including algal strain screening and growth optimization, extraction methods and scale-up feasibility, as well as metabolic engineering, which could enhance the production of target fuels and chemicals.

REFERENCES

Ahmad, A. L.; Mat-Yasin, N. H; Derek, C. J. C.; Lim, J. K. Optimization of microalgae coagulation process using chitosan. *Chem. Eng. J.,* 2011, 173, 879-882.

Amin, S. Review on biofuel oil and gas production processes from microalgae. *Energ. Convers. Manage.*, 2009, 50 (7), 1834-1840.

Antolin, G.; Tinaut, F. V.; Briceno, Y.; Castano, V.; Perez, C.; Ramirez, A. I. Optimization of biodiesel production by sunflower oil transesterification. *Bioresour. Technol.*, 2002, 83, 111-114.

Araujo, G.; Matos, L. J. B. L.; Fernandes, J. O.; Cartaxo, S. J. M.; Gonçalves, L. R. B.; Fernandes, F. A. N.; Fariasa, W. R. L. Extraction of lipids from microalgae by ultrasound application: Prospection of the optimal extraction method. *Ultrason. Sonochem.*, 2013, 20 (1), 95-98.

Bhave, R.; Kuritz, T.; Powell, L.; Adcock, D. Membrane-based energy efficient dewatering of microalgae in biofuels production and recovery of value added co-products. *Environ. Sci. Technol.*, 2012, 46, 5599-5606.

Biller P., Ross A. B. Potential yields and properties of oil from the hydrothermal liquefaction of microalgae with different biochemical content. *Bioresour. Technol.*, 2011, 102, 215-225.

Biller, P.; Riley, R.; Ross, A. B. Catalytic hydrothermal processing of microalgae decomposition and upgrading of lipids. *Bioresour. Technol.,* 2011,102, 4841-4848.

Biller P., Friedman C., Ross A. B. Hydrothermal microwave processing of microalgae as a pre-treatment and extraction technique for bio-fuels and bio-products. *Bioresour. Technol.*, 2013, 136, 188-195.

Brennan, L.; Owende, P. Biofuels from microalgae – A review of technologies for production, processing, and extractions of biofuels and co-products. *Renew. Sust. Energ. Rev.*, 2010, 14, 557-577.

Brown, T. M.; Duan, P.; Savage P. E. Hydrothermal Liquefaction and Gasification of *Nannochloropsis* sp. *Energ. Fuel.,* 2010, 24, 3639-3646.

Cabanelas, I. T. D.; Ruiz, J.; Arbib, Z.; Alexandre, C. F.; Garrido-Pérez, C.; Rogalla, F.; Andrade N. I.; Perales, J. A. Comparing the use of different domestic wastewaters for coupling microalgal production and nutrient removal. *Bioresour. Technol.*, 2013, 13, 429-436.

Campanella, A.; Muncrief, R.; Harold, M. P.; Griffith, D. C.; Whitton, N. M.; Weber, R. S. Thermolysis of microalgae and duckweed in a CO_2-swept fixed-bed reactor: bio-oil yield and compositional effects. *Bioresour. Technol.,* 2012, 109, 154-162.

Clark, J. H.; Deswarte E. I. 2008. The Biorefinery concept - An integrated concept. Introduction to Chemicals from Biomass (eds J. H. Clark and F. E. I. Deswarte), John Wiley & Sons, Ltd., Chichester, UK.

Clarke, A.; Prescott, T.; Khan, A.; Olabi, A. G. Causes of breakage and disruption in a homogenizer. *Appl. Energ.*, 2010, 87 (12), 3680-3690.

Chakraborty, M.: Miao, Ch.; McDonald, A.; Chen, Sh. Concomitant extraction of bio-oil and value added polysaccharides from *Chlorella sorokiniana* using a unique sequential hydrothermal extraction technology. *Fuel*, 2012, 95, 63-70.

Chen, Y. M.; Liu, J. C.; Ju, Y. H.; Flotation removal of algae from water. *Colloid. Surface. B.*, 1998, 12, 49-55.

Chen, C. Y.; Yeh, K. L.; Aisyah, R.; Lee, D. J.; Chang, J. S. Cultivation, photobioreactor design and harvesting of microalgae for biodiesel production: a critical review. *Bioresour. Technol.*, 2011, 102, 71-81.

Cherubini, F. The Biorefinery concept: using biomass instead of oil for producing energy and chemicals. *Energ. Convers. Manage*, 2010, 51, 1412-1421.

Chin L. Y.; Engel A. J. Hydrocarbon feedstocks from algae hydrogenation. *Biotechnol. Bioeng. Symp.,* 1981, 11, 171-86.

Chisti, Y. Biodiesel from microalgae. *Biotechnol. Adv.*, 2007, 25, 294-306.

Chiu, Y-W.; Wu, M. Considering water availability and wastewater resources in the development of algal bio-oil. *Biofuels Bioprod. Bioref.*, 2013, 7 (4), 406-415.

Cho, S.; Lee, N.; Park, S.; Yu, J.; Luong, T. T.; Oh, Y-K.; Lee, T. Microalgae cultivation for bioenergy production using wastewaters from a municipal WWTP as nutritional sources. *Bioresour. Technol.*, 2013, 131, 515-520.

Christenson, L.; Sims R. Production and harvesting of microalgae for wastewater treatment, biofuels, and bioproducts. *Biotechnol. Adv.*, 2011, 29 (6), 686-702.

Danquah, M. K.; Ang, L.; Uduman, N.; Moheimani, N.; Forde, G. M. Dewatering of microalgal culture for biodiesel production: Exploring polymer flocculation and tangential flow filtration. *J. Chem. Technol. Biotechnol.*, 2009a, 84, 1078–1083.

Danquah, M. K.; Gladman, B.; Moheimani, N.; Gorde, G. M. Microalgal growth characteristics and subsequent influence on dewatering efficiency. *Chem. En. J.*, 2009b, 151, 73-78.

de Boer, K.; Moheimani, N. R.; Borowitzka, M. A.; Bahri, P. A. Extraction and conversion pathways for microalgae to biodiesel: a review focused on energy consumption. *J. Appl. Phycol.*, 2012, 24 (6), 1681-1698.

de Fraiture, C.; Giordano, M.; Liao, Y. Biofuels and implications for agricultural water use: blue impacts of green energy. *Water Policy*, 2008, 10 (1): 67-91.

Demirbas, A. Oily products from mosses and algae via pyrolysis. *Energ. Source*, 2006, 28, 933-940.

Demirbas, A.; Demirbas, M. F. Algae energy: algae as a new source of biodiesel. London Springer-Verlag, 2010.

Dickinson, K. E.; Whitney, C. G.; McGinn, P. J. Nutrient remediation rates in municipal wastewater and their effect on biochemical composition of the microalga *Scenedesmus* sp. AMDD. *Algal Research*, 2013, 2, 127-134.

Divakaran, R.; Pillai, V. N. S. Flocculation of algae using Chitosan. *J. Appl. Phycol.*, 2002, 14, 419-422.

DOE http://www1.eere.energy.gov/bioenergy/integrated_biorefineries.html. June 2013.

Dote, Y.; Sawayama, S.; Inoue, S.; Minowa, T.; Yokoyama, S-Y. Recovery of liquid fuel from hydrocarbon-rich microalgae by thermochemical liquefaction. *Fuel*, 1994, 73, 1855-1887.

Du, Z.; Mohr, M.; Ma, X.; Cheng, Y.; Lin, X.; Liu, Y.; Zhou, W.; Chen, P.; Ruan, R. Hydrothermal pretreatment of microalgae for production of pyrolytic bio-oil with a low nitrogen content. *Bioresour. Technol.,* 2012, 120, 13-18.

Duan, P.; Savage, P. E. Hydrothermal liquefaction of a microalga with heterogeneous catalysts. *Ind. Eng. Chem. Res.,* 2010, 50, 52-61.

Duan, P.; Jin, B.; Xu, Y.; Yang, Y.; Bai, X.; Wang, F.; Zhang, L.; Miao, J. Thermo-chemical conversion of *Chlorella pyrenoidosa* to liquid biofuels. *Bioresour. Technol.*, 2013, 133, 197-205.

Encinar, J. M.; Beltran, F. J.; Ramiro, A. Pyrolysis/gasification of agricultural residues by carbon dioxide in the presence of different additives: influence of variables. *Fuel Process. Technol.,* 1998, 55, 219-233.

Ferreira, A. F.; Ribeiro, L. A.; Batista, A. P.; Marques, P. A.; Nobre, B. P.; Palavra, A. M. F.; Pereira, P. S.; Gouveia, L.; Silva, C. A biorefinery from *Nannochloropsis* sp. microalga–Energy and CO_2 emission and economic analyses. *Bioresour. Technol.*, 2013, 138, 235-244.

Flynn, K. J.; Mitra, A.; Greenwell, H. C.; Sui, J. Monster potential meets potential monster: pros and cons of deploying genetically modified microalgae for biofuels production. *Interface Focus,* 2013, 3(1) UNSP 20120037.

Gallagher, B. J. The economics of producing biodiesel from algae. *Renew. Energ.*, 2011, 36, 158-162.

García Alba, L.; Torri, C.; Samori, C.; van der Spek, J.; Fabbri, D.; Kersten, S. R. A.; Brilman, D. W. F. Hydrothermal treatment (HTT) of microalgae: evaluation of the process as conversion method in an algae biorefinery concept. *Energ. Fue*l, 2011, 26 (1), 642-657.

García, F.; Jawiarczyk, N.; González, C. V.; Fernández, J. M.; Acién Fernández, F. G. Valorization of microalgal biomass: Exploitation of proteins, carbohydrates and lipids. *Rev. Latinoam. Biotecnol. Amb. Algal,* 2013, 3,147-161.

Gendy, T. S.; El-Temtamy, S. A. Commercialization potential aspects of microalgae for biofuel production: An overview, *Egypt. J. Petrol.*, 2013, 22 (1), 43-51.

Greenwell, H. G.; Laurens, L. M. L.; Shields, R. J.; Lovitt, R. W.; Flynn, K. J. Placing microalgae on the biofuels priority list: a review of the technological challenges. *J. R. Soc. Interface*, 2010, 7, 703-726.

Grobbelaar, J. U. Algal nutrition. In: Richmond, A. Handbook of microalgal culture: biotechnology and applied phycology. USA, Blackwell. 2004, 97-115.

Gouveia, L.; Oliveira, A. C. Microalgae as a raw material for biofuels production. *J. Ind. Microbiol. Biotechnol.*, 2009, 36, 269-274.

Gouveia, L. Microalgae as a feedstock for Biofuels. Springer Briefs in Microbiology, London New York, Springer Heidelberg Dordrecht, 2011.

Halim, R.; Harun, R.; Danquah, M. K.; Webley, P. A. Microalgal cell disruption for biofuel development. *Appl. Energy*, 2012, 91, 116-121.

Harun, R.; Singh, M.; Forde, G. M.; Danquah, M. K. Bioprocess engineering of microalgae to produce a variety of consumer products. *Renew. Sust. Energ. Rev.*, 2010a, 14, 1037-1047.

Harun, R.; Danquah, M. K.; Forde, G. M. Microalgal biomass as a fermentation feedstock for bioethanol production. *J. Chem. Technol. Biotechnol.,* 2010b, 85 (2), 199-203.

Harun R., Danquah M. K. Influence of acid pre-treatment on microalgal biomass for ethanol production. *Process. Biochem.*, 2011, 46, 304-309.

Herrero, M.; Cifuentes, A.; Ibañez, E. Sub- and supercritical fluid extraction of functional ingredients from different natural sources: Plants, food-by-products, algae and microalgae: A review. *Food Chem.*, 2006, 98 (1), 136-148.

Ho, S-H.; Chen, C-Y.; Chang, J-S. Effect of light intensity and nitrogen starvation on CO_2 fixation and lipid/carbohydrate production of an indigenous microalga Scenedesmusobliquus CNW-N. *Bioresour. Technol.,* 2012, 113, 244-252.

Hu, Q.; Sommerfeld, M.; Jarvis, E.; Ghirardi, M.; Posewitz, M.; Seibert, M.; Darzins, A. Microalgal triacylglycerols as feedstocks for biofuel production: perspectives and advances. *Plant J.*, 2008, 54, 621-639.

Hu, Z.; Zheng, Y.; Yan, F.; Xiao, B.; Liu, Sh. Bio-oil production through pyrolysis of blue-green algae blooms (BGAB): Product distribution and bio-oil characterization. *Energy*, 2013, 52, 119-125.

Huang, G.; Chen, F.; Wei, D.; Zhang, X.; Chen, G. Biodiesel production by microalgal biotechnology. *Appl. Energy*, 2010, 87, 38-46.

Huntley, M. E.; Redalje, D. G. CO_2 mitigation and renewable oil from photosynthetic microbes: a new appraisal. *Mitig. Adapt. Strat. Glob. Change,* 2007, 12, 573-608.

IEA. 2007 Bioenergy Task 42- Countries Report Final. Available from: www.biorefinery.nl/fileadmin/biorefinery/docs/CountryReportsIEABioenergyTask42Final170809.pdf

Jena, U.; Das, K. C. Comparative Evaluation of Thermochemical Liquefaction and Pyrolysis for Bio-Oil Production from Microalgae. *Energ. Fuels*, 2011, 5472-5482.

Johnson, M. B.; Wen, Z. Y. Development of an attached microalgal growth system for biofuel production. *Appl. Microbiol. Biotechnol.*, 2010, 85, 525-534.

Jungmeler G., Hingsamer M., van Ree R. Biofuel-driven Biorefineries. A selction of the most promising biorefinery concepts to produce large volumes of road transportation biofuels by 2025. Report for the IEA-Bioenergy Task 42 Biorefinery. February 2013.

Kaufmann, B.; Christen, P.; Veuthey, J. L. Parameters Affecting Microwave-assisted Extraction of Withanolides. *Phytochem. Anal.*, 2001, 12, 327-331.

Kelly, W.; Muske, W. Optimal operation of high-pressure homogenization for intracellular product recovery. *Bioprocess Biosys. Eng.*, 2004, 27:25-37.

Klasson, K. T. Char from sugarcane bagase. In Bergeron, C.; Carrier, D. J.; Ramaswamy, S. H. Biorefinery Co-Products: Phytochemicals, Primary Metabolites and Value-Added Biomass Processing. John Wiley & Sons. United Kingdom. 2012, 327-345.

Kliphuis, A. M. J.; Winter, L.; Vejrazka, C.; Martens, D. E.; Janssen, M.; Wijffels, R. H. Photosynthetic efficiency of *Chlorella sorokiniana* in a turbulently mixed short light-path photobioreactor. *Biotechnol. Prog.*, 2010, 26 (3), 687-696.

Kong, Q.; Li, L.; Martinez, B.; Chen, P.; Ruan, R. Culture of microalgae *Chlamydomonas reinhardtii* in wastewater for biomass feedstock production. *Appl. Biochem. Biotechnol.*, 2010, 160, 9-18.

Kula, M-R.; Schütte, H. Purification of proteins and the disruption of microbial cells. *Biotechnol. Prog.*, 1987, 3 (1), 31-42.

Lakaniemi, A. M.; Tuovinen, O. H.; Puhakka J. A., Anaerobic conversion of microalgal biomass to sustainable energy carriers -*A review-Bioresour. Technol.*, 2013, 135, 222-231.

Lee, J-Y.; Yoo, C.; Jun, S-Y.; Ahn, C-Y.; Oh, H-M. Comparison of several methods for effective lipid extraction from microalgae. *Bioresour. Technol.,* 2010, 101, S75-S77.

Lee, A. K.; Lewis, D. M.; Ashman, P. J. Disruption of microalgal cells for the extraction of lipids for biofuels: Processes and specific energy requirements. *Biomass Bioenerg.*, 2012, 46, 89-101.

Li, Y. G.; Xu, L.; Huang, Y. M.; Wang, F.; Guo, C.; Liu C. Z. Microalgal biodiesel in China: opportunities and challenges. *Appl. Energ.*, 2011, 88 (10), 3432-3437.

López, B. D.; Prins, W.; Ronsse, F.; Brilman, W. Hydrothermal liquefaction (HTL) of microalgae for biofuel production: State of the art review and future prospects. *Biomass Bioenerg.*, 2013, 53, 113-127.

Matsui, T.; Nishihara, A.; Ueda, Ch.; Ohtsuki, M.; Ikenaga, N.; Suzuki, T. Liquefaction of microalgae with iron catalyst. *Fuel*, 1997, 76, 1043-1048.

Mata, T. M.; Martins, A. A.; Caetano, N. S. Microalgae for biodiesel production and other applications: a review. *Renew. Sust. Energ. Rev.*, 2010, 14, 217-232.

McGinn, P. J.; Dickinson, K. E.; Park, K. C.; Whitney, C. G.; MacQuarrie, S. P.; Black, F. J.; Frigon, J-C.; Guiot, S. R.; O'Leary, S. J. B. Assessment of the bioenergy and bioremediation potentials of the microalga *Scenedesmus* sp. AMDD cultivated in municipal waste water effluent in batch and continuous mode. *Algal Research*, 2012, 1 (2), 155-165.

Middelberg, A. P. J. Process-scale disruption of microorganisms. *Biotechnol. Adv.*, 1995, 13 (3), 491-551.

Minowa, T.; Yokoyama, S-Y.; Kishimoto, M.; Okakura T. Oil production from algal cells of Dunaliellatertiolecta by direct thermochemical liquefaction. *Fuel*, 1995, 74, 1735-1738.

Molina, G. E.; Robles, M. A.; Giménez, G. A.; Ibáñez, G. M. J. Gram-scale purification of eicosapentaenoic acid (EPA 20:5n3) from wet *Phaeodactylum tricornutum* UTEX 640 biomass. *J. Appl. Phycol.*, 1996, 8, 359-367.

Molina G. E.; Belarbia E. H.; Acien F. G.; Robles A., Chisti Y. Recovery of microalgal biomass and metabolites: process options and economics. *Biotechnol. Adv.*, 2003, 491-515.

Mollah, M. Y. A.; Morkovsky, P.; Gomes, J. A. G.; Kesmez, M.; Parga, J.; Cocke, D. L. Fundamentals, present and future perspectives of electrocoagulation. *J. Hazard. Mater.*, 2004, 114,199-210.

Murugesan, R.; Orsat, V. Spray Drying for the Production of Nutraceutical Ingredients—A Review. *Food Bioprocess Technol.*, 2012, 5, 3-14.

Mussgnug, J. H.; Klassen, V.; Schlüter, A.; Kruse, O. Microalgae as substrates for fermentative biogas production in a combined biorefinery concept. *J. Biotechnol.*, 2010, 150, 51–56.

Nobre, B. P.; Villalobos, F.; Barragán, B. E.; Oliveira, A. C.; Batista, A. P.; Marques, PASS; Mendes, R. L.; Sovová, H.; Palavra, A. F.; Gouveia, L., A biorefinery from *Nannochloropsis* sp. Microalga-extraction of oils and pigments. Production of biohydrogen from the left over biomass. *Bioresour. Technol.*, 2013, 135, 128-136.

Olguín, E. J. Phycoremediation: key issues for cost-effective nutrient removal processes. *Biotechnol. Adv.*, 2003, 22, 81–91.

Olguín, E. J. Dual purpose microalgae–bacteria-based systems that treat wastewater and produce biodiesel and chemical products within a Biorefinery. *Biotechnol. Adv.*, 2012, 30, 1031-1046.

Orset, S.; Leach, G. L.; Morais, R.; Young, A. J. Spray-Drying of the Microalga *Dunaliellasalina*: Effects on â-Carotene Content and Isomer Composition. *J. Agric. Food Chem.*, 1999, *47,* 4782-4790.

Patil, V.; Kallqvist, T.; Olsen, E.; Vogt, G.; Gislerod, H. R. Fatty acid composition of 12 microalgae for possible use in aquaculture feed. *Aquacult. Int.*, 2007, 15, 1-9.

Prakash, J.; Pushparaj, B.; Carlozzi, P.;.Torzillo, G.; Montaini, E. Materassi R. Microalgal biomass drying by a simple solar device. *Int. J. Sol. Energ.,* 1997, 18 (4), 303-311.

Qureshi, N.; Annous, B.; Ezeji, T.; Karcher, P.; Maddox, I. Biofilm reactors for industrial bioconversion processes: employing potential of enhanced reaction rates. *Microbiol. Cell Fact.*, 2005, 4, 24.

Ramzan, N.; Ashraf, A.; Naveed, S.; Malik, A. Simulation of hybrid biomass gasification using Aspen plus: a comparative performance analysis for food, municipal solid and poultry waste. *Biomass Bioenerg.*, 2011, 35 (9), 3962-3969.

Rashid, N.; Choi, W.; Lee, K. Optimization of two-staged bio-hydrogen production by immobilized *Microcystis aeruginosa*. *Biomass Bioenerg.*, 2012, 36, 241-249.

Rawat, I.; Kumar, R. R.; Mutanda, T.; Bux, F. Biodiesel from microalgae: A critical evaluation from laboratory to large scale production. *Appl. Energ.*, 2013, 103, 444-467.

Rawat, I.; Kumar, R. R., Mutanda T., Bux F. Dual role of microalgae: phycoremediation of domestic wastewater and biomass production for sustainable biofuels production. *Appl. Energ.*, 2011, 88, 3411-24.

Richmond A. 2004. Handbook of microalgal culture: biotechnology and applied phycology. USA, Blackwell. 2004.

Rodriguez, G.; Vasseur, J.; Courtois, F. Design and Control of Drum Dryers for the Food Industry. Part 1. Set-Up of a Moisture Sensor and an Inductive, *J. Food Eng.*, 1996, 28, 271-282.

Romano, R. T.; Zhang, R. Co-digestion of onion juice and wastewater sludge using an anaerobic mixed biofilm reactor. *Bioresour. Technol.*, 2008, 99 (3), 631–637.

Ross, A. B.; Biller, P.; Kubacki, M. L.; Li, H.; Lea-Langton, A.; Jones, J. M. Hydrothermal processing of microalgae using alkali and organic acids. *Fuel,* 2010, 89, 2234-2243.

Sánchez- Tuirán, E.; El-Halwagi M. M.; Kafarov, V. Integrated utilization of algae biomass in a biorefinery based on a biochemical processing platform. In Integrated Biorefineries: Design, Analysis and Optimization Edited by Stuart PR and El-Halwagi, M. 2013. CRC Press Boca Raton Florida. 707-726.

Sawayama, S.; Inoue, S.; Dote, Y.; Yokoyama, S-Y. CO_2 fixation and oil production through microalga. *Energ. Convers. Manage.*, 1995, 36, 729-731.

Sharar, M. A.; Clausen, E. C.; Carrier, D. J. An overview of biorefinery technology. In Bergeron, C; Carrier, DJ; Ramaswamy, SH. Biorefinery Co-Products: Phytochemicals, Primary Metabolites and Value-Added Biomass Processing. John Wiley & Sons. United Kingdom. 2012. 1-16.

Schavan, A.; Aigner, I. 2012. Biorefineries Roadmapas part of the German Federal Government action plans for the material and energetic utilisation of renewable raw materials. The German Federal Government.108 pp.

Sialve, B.; Bernet, N.; Bernard, O. Anaerobic digestion of microalgae as a necessary step to make microalgal biodiesel sustainable. *Biotechnol. Adv.*, 2009, 27, 409-416.

Singh, S.; Kate, B. N.; Banecjee, U. C. Bioactive compounds from cyanobacteria and microalgae: an overview. *Crit. Rev. Biotechnol.*, 2005, 25, 73-95.

Singh, J.; Gu, S. Commercialization potential of microalgae for biofuels production. *Renew Sust. Energ. Rev.*, 2010, 14, 2596-2610.

Sivakumar, G.; Xu, J.; Thompson, R. W.; Yang, Y., Randol-Smith, P.; Weathers, P. J. Integrated green algal technology for bioremediation and biofuel. *Bioresour. Technol.,* 2012, 107, 1-9.

Spolaore, P.; Joannis-Cassan, C.; Duran, E.; Isambert, A. Commercial applications of microalgae. *J. Biosci. Bioeng.*, 2006,101, 87-96.

Suali, E.; Sarbatly, R. Conversion of microalgae to biofuel. *Renew. Sust. Energ. Rev.*, 2012, 16, 4316-4342.

Sukenik, A.; Shelef, G. Algal autoflocculation—verification and proposed mechanism. *Biotechnol. Bioeng.*, 1984, 26, 142–147.

Sturm, B. S. M.; Lamer, S. L. An energy evaluation of coupling nutrient removal from wastewater with algal biomass production. *Appl. Energ.*, 2011, 88, 3499-3506.

Sun, Y.; Cheng, J. Hydrolysis of lignocellulosic materials for ethanol production: a review. *Bioresour. Technol.*, 2002, 83, 1-11.

Sheng, J.; Vannela, R.; Rittmann, B. E. Disruption of *Synechocystis* PCC 6803 for lipid extraction. *Water Sci. Technol.*, 2012, 65, 567-573.

Shuping, Z.; Yulong, W.; Mingde, Y.; Chun, L. Thermochemical catalytic liquefaction of the marine microalgae *Dunaliella tertiolecta* and characterization of Bio-oils. *Energ. Fuels,* 2009, 23, 3753-3758.

Uduman, N.; Qi, Y.; Danquah, M. K.; Forde, G. M.; Hoadley, A. Dewatering of microalgal cultures: A major bottleneck to algae-based fuels. *J. Renew. Sust. Energ.*, 2010, 2, 012701.

Valderrama, J. O.; Perrut, M.; Majewski W. Extraction of Astaxantine and Phycocyanine from Microalgae with Supercritical Carbon Dioxide. *Chem. Eng. Data,* 2003, 48, 827 830.

Vanthoor-Koopmans, M.; Wijffels, R.; Barbosa, M. J.; Eppink, M. Biorefinery of microalgae for food and fuel. *Bioresour. Technol.*, 2013, 135, 142–149.

Wang, B.; Li, N.; Lan, C. Q. CO_2 bio-mitigation using microalgae. *Appl. Microbiol. Biotechnol.*, 2008, 79, 707-718.

Wang, L.; Min, M.; Li, Y.; Chen, P.; Chen, Y.; Liu, Y.; Wang, Y.; Ruan, R. Cultivation of green algae *Chlorella* sp. in different wastewaters from municipal wastewater treatment plant. *Appl. Biochem. Biotechnol.,* 2009, 162 (4),1174-1186.

Wijffels, R. H.; Barbosa. An Outlook on Microalgal Biofuels. *Science,* 2010, 329: 796-799.

Wijffels, R. H.; Barbosa, M. J.; Eppink, M. H. M. Microalgae for the production of bulk chemicals and biofuel. *Biofuels, Bioprod. Bioref.*, 2010, 4, 287-295.

Wiltshire, K. H.; Boersma, M.; Möller, A.; Buhtz, H. Extraction of pigments and fatty acids from the green alga *Scenedesmus obliquus* (Chlorophyceae). *Aqua Ecol.*, 2000, 34 (2),119-126.

Yang, Y. F.; Feng, C. P.; Inamori, Y.; Maekawa, T. Analysis of energy conversion characteristics in liquefaction of algae. *Resour. Conserv. Recycl.*, 2004, 43, 21-33.

Yang, Z.; Guo, R.; Xu, X.; Fan, X.; Luo, Sh. Hydrogen and methane production from lipid-extracted microalgal biomass residues. *Int. J. Hydrogen Energ.*, 2011, 36, 3465–3470.

Yen, H-W.; Hu, I-C.; Chen, C-Y.; Ho, S-H.; Lee, D-J.; Chang, J-S. Microalgae-based biorefinery– From biofuels to natural products. *Bioresour. Technol.,* 2012, 135, 166-174.

Yoon, R. H.; Luttrell, G. H. The effect of bubble size on fine particle flotation. *Miner. Process Extract Metal Rev.: An Int. J.*, 1989, 5 (1-4), 101-122.

Yuan, X.; Wang, J.; Zeng, G.; Huang, H.; Pei, X.; Li, H.; Liu, Z.; Cong, M. Comparative studies of thermochemical liquefaction characteristics of microalgae using different organic solvents. *Energy*, 2011, 36, 6406-6412.

Zhang, J.; Chen, W. T.; Zhang, P.; Luo, Z.; Zhang, Y. Hydrothermal liquefaction of Chlorella pyrenoidosa in sub- and supercritical ethanol with heterogeneous catalysts. *Bioresour. Technol.*, 2013, 133, 389-397.

Zheng, H.; Gao, Z.; Yin, J.; Tang, X.; Ji, X.; Huang, H., Harvesting of microalgae by flocculation with poly (γ-glutamic acid). *Bioresour. Technol.*, 2012, 112, 212-220.

In: Energy and Environment Nowadays
Editors: Luis G. Torres and Erick R. Bandala
ISBN: 978-1-63117-398-1

Chapter 3

Environmental Impact of Biofuels

Eugenio Sánchez-Arreola[1]*, José D. Lozada-Ramírez[1], Luis R. Hernández[1] and Horacio Bach[2]
[1]Departamento de Ciencias Químico Biológicas
Universidad de las Américas Puebla, Cholula, Puebla, México
[2]Department of Medicine, Division of Infectious Diseases
University of British Columbia, Vancouver, Canada

Introduction

Problems associated with the use of fossil fuels have prompted a search for alternative fuel sources. Alternative sources for engine fuel mainly include the use of crops or crop-derived products, such as oil, and anthropogenic waste sources. In this chapter, the production of the three main biofuels currently used is described: biodiesel, biogas, and bioethanol. The environmental impact of these biofuels is also discussed.

1. Biodiesel

Initially, the use of pure vegetable oils or vegetable oils mixed with diesel were proposed, but their use was not satisfactory nor practical because of high viscosity, free acid content, and fatty acid and gum formation due to oxidation and polymerization during storage and combustion.

1.1. Production

To solve some of the above mentioned problems, modification of oils by a transesterification or alcoholysis reaction was proposed. In this reaction, a fatty acid or oil

* email: egugenio.sanchez@udlap.mx.

reacts with an alcohol to form esters and glycerol. A catalyst is used to improve the reaction rate and yield because the reaction is reversible, and by using an excess of alcohol the reaction is shifted to the product side. The alkyl esters obtained from this reaction are biodiesel (Fangrui, 1999).

1.1.1. Sources of Biomass

The main raw materials used for the production of biodiesels are listed in Table 1 (Dennis, 2010; Ghaly, 2010).

The advantages of biodiesel use are mainly related to its environmental benefits, because it can be obtained from renewable resources. Disadvantages include its high production cost and the limited availability of raw materials. Production costs include the production of raw material (fats and oils) and represents 60-75% of the total cost of biodiesel (Fangrui, 1999).

1.2. Biodiesel Environmental Impact

1.2.1. Positive Impacts

The most common positive impact of biodiesel use is the reduction of greenhouse gas emissions, particularly CO_2 emissions. This is because plants (biomass) absorb CO_2 in an equal amount to the CO_2 emitted when biodiesel is burned. However, the net positive impact from this relies on 100% renewal of the biomass, i.e.; the rate of biomass use must be equal to the new biomass cultivation rate.

Biodiesel can also contribute positively to regional development and sustainability since some plants suitable for biodiesel production can be grown on non-arable land. For example, in Asian countries like India and China they produce biodiesel from hardy inedible plants, like *Jatropha curcas,* which grown in marginal areas not suitable for food production (Soetaert, 2009).

One of the most important potential positive impacts of biodiesel use is low sulfur dioxide (SO_2) emissions. This is due to the low content of sulfur in biomass. However, in a complicated supply system of biodiesel production, this advantage is abolished because of fossil fuel usage involved in biomass cultivation, harvesting, and transportation stages.

Table 1. Major raw materials used in the production of biodiesel

Edible oils	Inedible oils	Others
Soybean	*Jatropha curcas*	Used cooking oil
Rapeseed	*Pongamia pinnata*	
Sunflower	*Calophyllum inophyllum*	
Palm	Sea mango	
Peanut	Palanga	
Corn	Tallow	
Cotton	Nile tilapia	
Pumpkin	Poultry	

1.2.1.1. Ecologic Efficiency

Ecological efficiency is defined as the effectiveness by which energy is transferred among trophic levels. Studies carried out to determine the ecologic efficiency of biodiesel and diesel in a thermoelectric power plant found efficiencies of 98.2 and 93.2% when pure biodiesel fuel (B100) and biodiesel blended with conventional diesel fuel (B20) (20% biodiesel: 80% diesel) were analyzed, respectively. Interestingly, a lower ecologic efficiency of 92.2% was found with conventional diesel (Rodriguez et al., 2010).

A study addressing the environmental impact of the use of gasoline, diesel and biodiesel was conducted in Greece. Results showed that although the use of biodiesel appears to be very attractive (reductions of greenhouse gas emissions), the logistics involved in the transportation of the fuel increases the emissions of particulate matter and NOx (Nanaki, 2012). Moreover, a study of biodiesel obtained from *Jatropha curcas* also showed that the emissions of NOx, SO_2, and particulate matter were higher than for diesel (Xing, 2010). However, another study compared the emissions of B100 vs. B0 and determined that significant decreases of 9-55, 50-67, 26-37, and 45-58% of smoke, CO, NOx, and unburned hydrocarbon emissions were obtained, respectively (Dayong, 2012). Moreover, smoke, HC, CO, and NOx emissions for B50 blend of biodiesel from *Thespesia populnea* at different loads were found higher or comparable to diesel fuel (Nanasaheb, 2012).

1.2.1.2. Net Energy Ratio

Biodiesel contribution to non-renewable source depletion (mainly fossil fuels) depends on their net energy ratio (NER). NER is equal to biofuel energy content/ energy consumed for its production and distribution. In most of the biofuel production cases, NER is greater than 1. However, in some cases, NER is lower than 1, which means that these production systems are unacceptable from an energy point of view. NER depends on the limits of the system and other parameters, such as cultivation techniques, production methods, etc. For example, the energy benefits of biodiesel produced from *Jatropha* cultivation on two different types of soil (poor and normal soils) were evaluated with and without irrigation. The study estimated the net energy balance and NER for both models and found that the energy values were high for both systems (Chatterjee, 2012).

In another study, the environmental impacts of biodiesel from *Jatropha* was compared to the impacts of a fossil fuel used as a reference system delivering the same amount of products. Results show that the production and use of *Jatropha* biodiesel decreases in a 82% the non-renewable energy requirement with a NER = 1.85 (Achten, 2010). However, another study reported a lower NER using the same plant biodiesel (Wang, 2011).

Studies conducted on oils obtained from different organisms report variable NER values. For example, biodiesel produced from canola and microalgae gave NER values of 1.72 and 0.93, respectively (Batan, 2010; Rustandi, 2010). However, best results were obtained with palm biodiesel with a NER of 3.53 (Yee, 2009). Another plant analyzed for the production of biodiesel was rapeseed (*Brassica napus*), which gave a NER= 1.78 (Fore et al., 2011).

1.2.1.3. Ozone Layer Depletion

Another positive factor of biodiesel use is the reduction of ozone layer depletion. For instance, the production of biodiesel derived from the transesterification of crude rapeseed oil, one of the most important sources of biodiesel in Europe, showed that using B100 derived from this oil instead of petroleum-based diesel would reduce ozone layer depletion by approximately 44% (González-García et al., 2013).This decrease can be explained because the use of biodiesel reduces NOx and SOx concentrations that are expelled into the atmosphere, since it has been shown that the NOx and SOx gases can deplete ozone layer. Sulfur trace gases and SO_2 as their main degradation product can potentially reach the stratosphere and can influence stratospheric ozone by enhancing its degradation through heterogeneous reactions (WMO, 2010).

1.2.2. Negative Impacts

Negative impacts from biodiesel are observed in a number of areas; such as ozone layer depletion, eutrophication, and acidification. In some cases the negative impacts of biodiesel use are worse than those corresponding to fossil fuels. These impacts vary from study to study and depend upon the definition of the limits in each system, the cultivation, production methods, etc. A study of the production of biodiesel derived from the transesterification of crude rapeseed oil, one of the most important sources of biodiesel in Europe, showed that using B100 derived from rapeseed oil instead of petroleum-based diesel increased acidification in a 59% and eutrophication in a 214 % (González-García et al., 2013).

Another study conducted in China studied eutrophication and acidification of biodiesel obtained from soybean, *Jatropha*, and microalgae were compared to fossil diesel, and the results indicated that the use of biodiesel increases both negative factors (Hou, 2011). In Denmark, the use of fossil diesel in the transport sector appeared to be environmentally preferable over biodiesel regarding their impact on acidification, aquatic eutrophication, and land occupation (Tonini, 2012). This can be explained when biodiesel used as transportation fuel increases the level of nutrients, such as nitrogen and phosphorous, which will increase the level of eutrophication (Nanaki, 2012). Similar results were reported in India when *Jatropha* biodiesel was compared to fossil fuel. This study reported increases of 49 and 430% for acidification and eutrophication, respectively (Achten, 2010).

One of the most serious problems arising from biodiesel production is the increase of food market prices resulting from arable land being converted from food production to biomass production. In addition, the agricultural commodity trading in the future market pushes their prices to rise in an unpredictable way. To reduce the impact on food prices, Kovacs (2012) recommends that feedstock for biodiesel must include waste and secondary oil sources with low impact on the food chain. There is also growing interest in using oils from inedible plants like *Jatropha curcas* as the feedstock for biodiesel production because it does not compromise the production of edible oils, which are mainly used for food consumption (Koh, 2011).

The mass production of biodiesel can lead to the increase of gases contributing to the greenhouse effect. This is due principally to the use of fossil transportation fuels in the complicated logistics needed for biomass collection, transportation, and final product distribution. Moreover, deforestation or clearing of grasslands used for biomass cultivation leads to emission of CO_2 captured in soil biomass into the atmosphere (Petrou, 2009).

Several studies in different countries have been conducted to determine the benefits of biodiesel use. For example, a study performed in Denmark about the use of biodiesel for heavy terrestrial transportation showed that greenhouse gas emissions could be significantly reduced (Tonini, 2012). Others reported that simulations of production of biodiesels of different origin, such as *Jatropha* oil methylester (JME), cottonseed oil methylester (COME), rapeseed oil methylester (RME), soybean oil methylester (SME), and castor oil methylester

(CAME) showed that CO_2 emissions were higher for SME, RME, and COME; and lower for JME and CAME (Nasim, 2012).

In China, the net energy, CO_2 emission, and cost efficiency of *Jatropha* biodiesel as a fuel substitute in Panzhihua region (Sichuan Province), which is well suited to growing *Jatropha*, was evaluated. In this report, low energy efficiency was obtained when compared to fossil fuels suggesting that biodiesel production from *Jatropha* should be improved (Deng, 2012).

Another negative impact of biodiesel use is the increase in N_2O (nitrous oxide) and NOx emissions due to the use of fertilizers on crops (No, 2010). It has generally been argued that greenhouse gas releases from land use changes and N_2O emissions from the use of fertilizers can potentially be significant enough to change the environmental profile of biodiesel (Acquaye, 2011). It has been demonstrated that NOx emissions was higher for *Jatropha*, cottonseed, rapeseed, soybean, and castor biodiesel compared to fossil diesel fuel (Nasim, 2012). Other studies also report similar results for the use of *Jatropha* biodiesel with a slightly increase of NOx emissions when compared to fossil diesel (Banerjee, 2009; Jayed, 2009).

Studies have reported that the current use of several agricultural crops for energy production can lead to an increase in the N_2O emissions that can be large enough to cause climate warming. However, it has been observed that low-temperature combustion effectively reduces NOx emissions, because less thermal NOx is formed (Crutzen et al., 2008). Although biodiesel combustion produces more NOx for both conventional and late-injection engines, with the latter leading to a low-temperature combustion mode, various mixtures of soy biodiesel and European low-sulfur diesel reduced the NOx emission considerably. For example, the levels of NOx emissions from mixtures containing B20, B50, and B100 of the above-mentioned combustibles, reduced the emissions in a 68, 67, and 64%, respectively, when compared to the use of pure European low-sulfur diesel (Fang, 2008). Other study with rapeseed biodiesel showed that NOx emissions were lower for B5, B20, B40, and B50, while higher for B80 and B100 (Saqib, 2012).

Use of biodiesel in some diesel engines can cause low efficiency and poor combustion for example, for engines using mechanical fuel injectors in a pump-line-nozzle style fuel injection system, the fuel pump produces pressure pulses which travel down individual fuel lines to each injector. The time at which the pressure wave reaches the fuel injectors directly influences the fuel injection time. For biodiesel, the pressure wave travels down the fuel lines more rapidly due to higher speed of sound and bulk modulus, thereby advancing the injection timing, in some cases by several crank angle degrees. As a result of earlier injection, residence times are longer for burned gases and more heat release occurs closer to top dead center when the cylinder volume is reduced. The cylinder temperatures are therefore increased, leading to higher NOx emissions (Adi, 2009).

Engine performance with biodiesel blends and conventional diesel fuels is generally similar except that brake-specific fuel consumption increases using biodiesel due to its lower energy content (Chin, 2012). In other hand several studies reported that to biodiesel contributing NOx increases, especially with newer engines (Hajbabaei, 2012).

When comparing the performance of biodiesel and fossil diesel regarding engine efficiencies, the relation between biodiesel, engine calibration, and turbocharger system has to be taken in account. For instance, results showed that biodiesel has a lower performance than diesel (Beatrice 2012), including the compression ignition of the engine (Özkan, 2007). Then, car companies should address these issues by adjusting the engines for a better performance. A summary of the some positive and negative effects of biodiesel compared with fossil diesel are shown in Table 2.

Table 2. Positive and negative effects of biodiesel

Positive effects	Negative effects
1) Lower emission of CO_2, CO, SO_2, hydrocarbons and particulate matter 2) Regional development 3) Production sustainable 4) Better ecologic efficiency 5) Good net energy ratio (NER)	1) Increase NOx emission 2) Ozone layer depletion 3) Eutrophication 4) Acidification 5) Competition with the food market*

*Only when edible oils were used

1.3. Future Directions and Perspectives

One of the main issues to be address in the future is to obtain a lower cost biodiesel. Raw materials are a key factor for the production of biodiesel, as they represent a 70-80% of the total cost of production (Qian, 2009). In general, the cost of biodiesel production varies considerably and ranges from $0.29 to over $9.00 per liter depending on the feedstock available for their production. In addition, not all of the countries are equally suited to large-scale biodiesel production (Johnston, 2007), therefore it is necessary to increase the efforts to reduce the cost of biodiesel. Some ways to reduce the cost of biodiesel production is the selection of inexpensive raw materials or conducting research in order to optimize the reaction process in its production.

Another aspect to consider is the biodiesel market. Currently, in some countries, biodiesel is mainly used in public transportation, but its use can be extended to private cars and property heating. Consequently, it is expected that the biodiesel demand would increase in the future encompassing a greater possibility of reducing costs (Santibañez-Aguilar, 2011).

When comparing the performance of biodiesel and fossil diesel regarding engine efficiencies, the relation between biodiesel, engine calibration, and turbocharger system has to be taken in account. For instance, results showed that biodiesel has a lower performance than diesel (Beatrice 2012), including the compression ignition of the engine (Özkan, 2007). Then, car companies should address these issues by adjusting the engines for a better performance.

Another important aspect to consider is the distribution and storage of biodiesel due to its stability as some studies have shown that it can be oxidized easily (McCormick, 2010).

2. Biogas

2.1. Production

Biogas is a gas generated from natural sources or specific devices called digesters (first generation) or bioreactors (second generation) (Persson et al., 1979). It comprises of a mixture of gases consisting mainly of methane (CH_4, 55-65%), carbon dioxide (CO_2, 35-45%), and low amounts of hydrogen (H_2, 0-1%), hydrogen sulfide (H_2S, 0-1%) and nitrogen (N_2, 0-3%). Biogas production occurs naturally mainly in wetlands, landfills and animal intestinal tracts, by the activity of methanogen microorganisms. It is estimated that the microbiological activity releases into the atmosphere between 590-880 million tons of CH_4 annually (Saleh et al., 2012), which can be collected and pressurized as an energy resource (Gayh, 2012).

The overall process of CH_4 production comprises of three main stages as illustrated in Figure 1:

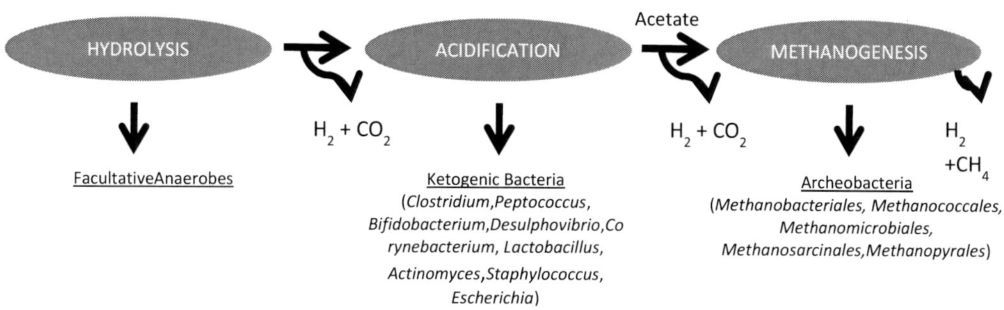

1. Hydrolysis: Bacteria decompose raw organic matter converting long chains of carbon into shorter and simpler chains (organic acids).
2. Acidification and Acetogenesis: During the acetogenesis phase, the products of the hydrolysis stage are converted to hydrogen, CO_2, formic acid and acetic acid. The step occurs for the action of microbes called obligatory hydrogen producing acetogens (OHPA). Degradation reaction of organic acids into acetate and release of H_2 and CO_2 occurs. This endergonic reaction is possible due to the close relationship between ketogenic and methanogenic bacteria, which remove the final products from the ketogenic bacteria, facilitating the flow of the process. The presence of low concentration of final products facilitates the reaction by causing activation of methanogenic bacteria enabling degradation and maintaining the energetic balance. Microorganisms participating in the ketogenic phase are composed of anaerobic bacteria or non-methanogenic bacteria, and include facultative and obligate anaerobes. These microorganisms are also identified as acid-forming bacteria.
3. Methanogenesis: Microorganisms involved in this stage belong to the Archaea domain termed methanogens, which possess unique features. These microorganisms are extremely sensitive to oxygen and to environmental changes. The final products accomplished at this stage is the microbial anaerobic respiration using as carbon source acetate and other short chain organic acids to generate the final products CH_4 and CO_2 (Perot and Amar, 1989; Verma, 2002). Methanogens are subdivided into two subcategories (Chantaraporn and Maneerat, 2012): a) Hydrogenotrophic methanogens ($CO_2 + 4H_2 \rightarrow CH_4 + 2H_2O$); and b) Acetotrophic methanogens ($CH_3COOH \rightarrow CH_4 + CO_2$).

Microorganisms involved in each phase have different properties that are very important and should be known in order to understand the balance and optimal operation of a digester (Mata-Álvarez et al., 2000). The different characteristics of each step are summarized in Table 3.

Table 3. Microbial decomposition of organic matter to CH_4 and CO_2

Hydrolytic and Ketogenic Phases	Methanogenic Phase
- Facultative bacteria	- Strict anaerobes
- Fast growing	- Slow growing
- No sensitivity to changes in acidity and temperature	- Sensitive to changes in acidity and temperature
- Major metabolites: organic acids	- Major end products: CH_4 and CO_2

During process for biogas production, the metabolic activity involved in the methanogenic phase is affected by several factors. Because each group of microorganisms involved in the various stages responds differently to environmental changes, it is impossible to give qualitative values of degree to each of them regarding to their impact on biogas production. Nevertheless, among the most important factors that should be taken into account are: type of substrate (available nutrients), substrate temperature, volumetric loading, retention time, level of acidity (pH), carbon : nitrogen ratio, substrate concentration, generation of inocula, degree of mixing, and presence of inhibitory compounds during the process (Jakobsen, 1992; Sreekrishnan et al., 2004; Funda, 2011).

Sources of Organic Materials for Biogas Production

Biogas production can be accomplished by using a variety of organic materials. For example, crop residues, manure, household waste, water hyacinth, algae, effluents from food industry, drinks, pulping and paper, and some chemicals (Stabnikova et al., 2010). Other raw materials for biogas production comprise of organic wastes from domestic, industrial and agricultural sources (Taherzadeh and Karimi, 2008; Martins das Neves et al., 2009). Domestic wastes include kitchen waste and fresh organic garbage, solid wastes, and sewage sludge (Verma, 2002; Zhang et al., 2005).

Several industrial wastes from agro-industry have been reported as suitable materials, such as the case of coffee, cherry, pulp juice, tapioca, pineapple, and other wastes from fruits for canning (Lee et al., 2003; Paepatung et al., 2009). The most important agricultural sources for biogas production are animal manure (cattle, swine, horse, elephant, and birds) mixed with water, and plant residues like fruit peels and straws (Van Horn and Hall, 1997; Sadi, 2010). Plant residues must be pretreated because their content of lignin, cellulose, and hemicellulose affects directly the biogas yield. This pre-treatment consists of chopping, grinding, and enzyme hydrolyzing processes (Carolan et al., 2007; Santos, 2011).

The stages of biogas production are characterized by different reaction velocities. For instance, cellulose degradation occurs in weeks, hemicelluloses and proteins in days, and small molecules, such as sugars, fatty acids and alcohols in hours (Lacourt, 2011). One of the advantages of this system is that the use of pure cultures of microorganisms is not necessary, and the various microorganism consortia capable of decomposing organic substances and produce biogas are ubiquitously distributed in nature including in human or animal excreta (Marchaim, 1992, Abbasi et al., 2012). These bacteria can be activated and maintained indefinitely with proper management. Insoluble materials such as paper, straw and other lignocellulosic material may require days of treatment (Kayhanian and Hardy, 1994).

2.2. Biogas Environmental Impact

2.2.1. Positive Impacts

The environmental advantages related to biogas production are: 1) Reduction of greenhouse gas emissions due to the capture of CH_4, which is a gas that contributes significantly to climate changes (Rao and Riahi, 2006; Bogner et al., 2008); 2) Obtaining a renewable energy from wastes by reducing dependence on fossil fuels, and preserving natural resources (Panwar et al., 2011); 3) Reduction of water and groundwater pollution by removing 70%-90% of the Biological Oxygen Demand (BOD) in wastewater treatment (Kivaisi, 2001; Renou et al., 2008); 4) Confinement and elimination of microbial pathogens (Oliver et al., 2007); 5) Obtaining of high quality organic fertilizers from wastes; 6) Reduction of CO_2 emissions as a result of a reduction of the demand for fossil fuels (Wyman, 1994; Gustavsson et al., 1995; Sims et al., 2003; Sims, 2003; Akella et al., 2009); 7) Effective reduction of odors caused mainly by NH_3, H_2S, and volatile fatty acids using sealed biodigesters equipped with odor dissipation systems (Welsh et al., 1977; Demirer and Chen, 2005).

Indirectly, biogas production can also contribute in reducing the use of forest resources, particularly for household energy purposes. In this way, biogas can contribute to diminish deforestation, soil degradation and natural-associated catastrophes like flooding or desertification (Heltberg et al., 2000). In addition, since most of the raw materials used for biogas production are produced from waste, there is no competition to arable lands (United Nations, 2009).

2.2.2. Negative Impacts

Implementation of biogas technology must be seen as an alternative to solve energetic concerns, but also as an alternative to reduce pollution, which implies the use of materials that can be a contamination source, i.e., manure (Holm-Nielsen et al., 2009).

There are potentially negative environmental impacts during the production of biogas, which include: the generation of NOx and NH_3emissions from anaerobic digesters (Oenema et al., 2005; Shih et al., 2006); the release of volatile solid matter from manure if not properly handed; and the use of raw-animal by products, such as skulls, brains, trigeminal ganglia, eyes, spinal cords that could lead to exposure of organisms to prions (Wong, 1990; Lansing et al., 2010).

2.3. Future Directions and Perspectives

The bioconversion of rural and urban organic wastes and wastewater into gas using anaerobic digestion causes an increase interest to international Governments. For this purpose, local and global networks have been arising, concentrating efforts to improve knowledge about fundamental research of anaerobic microorganisms, such as development of molecular biology techniques to improve microorganisms capacity to convert organic material into biogas; the use of agents and additives to increase anaerobic digestion; the use of granular anaerobic sludge; and development of novel processes and equipment for biogas production (Holm-Nielsen et al., 2009).

There are numerous programs financing biogas research projects around the globe. Those programs are mainly focused on the study of anaerobic microorganisms and their impact on enhancement of anaerobic digestion technology. The most important advances in these subject matter lies on modification of bacteria using molecular biology techniques in order to improve bacterial capacity to convert substrates into biogas more efficiently and developing stability and tolerance of microorganisms to a wide range of environments.

3. Bio-hydrogen

For the past years researchers has been exploring new sources of energy due to the fact that the sources of fossil fuel has decreased globally. An important concern is that emissions of CO_2 into the atmosphere have reached high levels, thus compromising population's health. In recent research, biological hydrogen or bio-hydrogen (bioH_2) has been studied because it is considered a viable alternative of fuel for the future (Karapinar, 2006).

3.1. Production

BioH_2 is a clean fuel with no CO_2 emissions and with a very high-energy yield. In addition, there are many processes in which it can be obtained from raw materials (Kelly-Yong et al., 2007). Although H_2 can be obtained from fossil fuel or electrolysis of water, these methods cannot be used for bioH_2production, because the processes involved in its production result in emission of CO_2 to atmosphere (Govoni et al., 2008; Lee et al., 2010).

Production of bio$H_2$is a sustainable and waste minimization process that can be performed by anaerobic and photosynthetic microorganisms using carbohydrates and non-toxic raw materials as carbon source (Mudhoo et al., 2011). Anaerobic process consists on conversion of organic wastes into organic acids for CH_4 generation. Photosynthetic step include algae and/or photo-heterotrophic bacteria, which are able to use CO_2 (some microorganisms use organic acids) and H_2O for H_2 production. However, the rate of bioH_2 production is low and technology for this process needs to be improved (Bae et al., 2005; Turker et al., 2008).

Biomass can be used to produce bio$H_2$from non-thermal processes, using bacteria by photo-fermentation. This microbial process is based on the activity of hydrogenases (enzymes that catalyze bioH_2 production) (Bartacek et al., 2007; Patel et al., 2012). In the past years, different anaerobic and photo-fermentation processes were tested for thebioH_2production (Wukovits et al., 2007). These studies revealed that a promising way for the production of bio$H_2$from biomass consists in a two-step path, in which the first step consists of a thermophilic fermentation to produce intermediates that are sequentially used to produce H_2 and CO_2 (Sorensen, 2011).

Several metabolic pathways of non-sulfur purple photosynthetic bacteria have been shown to lead to the production of bioH_2, especially in an anaerobic atmosphere (Franchi et al., 2005). Although this study was conducted using genetically modified bacteria, the over-expression of enzymes involved in the production of H_2 can increment the production of this biofuel.

BioH_2 can be produced either by algae, phototropic bacteria or bacterial anaerobic fermentation (Fakhru'l-Razi et al., 2006). Photo-fermentation is a photosynthetic process (Loubette and Junker, 2006) that represents an easy way to obtain H_2 because it uses light as the energy source. Although bacteria are able to ferment, the levels of O_2 produced during the reaction can quantitatively reduce the H_2levels. Organic matter can be converted to H_2 and CO_2 with an efficiency of 75% (Wukovits et al., 2007).

Photosynthetic microorganisms (algae and cyanobacteria) can produce H_2 from organic acids and water using sunlight (Fakhru'l-Razi et al., 2006), but when sunlight is not available, the production of H_2 decrease considerably. Although light intensity has a stimulatory effect on H_2 yield, it has also an adverse effect on light conversion efficiency (Karapinar., 2006). Bacteria participating in this process are purple bacteria, which contain the enzyme nitrogenase capable of producing H_2 as a result of N_2 fixation (Bazac, 2007).

Although anaerobic microorganisms are able to produce H_2 from endogenous anaerobic metabolic pathways (Fakhru'l-Razi et al., 2006), there is also a dark fermentation process, which differs from photo-fermentation by the fact that it does not require sunlight to produce H_2. Advantages of this metabolic pathway are that energy does not come from converting light into energy, and then the reaction is not limited to the rate of O_2 production. Therefore, high yields of H_2 can be obtained as compared to the photo-fermentation process, but with the formation of by products such as NH_3. Products obtained from dark fermentation can provide enough amount of organic acids acting as precursors for photo-fermentation (Karapinar, 2006). In this sense, for best results in production of H_2, both processes are needed, because residual products from one process can be reutilized, and then an increase in the yield of H_2 is obtained.

Dark fermentation is highly dependent on the process conditions e.g., temperature, pH, mineral medium formulation, type of organic acids, hydraulic residence time, type of substrate and concentration, hydrogen partial pressure, and reactor configuration (Moreno-Dávila, 2011). Dark fermentation is an anaerobic system that needs a perfect balance of all of the components in the system. If any of them is missing or present in low amounts, the production of H_2 can be drastically reduced.

3.2. Environmental Concerns

The production of bioH_2 does not have environmental concerns *per se*. BioH_2offers the source of the most efficient energy and emission-free due to the release of water during its combustion.

Problems associated to bioH_2production are related to infrastructure, because of the low rates and yields of H_2 formation. Nevertheless, some research establish that emissions from a global change to H_2 will produce an increase in CH_4 and ozone, which are considered global warming gases, bringing a potential negative effect of H_2 to the environment. There has been a rising concern about H_2 leakage. When H_2 is accumulated in the stratosphere, a decrease in total ozone is observed, similar to those generated by chlorofluorocarbons.H_2 in the stratosphere increases water vapor in the ozone layer, because it can form H_2O by the reaction of H_2 with OH (Khalil and Rasmussen, 1990; Dessler et al., 1994; Hurst et al., 1999; Tromp et al., 2003; Jacobson and Golden, 2004).

3.3. Perspectives

Requirements for new sources of energy, increases in combustible prices, and environmental concerns intensify the need to obtain H_2 from biological sources. For that purpose and in order to accelerate the development of new technologies and infrastructures, the collaboration of centers of investigation, universities, and industry is needed to develop efficient processes for an optimal commercialization of bioH_2.

BioH_2must be produced under a sustainability frame, working on the conversion of solar energy into chemical energy. The fields in which that work must be addressed include molecular biology to improve the efficiency of the enzymatic reactions, light collectors and the design of photobioreactors (Akkerman et al. 2003).

4. Bioethanol

The global trend of reducing greenhouse gas emissions and dependence on fossil fuels, challenges the scientific community to develop new combustible alternatives for vehicle fuels. One such alternative is bioethanol or ethanol produced by fermentation processes.

Ethanol or ethyl alcohol is a chemical compound obtained industrially from the fermentation of sugars in plants, plant derived products or wastes known as biomass. Thus, biomass can be defined as the organic matter originated during a biological process and can be used as an energy source.

Ethanol produced from biomass is a very attractive alternative to fossil fuels because of its environmental and economic benefits. Then, the use of bioethanol as an energy source should not be seen as a complement to solar, wind or other sources of energy (Lin and Tanaka, 2006).

The main source of sugars for the production of bioethanol is found in the supporting tissue of plants, which consists mainly of cellulose, a polymer of β-glucose. Other plant parts, such as seeds, roots, tubers and fruits, contain other types of sugars, such as starch, sucrose, fructose, and free glucose. This richness in carbohydrates is the main source of materials for the production of bioethanol.

4.1. Production of Bioethanol

The production of bioethanol consists mainly in four steps (Figure 2):

1. Conversion of biomass to fermentable sugars.
2. Adjustment of the sugar concentration.
3. Sugar fermentation carried out by microorganisms, such as yeast.
4. Water elimination.

These steps correspond strictly to the production of bioethanol and do not include previous steps, such as land preparation, cultivation, harvesting, packing, transport, distribution, and waste treatment and management. The relevant importance of each of these steps will depend on the source used for the production of bioethanol and thus, they have to be carefully taken in account to develop an efficient process for bioethanol production with the lowest environmental impact.

Sources of Biomass

There are several sources of biomass that can be grouped as follows:

1. Wood waste, mainly from paper industries, furniture and woodworking.
2. Municipal solid waste.
3. Agricultural residues.
4. Crops for bioethanol production.

In order to be suitable for an efficient fermentation in the bioethanol process, biomasses have to undergo different types of pre-treatments in order to make fermentable sugars available to fermentation. For example, sugars from fruit, sugar cane, and sugar beet are directly converted into ethanol by fermentation and do not require a complex pre-treatment, at most an adjustment in the concentration of sugars and the pH. When using starches from different sources, such as corn, wheat, oat, rice, potato, cassava, or sorghum; a pre-treatment is required. This pre-treatment is performed in a dual-step process consisting in liquefying the starch at 150°C in presence of α-amylase, and then, the products of this reaction are treated with glucoamylase in order to convert them to glucose. This process is very efficient because high yields of ethanol are obtained; however, production costs are high due to high working temperatures. Nevertheless, a pre-treatment at low temperatures have been developed to reduce the high costs associated to the high temperatures (Matsumoto et al., 1985).

Biomass from agricultural waste and wood contains mainly cellulose. This polymer has to be depolymerized with inorganic acids (hydrolysis) to obtain glucose, which is rapidly converted to ethanol by microbial fermentation. However, the major problem affecting this source of biomass is the presence of lignocellulose, another major polymer consisting of cellulose, hemicellulose and lignin. Hemicellulose is a complex structure, consisting of various carbohydrates, such as pentoses (xylose and arabinose), hexoses (glucose, mannose and galactose), and uronic acids. The main hemicellulosic components of hardwoods are xylans, whereas glucomannan is in softwoods. Hemicellulose is of lower molecular weight compared to cellulose and contains short side chain branches with different easily hydrolyzable sugars (Hendriks and Zeeman, 2009).

Lignin is a hydrophobic polymer biosynthesized by the polymerization of three types of irregular phenylpropanoids that includes *p*-coumaryl, coniferyl, and sinapyl alcohols. Lignin is intercalated between the cellulosic fibers and hemicellulose conferring rigidity and protection from physical and chemical damages, and giving resistance to the acidic hydrolysis. For this reason, in the process of ethanol production from lignocellulosic materials, an additional step called delignification or release of the hemicellulose and cellulose from lignin is necessary. As expected, this delignification step in this process consumes energy, and thus raises production costs and environmental impacts. Moreover, the delignification process is also a critical step that must be executed under controlled conditions in order to favor a high sugar concentration to produce large amounts of ethanol. If the process is not controlled properly, several factors can reduce or even stop the production of ethanol. These factors include: (1) High concentrations of alcohol that are detrimental for the microbial fermentation; (2) High concentrations of salts in the starting lignocellulosic materials can generate an osmotic stress to the microorganism; and (3) Inhibitors of the enzyme cellulase generated from lignin decomposition (Taherzadeh and Karimi, 2011; Lin and Tanaka, 2006).

As a result of these potential concerns, there have been and continue to be many studies on improvement of conversion of lignocellulosic materials into ethanol, which are summarized in Table 4 (Karimi et al. 2006; Sanchez et al. 2004; Schell et al. 2003; Tucker et al. 2003, Nguyen et al. 2000, Lee et al. 1999; Barl et al. 1991; Arato et al. 2005; Sidiras and Koukios 2004, Alizadeh et al. 2005; Sassner et al. 2005; Ohgren et al. 2005; Berlin et al. 2006).

Table 4. Biomass pre-treatments used in lignocellulosic materials for bioethanol production*

Method	Change in Biomass	Advantages and Disadvantages	Examples
Physical	Increases surface accessibility and pore size Decreases crystallinity of cellulose Reduces the polymer size Lignin is not eliminated completely	High energy demand Not recommended for industrial applications No chemicals involved	Grinding Irradiation
Chemical and physico-chemical	Increases surface accessibility Decreases crystallinity of cellulose Reduces the polymer size Lignin is eliminated almost completely Partial to complete elimination of hemicellulose	Recommended for industrial applications Fast process Harsh conditions Chemicals involved	Alkalis Acids Explosion Gases Lignin dissolvent Oxidant agents
Biologic	Lignin eliminated Reduces the polymer size Partial elimination of hemicellulose	Low energy demand No chemicals involved Mild conditions Treatment velocity low Not for commercial applications	Fungi Actinomycetes

* Adapted from Taherzadeh and Karimi, 2011.

Of all the possible raw material sources for bioethanol production, lignocellulose is of high interest due to their abundance worldwide. The global production of plant biomass is estimated at approximately $2x10^{11}$ tons/year, of which about 90% is lignocellulose, which makes approximately $2x10^{10}$ tons of plant biomass potentially available for the production of bioethanol (Lin and Tanaka, 2006).

United States and Brazil are the largest countries producing bioethanol from corn and sugarcane, respectively, leading with a 60% of the worldwide bioethanol production (Chandel et al., 2007). Another raw material used for biethanol production is cassava led by China and Thailand (Botero et al., 2011).

4.2. Environmental Impact in the Production of Bioethanol

One of the tools used to calculate the environmental impact of biofuels is the life cycle analysis (LCA). This tool can quantify the environmental impacts at different stages of the production of bioethanol, such as soil preparation, crop harvesting, transportation, production, and use of biofuels. This analysis is based on the ISO 14040 and 14044 standards, allowing not only the identification of critical points in the production of biofuels, but also can be used to compare to other sources of energy (Botero et al., 2011).

LCA results are based on ethanol formulations or gasohol (ethanol and gasoline mixtures) and are expressed with a capital letter and a number, which indicates the percentage of ethanol in the mixture. For instance, formulations E10 or E85 indicate that 10 or 85% of the mixture is ethanol, while the remaining percentage corresponds to conventional gasoline.

It has been reported that in the LCA of ethanol produced from eucalyptus in three different formulations (E10, E85 and E100), a reduction in the environmental impact in most of the categories studied was observed. However, these impact reductions did not apply to the formation of compounds photochemically oxidants, eutrophication, and ecotoxicity in land and sea (González-García et al., 2012). It means, among other things, that biomass production from forests should be oriented to the use of more efficient machinery, a reduction in the dose of fertilizers and pesticides, and to control the diffuse emission or foci not defined.

Another factor to be measured is the efficiency of energy conversion or energetic balance, which measures the relationship between the inputs vs. the outputs of energy. Although the optimal value is still under discussion, it is desired to be >1. For example, a study indicated that the effective value is from 1.28, but another study reported effective values greater than 1.43 (Yu and Tao, 2009). In general, studies reported have calculated positive energy balances. For instance, the use of cassava reported an energetic balance of 1.28 (Leng et al., 2008), whereas energetic balances of 1.43 and 1.34 were reported in China and Colombia, respectively (Yu and Tao, 2009; Botero Agudelo et al., 2011).Furthermore, studies of the environmental impact of bioethanol produced from paper waste showed that the profile of the impact of bioethanol produced from newsprint, office paper, magazines, and cardboard on the environment is favorable compared to gasoline (Wang et al., 2012).

Lastly, another important indicator of the impact to the environment is the carbon footprint, which is the total amount of greenhouse gases emitted directly or indirectly by any entity or product, indicated as the mass of equivalent CO_2(CO_2e). At present, there is no agreement in studies if the carbon balance for bioethanol is positive or negative (Johnson, 2008; Botero, 2011; Dias de Oliveira et al., 2005).Then, energetic analyses for future studies would depend on how the study is conducted, its limitations and scope, raw materials used, production and processing, and the region where the study is conducted.

4.3. Future Directions and Perspectives

Bioethanol production has its advantages and disadvantages. On one hand, the feasibility of using renewable waste materials makes it attractive to replace the use of fossil fuels, but on the other hand, further studies have to be conducted to achieve an efficient environmentally friendly process. Paradoxically, many of the benefits obtained by the use of bioethanol are lost by the processes involved in its production that are often not ecologically compatible.

Also, the replacement of cultivated land dedicated to produce food for humans generates economic conflicts that, in a large-scale production could mean shortages of food for the population, as farmers would prefer to use their crops for the production of ethanol, due to better economic benefits. Therefore, the production of bioethanol from biomass originated from crops has to be tight regulated to avoid these conflicts. To minimize these conflicts, the growth of crops that do not compete with those intended for human consumption is encouraged. Given the urgent energy situation facing humanity that will worsen with the passage of time, it is urgent to conduct studies to improve technologies, processes, and methods of producing ethanol and biofuels in general.

A comparative summary of the four biofuels described is shown in table 5.

Table 5. Biofuel characteristics

Biofuel	Source of biomass	% Yield	Positive effects	Negative effects
Biodiesel	Edible oil. Inedible oil. Used cooking oil.	80-99	Lower emission of CO_2, CO, SO_2, hydrocarbons and particulate matter. Regional development. Production sustainable. Better ecologic efficiency. Good net energy ratio (NER).	Increase NOx emission. Ozone layer depletion. Eutrophication. Acidification. Competition with the food market*.
Biohydrogen	Carbohydrates and non-toxic raw materials (manure slurry, crop straw, solid wastes).	35-40	Clean fuel with no CO_2 emissions. High energy yield. Renewable energy from wastes.	Increase in CH_4 and ozone.
Biogas	Crop residues, manure, household waste, water hyacinth, algae, effluents from food industry, drinks, pulping and paper, and some chemicals.	18-25	Reduction of greenhouse gas emissions (capture of CH_4). Renewable energy from wastes. Reduction of water and groundwater pollution. Confinement and elimination of microbial pathogens. Effective reduction of odors.	Generation of NOx and NH_3 emissions from anaerobic digesters. Release of volatile solid matter from manure if not properly handed. Exposure to prions, depending on the carbon source (biologic material).
Bioethanol	Sugar from fruit, cane, or beet.	High	Do not require a complex pretreatment.	Raw material is used as a human food source.
	Starch.	High	May be used some raw material not suitable for human food.	Pretreatment is required. High production costs. Many raw material is used as a human food source.
	Agricultural and wood wastes.	Medium to high.	Raw material is not used as a human food source.	Pretreatment is required. Many pollutants from the pretreatment. High production costs.

*Only when edible oils were used

ACKNOWLEDGMENT

We thank Jeffrey Helm for helpful discussions.

REFERENCES

Abbasi T., Tauseef S. M., Abbasi S. A. (2012). *Biogas energy* (Vol. 2). Springer.

Achten W. M. J., Almeida J., Fobelets V., Bolle E., Mathijs E., Singh V. P., Tewari D. N., Verchot L. V., Muys B. (2010). Life cycle assessment of Jatropha biodiesel as transportation fuel in rural India. *Applied Energy*, 87(12), 3652-3660.

Acquaye A. A., Wiedmann T., Feng K., Crawford R. H., Barrett J., Kuylenstierna J., Duffy, A. P., Koh, S. C. L., McQueen-Mason S., (2011). Identification of Carbon Hot-Spots' and Quantification of GHG Intensities in the Biodiesel Supply Chain Using Hybrid LCA and Structural Path Analysis. *Environmental Science & Technology*, 45(6), 2471-2478.

Adi G., Hall C., Snyder D., Bunce M., Satkoski C., Kumar S., Garimella P., Stanton D., and Shaver G. (2009). Soy-Biodiesel Impact on NOx Emissions and Fuel Economy for Diffusion-Dominated Combustion in a Turbo-Diesel Engine Incorporating Exhaust Gas Recirculation and Common Rail Fuel Injection. *Energy Fuels*, 23, 5821–5829.

Akella A. K., Saini R. P., Sharma M. P. (2009). Social, economical and environmental impacts of renewable energy systems. *Renewable Energy*, *34*(2), 390-396.

Akkerman, M., Janssen J. M., Rocha S., Reith J. H., Wijffels R. H. (2003). Photobiological hydrogen production: Photochemical efficiency and bioreactor design, In Biomethane and Biohydrogen Status and perspectives of biological methane and hydrogen production Edited by J. H. Reith, R. H. Wijffels and H. Barten.

Alizadeh, H.; Teymouri, F.; Gilbert, T. I.; Dale, B. E. (2005). Pretreatment of switchgrass by ammonia fiber explosion (AFEX). *Appl. Biochem. Biotechnol.*124, 1133-41.

Arato, C.; Pye, E. K.; Gjennestad, G. (2005). The lignol approach to biorefining of woody biomass to produce ethanol and chemicals. *Appl. Biochem. Biotechnol.*123, 871-882.

Bae J. H., Bardiya N., Reddy M. P. (2005). Bio-hydrogen: technology status and future prospects. *Wealth from Waste: Trends and Technologies*, 87.

Banerjee T., Bhattacharya T. K., Gupta, R. K. (2009). Assessment of fuel characteristics and emissions of biodiesel from jatropha. *Pollution Research*, 28(2), 143-148.

Barl, B.; Biliaderis, C. G.; Murray, E. D.; Macgregor, A. W. (1991). Combined chemical and enzymatic treatments of corn husk lignocellulosics. *J. Sci. Food Agric.*56, 195-214.

Bartacek J., Zabranska J., Lens P. N. (2007). Developments and constraints in fermentative hydrogen production. *Biofuels, Bioproducts and Biorefining*, *1*(3), 201-214.

Batan L., Quinn J., Willson B., Bradley T. (2010). Net Energy and Greenhouse Gas Emission Evaluation of Biodiesel Derived from Microalgae. *Environmental Science & Technology*, 44(20), 7975-7980.

Basak N., Das D. (2007). The prospect of purple non-sulfur (PNS) photosynthetic bacteria for hydrogen production: the present state of the art. World *J Microbiol Biotechnol*, 23:31-42.

Beatrice C, Guido C., Napolitano P. (2012). Implementation of the Closed-Loop Combustion Control Methodology in Modern Automotive Diesel Engines for Low-End Torque Increment Burning Biodiesel, Energy & Fuels, 26, 1305−1314.

Berlin, A.; Balakshin, M.; Gilkes, N.; Kadla, J.; Maximenko, V.; Kubo, S.; Saddler, J. (2006). Inhibition of cellulase, xylanase and beta-glucosidase activities by softwood lignin preparations. *J. Biotechnol.*125, 198-209.

Bogner J., Pipatti R., Hashimoto S., Diaz C., Mareckova K., Diaz L., Gregory, R. (2008). Mitigation of global greenhouse gas emissions from waste: conclusions and strategies from the Intergovernmental Panel on Climate Change (IPCC) Fourth Assessment Report. Working Group III (Mitigation). *Waste Management & Research*, *26*(1), 11-32.

Botero Agudelo, J.; Castaño Peláez, H.; Naranjo Merino, C. (2011). Life Cycle Assessment for bioethanol produced from cassava in Colombia. Producción + Limpia 6, 69-77.

Carolan J. E., Joshi S. V., Dale B. E. (2007). Technical and financial feasibility analysis of distributed bioprocessing using regional biomass pre-processing centers. *Journal of Agricultural & Food Industrial Organization*, *5*(2).

Chandel, A. K.; Chan, E. S.; Ravinder Rudravaram, M. Lakshmi Narasu, L. Venkateswar Rao and Pogaku Ravindra (2007). Economics and environmental impact of bioethanol production technologies: an appraisal Biotechnology and Molecular Biology Review. 2, 14-32.

Chantaraporn Phalakornkule and Maneerat Khemkhao (2012). Enhancing Biogas Production and UASB Start-Up by Chitosan Addition, Biogas, Dr. Sunil Kumar (Ed.), ISBN: 978-953-51-0204-5, InTech, Available from: http://www.intechopen.com/books/biogas/enhancing-biogas-production-and-uasb-start-up-by-chitosan-addition

Chatterjee R., Sharma V. Kumar S. (2012). Life cycle assessment of energy performance of biodiesel produced from Jatropha curcas. *Journal of Renewable and Sustainable Energy*, 4(5), 053110/1-053110/13.

Chin J., Batterman S. A., Northrop W. F., Bohac S. V. and Assanis D. N. (2012). Gaseous and Particulate Emissions from Diesel Engines at Idle and under Load: Comparison of Biodiesel Blend and Ultralow Sulfur. *Diesel Fuels Energy Fuels*, 26, 6737−6748.

Crutzen P. J., Mosier A. R., Smith K. A., and Winiwarter W. (2008). N_2O release from agro-biofuel production negates global warming reduction by replacing fossil fuels. *Atmospheric Chemistry and Physics*, 8, 389–395.

Dayong A. J., Yun B. B. (2012). Performances on engine-out emissions and combustion of ethylene glycol monobutyl ether palm oil monoester as a newtype biodiesel. *Journal of Renewable and Sustainable Energy*, 4(5).

Demirer G. N., Chen S. (2005). Two-phase anaerobic digestion of unscreened dairy manure. *Process Biochemistry*, *40*(11), 3542-3549.

Deng X., Han J., Yin F. (2012). Net energy, CO_2 emission and land-based cost-benefit analyses of Jatropha biodiesel: a case study of the Panzhihua region of Sichuan Province in China. *Energies* (Basel, Switzerland), 5, 2150-2164.

Dennis Y.C. Leung , Xuan Wu, Leung M. K. H., (2010). A review on biodiesel production using catalyzed transesterification. *Applied Energy* 87 1083–1095.

Dessler A. E., Weinstock E. M., Hintsa E. J., Anderson J. G., Webster C. R., May R. D., Elkins J. W., Dutton G. S. (1994). An examination of the total hydrogen budget of the lower stratosphere, *Geophys. Res. Lett.*, 21, 2563-2566.

Dias De Oliveira, M. E.; Vaughan, B. E.; Rykiel Jr., E. J. (2005). *Ethanol* as Fuel: Energy, Carbon Dioxide Balances, and Ecological Footprint. *Bioscience*. 55, 593-602.

Fang T., Lin Y. C., Foong T. M., Lee C. (2008). Reducing NOx Emissions from a Biodiesel-Fueled Engine by Use of Low-Temperature Combustion. *Environmental Science & Technology*, 42(23), 8865-8870.

Fangrui M., Milford A. H., (1999). Biodiesel production: a review. *Bioresource Technology* 70 1-15.

Fakhru'l-Razi, A., El-Mahdi, A.H., Luqman Chuah, T.G.A. (2006). State of the Art of Bio-Hydrogen Production from Biomass. *Jurutera – IEM Bulletin*, 2006 (11) 8-14.

Fore, S. R.; Porter, P.; Lazarus, W. (2011). Net energy balance of small-scale on-farm biodiesel production from canola and soybean, *Biomass and Bioenergy*. 35 85), 2234-2244.

Funda Cansu Ertem (2011). Improving biogas production by anaerobic digestion of different substrates. Calculation of Potential Energy Outcomes. Halmstad University. *Master Thesis*. Pp- 8-10.

Gayh (2012). Process intensification of biological desulphurisation of biogas. Thesis Dissertation, Technische Universität Hamburg-Harburg, Germany. Institut für Abwasserwirtschaft und Gewässerschutz.

Ghaly A.E., Dave D., Brooks M.S. and Budge S. (2010). Production of Biodiesel by Enzymatic Transesterification: Review. *American Journal of Biochemistry and Biotechnology* 6 (2): 54-76.

González-García S., García-Rey D., Hospido A. (2013). Environmental life cycle assessment for rapeseed-derived biodiesel. *International Journal Life Cycle Assess*, 18:61–76.

González-García, S.; Moreira, T.; Feijoo, G. (2012). Environmental aspects of eucalyptus based ethanol production and use. *Science of the Total Environment* 438, 1-8. DOI: 10.1016/j.scitotenv.2012.07.044

Govoni C., Morosinotto T., Giuliano G., Bassi R. (2008). Exploiting Photosynthesis for Biofuel Production. In *Biophotonics* (pp. 15-28). Springer Berlin Heidelberg.

Gustavsson L., Börjesson P., Johansson B., Svenningsson P. (1995). Reducing CO_2 emissions by substituting biomass for fossil fuels. *Energy*, *20*(11), 1097-1113.

Hajbabaei M., Johnson K. C., Okamoto R. A., Mitchell A., Pullman M. and Durbin T. D. (2012). Evaluation of the Impacts of Biodiesel and Second Generation Biofuels on NOx Emissions for CARB Diesel Fuels. *Environ. Sci. Technol.* 46, 9163−9173.

Heltberg R., Arndt T. C., Sekhar N. U. (2000). Fuelwood consumption and forest degradation: a household model for domestic energy substitution in rural India. *Land Economics*, 213-232.

Hendriks, A. T. W. M., Zeeman, G. (2009). Pretreatments to enhance digestibility of lignocellulosic biomass. *Bioresource Technology*. 100, 10-18.

Hilkiah Igoni A., Ayotamuno M. J., Eze C. L., Ogaji S. O. T., Probert S. D. (2008). Designs of anaerobic digesters for producing biogas from municipal solid-waste. *Applied energy*, *85*(6), 430-438.

Holm-Nielsen J. B., Al Seadi T., Oleskowicz-Popiel P. (2009). The future of anaerobic digestion and biogas utilization. *Bioresource Technology*, *100*(22), 5478-5484.

Hou J., Zhang P., Yuan X., Zheng Y. (2011). Life cycle assessment of biodiesel from soybean, jatropha and microalgae in China conditions. *Renewable & Sustainable Energy Reviews*, 15(9), 5081-5091.

Hurst D. F., et al., (1999). Closure of the total hydrogen budget of the northern extratropical lower stratosphere, *J. Geophys. Res.*, 104, 8191-8200.

Jakobsen S. T. (1992). Chemical reactions and air change during decomposition of organic matters. *Resources, conservation and recycling*, 6(3), 259-266.

Jayed, M. H.; Masjuki, H. H.; Saidur, R.; Kalam, M. A.; Jahirul, M. I. (2009). Environmental aspects and challenges of oilseed produced biodiesel in Southeast Asia. *Renewable & Sustainable Energy Reviews,* 13(9), 2452-2462.

Johnson, E. (2008). Goodbye to carbon neutral: Getting biomass footprints right. *Environmental Impact Assessment Review.* 29, 165–168.

Johnston M., Hollowayt. (2007). A Global Comparison of National Biodiesel Production Potentials. *Environmental Science & Technology*, 41(23) 7967-7973.

Karapinar K.I., Kargi F., (2006). Bio-hydrogen production from waste materials. *Enzyme and Microbial Technology* 38 569–582.

Karimi, K.; Kheradmandinia, S.; Taherzadeh, M. J. (2006). Conversion of rice straw to sugars by dilute acid hydrolysis. *Biomass Bioenerg.*30, 247-253.

Kayhanian M. and Hardy S. (1994). The impact of four design parameters on the performance of a high-solids anaerobic digestion of municipal solid waste for fuel gas production. *Environmental technology*, *15*(6), 557-567.

Kelly-Yong T. L., Lee K. T., Mohamed A. R., Bhatia S. (2007). Potential of hydrogen from oil palm biomass as a source of renewable energy worldwide. *Energy Policy*, *35*(11), 5692-5701.

Khalil M. A. K. and Rasmussen R. A. (1990). Global increase of atmospheric molecular hydrogen, *Nature*, 347, 743-745, 1990.

Kivaisi A. K. (2001). The potential for constructed wetlands for wastewater treatment and reuse in developing countries: a review. *Ecological Engineering*, *16*(4), 545-560.

Koh M. Y.; Ghazi T. I. M. (2011). A review of biodiesel production from Jatropha curcas L. oil. *Renewable & Sustainable Energy Reviews*, 15(5), 2240-2251.

Kovacs A. J. (2012). Capacity and efficiency in small- to medium-sized biodiesel production systems: increasing profitability through agroindustrial ecology principles. *Journal of Industrial Ecology*, 16(1), 153-162.

Lacourt W. (2011). Enrichment of Methanogenic Microcosms on Recalcitrant Lignocellulosic Biomass (Doctoral dissertation, University of Toronto, Canada).

Lansing S., Martin J. F., Botero R. B., Da Silva T. N., Da Silva E. D. (2010). Methane production in low-cost, unheated, plug-flow digesters treating swine manure and used cooking grease. *Bioresource technology*, *101*(12), 4362-4370.

Lee H. S., Vermaas W. F., Rittmann B. E. (2010). Biological hydrogen production: prospects and challenges. *Trends in biotechnology*, *28*(5), 262-271.

Lee S., Park J., Brissonneau D. (2003). Biogas Generation and Recovery Potential within Selected Agro-Industries and Solid Waste Management Sector in Thailand. E*nvironmental Engineering Research-Seoul*, *8*(3), 107-115.

Lee, Y. Y.; Iyer, P.; Torget, R. W. (1999). Dilute-acid hydrolysis of lignocellulosic biomass. *Adv. Biochem. Eng. Biotechnol.*65, 93-115.

Leng, R.; Wang, C.; Zhang, C.; Dai D, P. G. (2008). Life cycle inventory and energy analysis of cassava-based Fuel ethanol in China. *J. Clean. Prod.* 16, 374-384.

Lin, Y.; Tanaka, S. (2006). Ethanol fermentation from biomass resources: Current state and prospects. *Appl. Microbiol. Biotechnol.* 69, 627-642.

Marchaim U. (1992). Biogas processes for sustainable development (Vol. 95). Food & Agriculture Org.

Martins das Neves L. C., Converti A., Vessoni Penna T. C. (2009). Biogas production: new trends for alternative energy sources in rural and urban zones. *Chemical engineering & technology*, *32*(8), 1147-1153.

Mata-Álvarez J., Macé S., Llabrés, P. (2000). Anaerobic digestion of organic solid wastes. An overview of research achievements and perspectives. Bioresource Technology, 74(3), 16.

Matsumoto, N.; Yoshizumi, H.; Miyata, S.; Inoue, S. (1985). Development of the non-cooking and low temperature cooking systems for alcoholic fermentation of grains. Nippon Nogeikagaku Kaishi. 59, 291–299.

McCormick R. L., Westbrook S. R. (2010). Storage Stability of Biodiesel and Biodiesel Blends, *Energy Fuels*, 24, 690–698.

Moreno-Dávila I.M.M., Ríos-González L.J., Garza-García Y., Rodríguez-de la Garza J.A., Rodríguez-Martínez J. (2011). Biohydrogen production from diary processing wastewater by anaerobic biofilm reactors. African *Journal of Biotechnology,* 10 (27), 5320-5326.

Mudhoo A., Forster-Carneiro T., Sánchez A. (2011). Biohydrogen production and bioprocess enhancement: a review. *Critical reviews in biotechnology*, *31*(3), 250-263.

Nanaki E. A., Koroneos C. J. (2012). Comparative LCA of the use of biodiesel, diesel and gasoline for transportation, *Journal of Cleaner Production*, 20(1), 14-19.

Nanasaheb D. S., Chandrakant M. K. (2012). Preparation of Methyl Esters from Thespesia populnea L. Oil and its Engine Exhausts Studies. *International Journal of Green Energy*, 9(2), 130-138.

Nasim M. N. (2012). Simulation and environmental assessment of compression ignition engine powered by neat biodiesels of different origin. *Indian Journal of Science and Technology*, 5(7), 3017-3021.

Nematullah N. M. (2012). Simulation and environmental assessment of compression ignition engine powered by neat biodiesels of different origin. *Indian Journal of Science and Technology*, 5(7), 3017-3021.

Nguyen, Q. A.; Tucker, M. P.; Keller, F. A.; Eddy, F. P. (2000). Two-stage dilute-acid pretreatment of softwoods. *Appl. Biochem. Biotechnol.*84-86, 561-576.

No, Soo-Young (2010). Inedible vegetable oils and their derivatives for alternative diesel fuels in CI engines: A review. *Renewable & Sustainable Energy Reviews*, 15(1), 131-149.

Oenema O., Wrage N., Velthof G. L., van Groenigen J. W., Dolfing J., Kuikman, P. J. (2005). Trends in global nitrous oxide emissions from animal production systems. *Nutrient Cycling in Agroecosystems*, *72*(1), 51-65.

Ohgren, K.; Galbe, M.; Zacchi, G. (2005). Optimization of steam pretreatment of SO_2-impregnated corn stover for fuel ethanol production. *Appl. Biochem. Biotechnol.*121, 1055-1067.

Oliver D. M., Heathwaite A. L., Hodgson C. J., Chadwick D. R. (2007). Mitigation and current management attempts to limit pathogen survival and movement within farmed grassland. *Advances in agronomy*, *93*, 95-152.

Özkan M. (2007). Comparative Study of the Effect of Biodiesel and Diesel Fuel on a Compression Ignition Engine's Performance, Emissions, and Its Cycle by Cycle Variations, *Energy & Fuels*, 21, 3627–3636.

Paepatung N., Nopharatana A., Songkasiri W. (2009). Bio-methane potential of biological solid materials and agricultural wastes. *Asian Journal of Energy and Environment*, *10*(1), 19-27.

Panwar N. L., Kaushik S. C., Kothari S. (2011). Role of renewable energy sources in environmental protection: A review. *Renewable and Sustainable Energy Reviews*, *15*(3), 1513-1524.

Patel S. K., Kumar P., Kalia V. C. (2012). Enhancing biological hydrogen production through complementary microbial metabolisms. *International Journal of Hydrogen Energy*.

Perot C. and Amar, D. (1989). Optimisation of sludge anaerobic digestion by separation of hydrolysis-acidification and methanogenesis. *Environmental Technology*, 10(7), 633-644.

Persson S. P. E., Bartlett H. D., Branding A. E., Regan R. W. (1979). Agricultural Anaerobic Digesters Design and Operation, Bulletin 827, November 1979, The Pennsylvania State University, College of Agriculture, Agricultural Experiment Station, University Park, Pennsylvania

Petrou E. C and Pappis C. P. (2009). Biofuels: A Survey on Pros and Cons. *Energy & Fuels*, 23, 1055–1066.

Qian J., Yun Z. (2009). Cogeneration of Biodiesel and Nontoxic Cottonseed Meal from Cottonseed through in Situ Alkaline Transesterification. *Energy & Fuels*, 23, 507–512.

Rao, S. and Riahi, K. (2006) The role of non-CO_2 greenhouse gases in climate change mitigation: long-term scenarios for the 21st century. *Energy Journal*, *27*(3), 177-200.

Renou S., Givaudan J. G., Poulain S., Dirassouyan F., Moulin P. (2008). Landfill leachate treatment: Review and opportunity. *Journal of hazardous materials*, *150*(3), 468-493.

Rodriguez C. C., Villela A. C., Silveira J. L. (2010). Ecological efficiency in CHP: biodiesel case. *Applied Thermal Engineering*, 30(5), 458-463.

Rustandi F., Wu H. (2010). Biodiesel Production from Canola in Western Australia: Energy and Carbon Footprints and Land, Water, and Labour Requirements. *Industrial & Engineering Chemistry Research,* 49(22), 11785-11796.

Sadi M. A. (2010). *Design and Building of Biogas Digester for Organic Materials Gained from Solid waste* (Doctoral dissertation, National University, Palestine).

Saleh A. F., Kamarudin E., Yaacob A. B., Yussof A. W., Abdullah M. A. (2012). Optimization of biomethane production by anaerobic digestion of palm oil mill effluent using response surface methodology. *Asia-Pacific Journal of Chemical Engineering*, 7(3), 353-360.

Sánchez, G.; Pilcher, L.; Roslander, C.; Modig, T.; Galbe, M.; Liden, G. (2004). Dilute-acid hydrolysis for fermentation of the Bolivian straw material Paja Brava. *Bioresource Technol.* 93, 249-256.

Santibañez-Aguilar J. E, González-Campos J. B., Ponce-Ortega J. M., Serna-González M., and El-Halwagi M. M. (2011). Optimal Planning of a Biomass Conversion System Considering Economic and Environmental Aspects, *Ind. Eng. Chem. Res.*, 50, 8558–8570.

Santos A. M. S. D. (2011). Anaerobic digestion of organic matter. Biogas production from energy crop residues (Maize stalks).

Saqib M., Mumtaz M. W., Mahmood A., Abdullah M. I. (2012). Optimized biodiesel production and environmental assessment of produced biodiesel. *Biotechnology and Bioprocess Engineering*, 17(3), 617-623.

Sassner, P.; Galbe, M.; Zacchi, G. (2005). Steam pretreatment of Salix with and without SO_2 impregnation for production of bioethanol. *Appl. Biochem. Biotechnol.*121, 1101-1117.

Schell, D. J.; Farmer, J.; Newman, M.; McMillan, J. D. (2003). Dilute-sulfuric acid pretreatment of corn stover in pilot-scale reactor: Investigation of yields, kinetics, and enzymatic digestibilities of solids. *Appl. Biochem. Biotechnol.*105, 69-85.

Schmidt U. (1974) Molecular hydrogen in the atmosphere, *Tellus*, 26, 78-90, 1974.

Shih J. S., Burtraw D., Palmer K., Siikamäki J. (2006). Air Emissions of Ammonia and Methane from Livestock Operations. *RFF DP 06, 11.*

Sidiras, D.; Koukios, E. (2004). Simulation of acid-catalysed organosolv fractionation of wheat straw. *Bioresource Technol.*94, 91-98.

Sims R. E. (2003). Bioenergy to mitigate for climate change and meet the needs of society, the economy and the environment. *Mitigation and Adaptation Strategies for Global Change*, *8*(4), 349-370.

Sims R. E., Rogner H. H., Gregory K. (2003). Carbon emission and mitigation cost comparisons between fossil fuel, nuclear and renewable energy resources for electricity generation. *Energy Policy*, *31*(13), 1315-1326.

Soetaert W. and Vandamme E. J., (2009). Biofuels, Wiley Series in Renewable Resources, ISBN 978-0-470-02674-8.

Sorensen B. (2011) *Hydrogen and fuel cells: emerging technologies and applications.* Academic Press.

Sreekrishnan T. R., Kohli S., Rana V. (2004). Enhancement of biogas production from solid substrates using different techniques—a review. *Bioresource Technology*, *95*(1), 1-10.

Stabnikova O., Wang J. Y., Ivanov V. (2010). Value-added biotechnological products from organic wastes. In *Environmental Biotechnology* (pp. 343-394). Humana Press.

Taherzadeh M. J., Karimi K. (2008). Pretreatment of lignocellulosic wastes to improve ethanol and biogas production: a review. *International Journal of Molecular Sciences*, *9*(9), 1621-1651.

Taherzadeh, M. J.; Karimi, K. (2011). Fermentation Inhibitors in ethanol processes and different strategies to reduce their effects. In Biofuels, Alternative Feedstocks and Conversion Processes. pp: 287- 311. ISBN 978-0-12-385099-7. Elsevier Science & Technology.

Tonini D., Astrup T. (2012). LCA of biomass-based energy systems: A case study for Denmark. *Applied Energy*, 99, 234-246.

Tromp T. K., Run-Lie S., Allen M., Eiler J. M., Yung Y. L. (2003). Potential Environmental Impact of a Hydrogen Economy on the Stratosphere, *Science* 300:5626, 1740-1742

Tucker, M. P.; Kim, K. H.; Newman, M. M.; Nguyen, Q. A. (2003). Effects of temperature and moisture on dilute-acid steam explosion pretreatment of corn stover and cellulase enzyme digestibility. *Appl. Biochem. Biotechnol.*105, 165-177.

Turker L., Gumus S., Tapan A. (2008). Biohydrogen production: molecular aspects. *Journal of Scientific and Industrial Research*, *67*(11), 994.

United Nations Environment Programme. Biofuels Working Group, & United Nations Environment Programme. International Panel for Sustainable Resource Management. (2009). *Towards sustainable production and use of resources: Assessing biofuels.* United Nations Envir Programme.

Van Horn H. H. and Hall M. B. (1997). Agricultural and environmental issues in the management of cattle manure. In *ACS Symposium Series* (Vol. 668, pp. 91-109). American Chemical Society.

Verma S. (2002). *Anaerobic digestion of biodegradable organics in municipal solid wastes* (Doctoral dissertation, Columbia University, United States).

Wang, L.; Templer, R.; Murphy, R. J. (2012). Environmental sustainability of bioethanol production from waste papers: sensitivity to the system boundary. *Energy & Environmental Science*. 5, 8281-8293. DOI: 10.1039/c2ee21550k

Wang Z., Calderon M. M., Lu Y. (2011). Lifecycle assessment of the economic, environmental and energy performance of Jatropha curcas L. biodiesel in China. *Biomass and Bioenergy*, 35(7), 2893-2902.

Welsh F. W., Schulte D. D., Kroeker E. J., Lapp H. M. (1977). The effect of anaerobic digestion upon swine manure odors. *Canadian Agricultural Engineering*, *19*(2), 122-126.

WMO (World Meteorological Organization), Atmospheric Ozone 2010. Scientific Assessment of Ozone Depletion, Global Ozone Research and Monitoring Project–Report No. 52, Geneva, Switzerland.

Wong M. H. (1990). Anaerobic digestion of pig manure mixed with sewage sludge. *Biological wastes*, *31*(3), 223-230.

Wukovits W., Friedl A., Markowski M., Urbaniec K., Ljunggren M., Schumacher M., Zacchi G., Modigell M., (2007). Identification of a Suitable Process Scheme for the Non-Thermal Production of Biohydrogen. *Chem. Eng. Trans.* 12, 315-320.

Wyman C. E. (1994). Alternative fuels from biomass and their impact on carbon dioxide accumulation. *Applied biochemistry and biotechnology*, *45*(1), 897-915.

Xing A., Ma J., Zhang Y., Wang Y., Jin Y. (2010). Life cycle assessment of biodiesel environmental effects. Qinghua Daxue Xuebao, Ziran Kexueban, 50(6), 917-922.

Yee K. F., Tan K. T., Abdullah A. Z., Lee K. T. (2009) Life cycle assessment of palm biodiesel: Revealing facts and benefits for sustainability. *Applied Energy*, 86(1), S189-S196.

Yu, S.; Tao, J. (2009). Energy efficiency assessment by life cycle simulation of cassava-based fuel ethanol for automotive use in Chinese Guangxi context. *Energy*. 34, 22-31

Zhang B., Zhang L. L., Zhang S. C., Shi H. Z., Cai W. M. (2005) The influence of pH on hydrolysis and acidogenesis of kitchen wastes in two-phase anaerobic digestion. *Environmental technology*, *26*(3), 329-340.

In: Energy and Environment Nowadays
Editors: Luis G. Torres and Erick R. Bandala
ISBN: 978-1-63117-398-1

Chapter 4

LIFE CYCLE ANALYSIS AND GHG EMISSIONS ASSESSMENT IN BIOFUELS PRODUCTION

Gemma Cervantes and Mariana Ortega
UPUBI - Instituto Politécnico Nacional, Mexico

1. BIOFUELS AND NATURAL RESOURCES

1.1. Resources and Renewable and Alternative Energy

Alternative energies are those that propose different options to fossil fuels. Examples of these include solar, wind, hydraulic, tidal, geothermal, hydrogen, and biomass based. Among these are those classified as renewable, which are the ones that use such natural resources.

Indeed, natural resources are classified as nonrenewable, potentially renewable, and renewable (Xercavins et al. 2005). The non-renewable are those in which the consumption of the resource represents its depletion during the human era, those include, for example, oil and coal. Potentially renewable resources are those that can regenerate if the rate at which they are used is less than the natural rate of recovery. Examples of these are forests, water, and soil while renewable resources are those that are considered inexhaustible during the human era, such as the sun and the wind (Daly & Cobb 1989). The only authentically renewable resources are the sun, the wind, the tides and the geothermal activity (Costanza & Daly 1992). Biomass energy is not considered a renewable energy but it is potentially renewable. Furthermore, it is important to note that alternative energies may become polluting. Nuclear energy can be an alternative energy, but generates waste with long-term persistence radioactive contamination.

1.2. Types of Biofuels

The term biofuel includes all fuels generated from potentially renewable resources, which can be derived from forest products, agriculture, aquaculture, and biowaste. They are classified according to their physical state as: solid, such as wood pellets or charcoal; liquids such as biodiesel and ethanol; or as biogas. Another classification divides them into primary or secondary and unprocessed or processed. In turn, processed fuels are classified as first, second, and third generation.

Figure 1 shows such a classification according to raw materials and production process.

1.2.1. First Generation Biofuels

First generation biofuels are produced using conventional fermentation technology and transesterification, and basic raw materials used are seeds, whole grains, and crop plants such as corn, sugarcane, rapeseed, wheat, sunflower seeds, among others.

The main disadvantage of using first generation crops in biofuel production is that they probably compete with the demand for food (Timilsina 2010). Competition for food raw materials by biofuels production could lead to significant increases in the cost of food and of biofuels (Mata et al., 2010). As an example one can take the case of corn. The United States of America is the country that produces most corn worldwide (FAO 2013). Since 2000, that country increased the use of this crop for ethanol production from 5% of the total grain production to almost 40% (15% of global production) in 2010. The expansion in ethanol production in the USA since 2005 has cost Mexico 250-500 million dollars more in annual importation bills. Corn and tortilla prices went up by 69%, between 2005 and 2011. With an increased price of food and the cost of basic goods increased by 53%, the family budget was diminished. In fact, there was an increase of poverty levels and an impact on food security in Mexico (ActionAid 2012). On the other hand, it has been reported that first-generation biofuels result in low reduction levels of greenhouse gases (GHG) emission compared with fossil fuels. They also have high production costs and in the long run could replace fossil fuels only to a modest level due to the high requirement of crop land (Cherubini et al., 2009); thus, it is considered that raw materials of second and third generation represent better options for biofuel production.

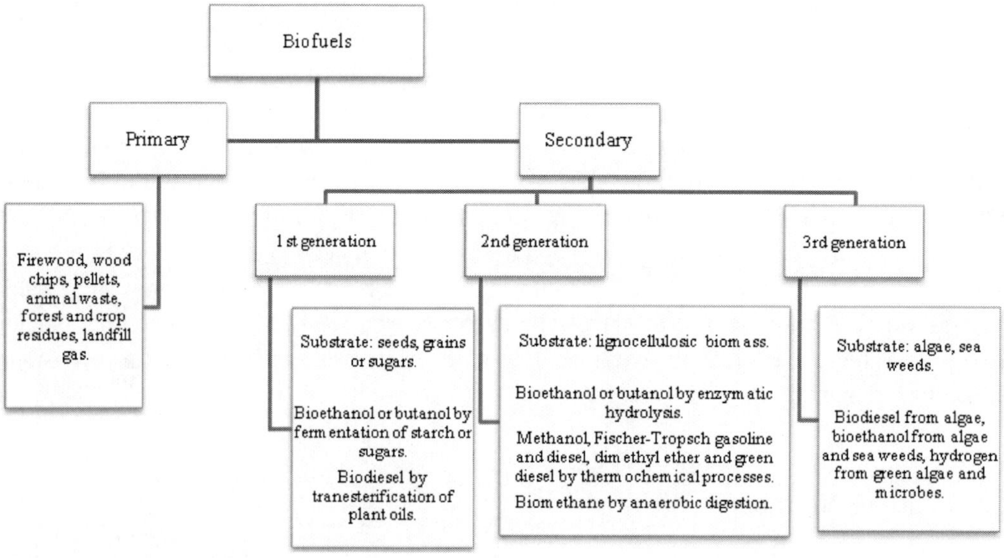

Source: Singh Nigam, 2010.

Figure 1. Classification of biofuels.

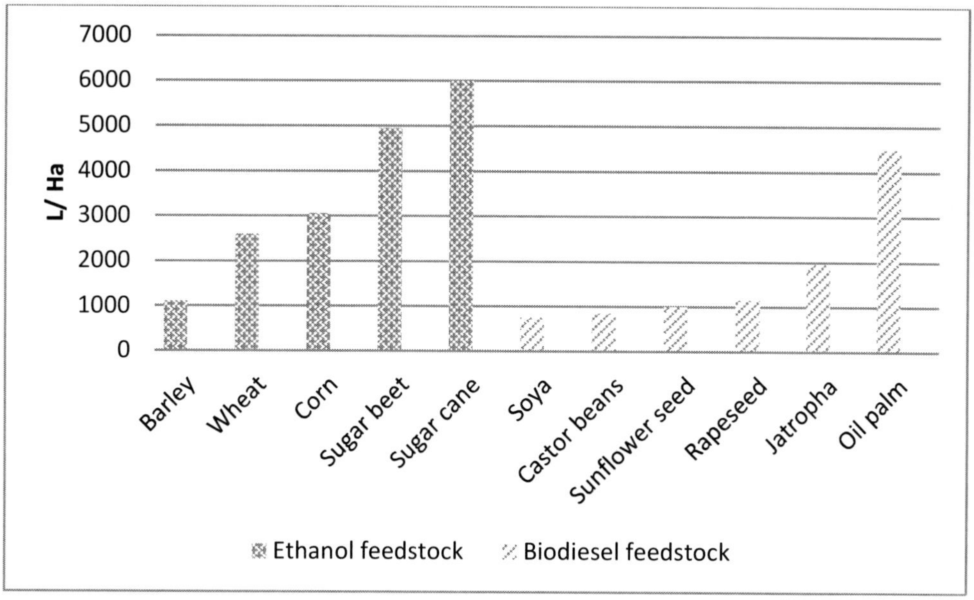

Adapted from CIFOR 2011.

Figure 2. Efficiency from the production of bioethanol and biodiesel according to the involved raw material.

Figure 2 shows observed yields of different raw materials from first generation pathways to produce biofuels.

1.2.2. Second Generation Biofuels

Second generation biofuels are produced from a variety of sources and inedible lignocellulosic biomass waste such as stalks of wheat, corn stover, wood, jatropha, and castor, among others. Lignocellulose (cellulose, hemicellulose and lignin) is the main component and more abundant in the biomass produced by photosynthesis. Lignocellulosic materials have a complex structure, and unlike the first generation, they require a special treatment before becoming biofuels. The technologies involved in the production of biofuels are the same as those used in the first generation, that is, biochemical conversion routes as hydrolysis and anaerobic digestion, but also include thermochemical routes such as gasification and pyrolysis (Singh 2010). Some advantages of the second generation materials with respect to the first are:

- They do not compete with land use for food production because with the applied technology wastes of lignocelullosic food crops are used.
- Increased efficiency since they use the complete product and not just the seeds.
- A flexible technology can be used that allows processing more sources of raw materials, and many of these are not dependent on climate conditions as is the case of first-generation sugar crops.

1.2.3. Third Generation Biofuels

Finally, the third generation energy sources are products specifically designed or adapted by various techniques to improve the conversion of biomass. Microalgae are an example of third generation raw materials. From these microorganisms, biodiesel, ethanol, hydrogen, and biogas can be readily produced (Vieira Costa 2010).

Microalgae are autotrophic eukaryotic microorganisms. They inhabit marine or fresh water environments and can grow autotrophically or heterotrophically. They have a tolerance to a wide range of temperature, salinity, and nutrient availability. As more evolved organisms they can produce lipids as triglycerides reserves (TAG's), even though only some species have the ability to accumulate them in large concentrations. Microbial species that accumulate over 20% of lipids in their cell biomass are considered oleaginous species (Stoycheva et al. 2011). A distinguishing characteristic of some of these oleaginous strains of algae is their ability to accumulate amounts of lipids, sometimes greater than 60% of their dry weight.

It has been reported that a given area of cultivation of microalgae can produce about 10 to 100 times more lipids compared to any other oilseed crop. Whereas a terrestrial crop cycle takes 3 months to 3 years to be exploited, algae begin to produce lipids between days 3 and 5 of culture, and therefore can be harvested daily. It has been reported that microalgae are the only raw material for generation of oils that could, according to its productivity, replace fossil-based diesel (Chisti, 2008).

The algal biomass is processed by biological and thermochemical methods.

Thermochemical methods include direct combustion to generate electricity, heat, and mechanical power. Biological methods include fermentation to produce energy carriers such as hydrogen, ethanol, and biogas, or the extraction of oils to produce biodiesel (Vieira 2010).

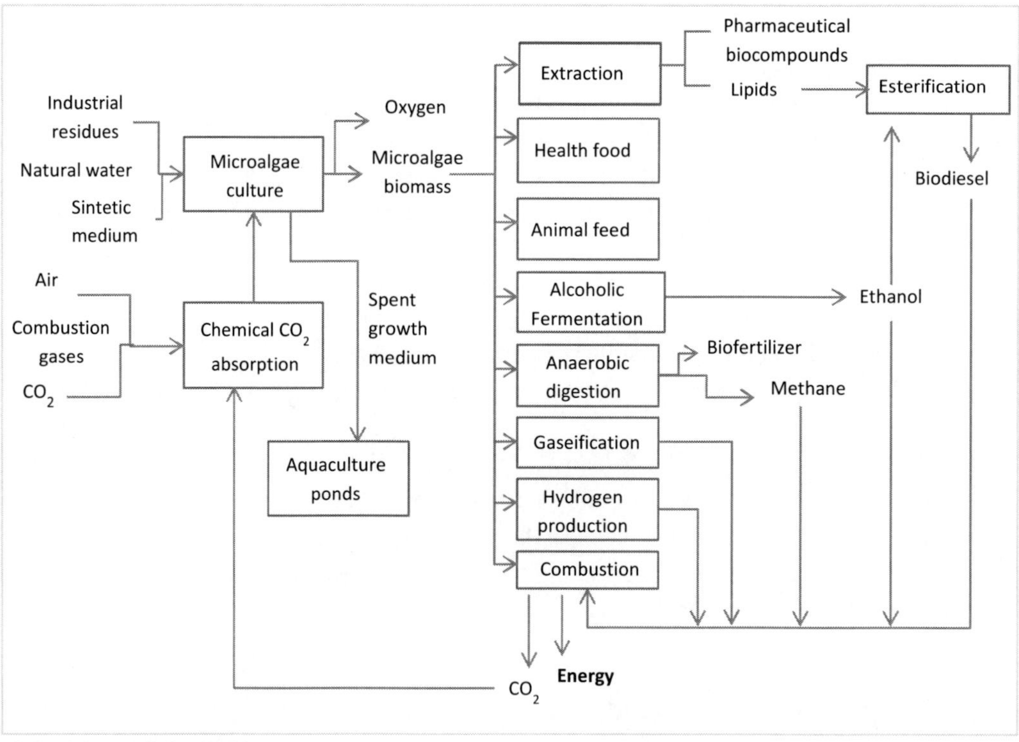

Adapted from Vieira 2010.

Figure 3. Diagram of potential processes generating bioenergy from microalgae.

Figure 3 shows a diagram of the potential production processes to obtain biofuels from microalgae. The final stage is, of course, combustion, which generates usable energy and CO_2, the latter to be again absorbed in the biomass growth cycle.

Several authors (Chisti 2007, Lardon et al. 2009, Stephenson et al. 2010, Campbell et al. 2011, etc.) have proposed the cultivation of microalgae as a system for CO_2 uptake as well as for the production of biodiesel and co-products based on several assertions about these microorganisms: a) photosynthetic efficiency greater than that of terrestrial plants, b) high growth rate, biomass doubling between 8 and 24 h, c) high lipid content of 20-70% of certain strains, d) high performance in lipid production per hectare, e) direct CO_2 uptake (100 tons of algae can bind nearly 183 tons of CO_2), f) potential large-scale production, g) no terrestrial plants competing with food production, h) production of valuable products, i) possibility of using waste water to provide nutrients for the crop, reducing process consumables and helping to treat domestic or industrial wastewater.

Algae can be grown both in salt water and fresh water in non-productive land where nothing else is produced, so there is great interest in promoting the use of algae for biofuel generation and provide an end to the debate on the cultivation of food and energy crops (Demirbas, 2010). Microalgae are emerging as one of the more promising sources for biodiesel production given several of their properties such as photosynthetic efficiency, their lack of competition with food crops, and their capacity to use multiple sources of carbon such as the CO_2 from industrial emissions (Collet et al. 2010).

The use of microalgae as raw material for large scale biodiesel production is still under development (Sander et al. 2010, Campbell et al. 2010, IPCC 2011). Biodiesel production process from microalgae implies their cultivation and extraction of their oils, both processes are currently under research at a worldwide level. Preliminary laboratory tests show high ratios of energy use in several stages, specially cultivation and extraction (Ortega 2012). This fact implies also high carbon footprint for the production process. Also an important challenge is to reduce the cost per unit area (Mata, 2010).

The difficulties in efficient biodiesel production from microalgae also relate to finding a strain with high lipid content, high speeds of growth, easy to harvest, and with a cost-effective cultivation system (Demirbas 2010).

2. Importance of Biofuels and Their Global Situation

Oil represented in 2009 more than 30% of the primary energy supply worldwide (IEA 2011) and most of it is used nowadays to produce transportation fuels: gasoline, diesel, and kerosene. The biomass energy is derived from burning it directly and from the processed biofuels. Biofuels can be used to generate heat, electricity, and liquid fuel.

Because in the field of alternative energy biomass is relevant to the generation of liquid fuels for transportation, therefore it has an important role in global energy supply. It has been reported that liquid biofuels could, to some extent, replace the corresponding fossil fuels (Singh 2010). Therefore, aside from the possible environmental benefits that will be discussed in sections ahead, biofuels have a strategic advantage to reduce fossil fuel dependence. The largest worldwide biofuel producer is the USA, followed by Brazil, and the third is the

European Union. These countries produce nearly 90% of world production, according to 2009 data (Figure 4).

According to the Biofuels Platform, their production in 2009 amounted to 74.0 billion liters (Bl) of bioethanol (approximately 37.7 Mtoe, 73%) and 17.9 Bl of biodiesel (14.1 Mtoe, 27%). In terms of production location, USA and Brazil are the main bioethanol producers whereas the EU is the first biodiesel producer (Guyomard et al. 2011).

Forecasts for biodiesel production in the upcoming years show an increasing tendency as it has been proposed to contribute to the following strategies: a) Reducing dependence on oil and improving energy security, b) reducing GHG emissions in order to mitigate climate change, and c) promoting agricultural development.

Indeed many governments around the world have developed policies that promote implementation of bioenergetics crops production. According to the European Commission, by 2030, biofuels will represent already 25% of the fuel for road transport in the European Union. The fuel would be provided mostly by an efficient European industry and would aim at substituting imported fossil fuel using innovative and sustainable technologies, and creating opportunities for biomass providers, biofuel industry, and the automotive industry (CEPAL, 2008).

In the USA, the government has promoted ethanol production. In 2008, the increase in ethanol consumption reduced gasoline demand by 5%. On the other hand, it has been reported that, as a consecuence of biofuel production, 1.2 million new green jobs will be created by 2038, assuming that biofuels will meet 30% of the fuel demand (Earley y Mceown, 2009).

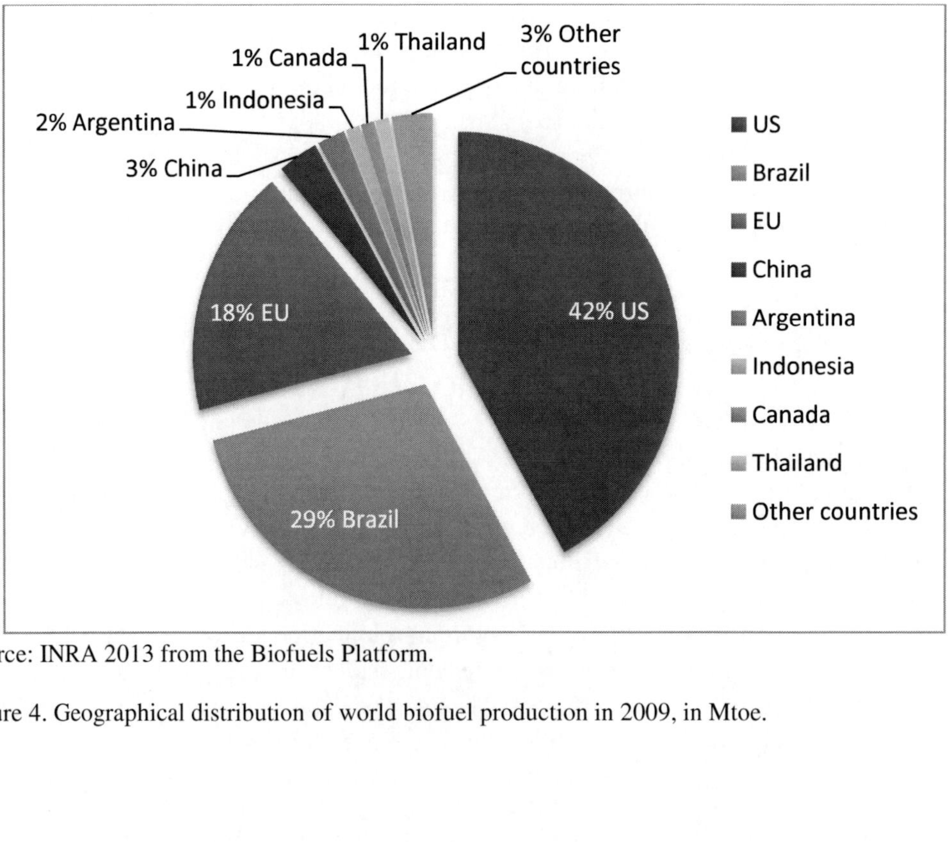

Source: INRA 2013 from the Biofuels Platform.

Figure 4. Geographical distribution of world biofuel production in 2009, in Mtoe.

Table 1. Policy goals planned for biofuels by governments

Country	Bioethanol	Biodiesel
USA	Standard and Alternative Fuels Renewable Fuel Standard: 28,000 million liters. Renewable fuels in 2012, 132,000 million liters. Of renewable and alternative fuels in 2017 (15% of projected gasoline use by 2017)	
Canada	5% in 2010	2% renewable content in diesel oil and fuel oil in 2012
EU	5.75% in 2010, 8% in 2015 and 10% in 2020 to biofuels to replace diesel and gasoline transport oil (computed based on energy usage)	
Japan	Replacing gasoline to transport 500,000 per year by 2010 (1.8 million liters / year of bioethanol in the short term, 6 million bioethanol produced locally, by 2030 representing 10% of the current demand for gasoline)	
China	15% of consumption for transport by 2020	
India	5% to 2012, 10% by 2017	
Australia	350 million liters of biodiesel-bioethanol by 2010	
Argentina	5% of the final product by 2010	5% of the final product by 2010
Bolivia		2.5% from 2007 to reach 20% in 2015
Brazil	22% since 2001	2% in 2008, and 5% from 2013 and 20% by 2020
Colombia		5% from 2008
Paraguay	18% minimum	1% in 2007, 3% in 2008, 5% in 2009
Peru	7.8% from 2006 and progressively by region	5% from 2008 and progressively by region

Adapted from CEPAL 2008.

Table 1 presents the goals set by some countries for replacing gasoline with ethanol and diesel with biodiesel. It is noted that Canada, the European Union, and Argentina intended replacing 5% gasoline (5.75 in the case of the EU) with bioethanol in 2010. In the case of Brazil, since 2001, bioethanol has been used in 22% of the consumed fuel volume, and for biodiesel it was planned to substitute diesel in 5% by 2010 and 20% by 2020.

In Brazil 57% of the vehicle fleet presents flex engines that can run on ethanol or gasoline. Between 80 and 90% of the cars leave the factory with this type of engine. So consumers choose what fuel to use and it has been reported that when ethanol has a lower cost, it fails to meet the demand (Caulyt, 2013).

In the case of Mexico, 88% of primary energy derived from petroleum (INEGI 2012). Energy consumption has grown steadily over recent years and the transport sector is the largest component, with over 2000 petajoules for 2010, a 48 % of the total energy compsumption. Regarding biofuels, in 2008, Mexico issued the law for the promotion and development of bioenergy (LPDB, acronyms in Spanish), whose purpose is to contribute to energy diversification and sustainable development (LPDB 2008). Indeed, it establishes the basis for promoting and marketing the production of raw materials for biofuels, while using them efficiently.

3. Environmental Impacts Associated with Biofuels

Evaluating the environmental impact of biofuels usually emphasizes the climate change issue, considering the reduction in GHG emissions and energy consumed for production. It is these two parameters that are mostly focused on when evaluating environmental benefits of biofuels. There are other relevant parameters which complete the environmental profile of biofuels. In the case of first generation biofuels, such as food crops, there is the risk of food cost increases and endangering food availability for the world's poorest people, according to the Food and Agriculture Organization of the United Nations (FAO, 2012).

Other important impacts of biofuels are declining biodiversity, land degradation, increasing pressure on water resources and effects on the carbon footprint.

3.1. Biodiversity

In terms of biodiversity, if the expansion of bioenergy crops is not planned and managed properly, it could trigger a change in land use both indirectly and directly, potentially promoting loss or deterioration of certain ecosystems and their resources as well as the corresponding services they provide. The establishment of large-scale monocultures for biofuel production would affect agrobiodiversity and traditional knowledge associated with their production and use.

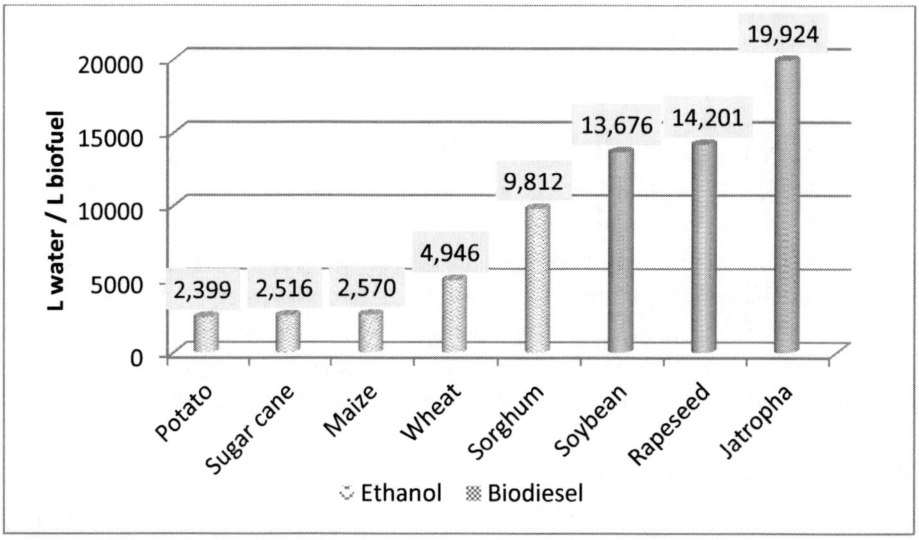

Adapted from Gerbens-Leenes (2008).

Figure 5. Water needed for a specific crop to produce one litre of ethanol or one litre of biodiesel (l water/l biofuel).

3.2. Water Consumption

Regarding the availability and quality of water, bioenergy crops can exert pressure on them and create competition with other uses, including production and preparation of food, also potentially contributing to water scarcity. In terms of water compsumption and efficiency in the production of the biofuel, figure 5 shows the amount of water needed for a specific crop to produce one litre of ethanol or one litre of biodiesel (l/l).

In terms of quality, leaching of fertilizers and pesticides used in crop production could have negative impacts on ground water. However, production of bioenergy projects could also contribute to investment in infrastructure related to the distribution and use of water.

3.3. The Use of Fertile Land

In terms of land use, if the operators do not implement appropriate management practices, the production of bioenergy crops could contribute to soil degradation through erosion, compacting, loss of organic matter and nutrients, and salinization. Some of these effects could result from excessive removal of crop residues and forestry, as these acts as soil fertilizers. In certain cases, cropland for bioenergy production can be grown on degraded land, and this can help to restore soil quality through the implementation of specific practices (Lal, 2005).

3.4. The Carbon Footprint

A portion of GHG emissions of biofuels is part of the carbon cycle, where carbon taken up by plants during photosynthesis is emitted during combustion to return to be absorbed into the bioenergetic system (Cherubini et al., 2009). Whereas carbon emitted during the combustion of fossil fuels cannot return to its source, causing an increase in its atmospheric concentration. Although there is not a general agreement that the global warming potential of CO_2 from biofuels is zero, the European Renewable Energy Directive (RED) recommends that this be considered (García et al. 2011).

GHG emissions balances in the life cycle of the biofuel or carbon footprint is a factor in the competitiveness of a biofuel supply chain. Indeed, the reduction in emissions due to the use of these sources is particularly important for their promotion.

Different authors have conducted comparative studies on the distribution of GHG emissions along the production chain of biofuels, such as bioethanol and biodiesel from different raw materials, by observing a decrease in emissions of up to 80% compared to fossil fuels (FAO 2008). However, technical efficiency of the ability of biofuels to help reduce GHG emissions is questioned when land use change is taken into account. In fact, some studies indicate that the production of biofuels on a large scale can lead to an increase in GHG emissions. For example, the conversion of a peat forest to palm oil cropland releases 3452 tons CO_2/ha and requires 423 years to pay off the "carbon debt" (Fargione et al. 2008).

The environmental impact of transportation fuels is such that, for example in Mexico, according to the 2009 national inventory of GHG emissions, GHG emissions from this sector, in 2006, contributed 38% of the national total, of which diesel and gasoline contributed almost 95% (SEMARNAT 2009).

4. Environmental Impact Assessment of Biofuel Production

It is critical to assess the potential environmental impacts of biofuel production, especially considering the importance that the world's governments are giving to the increased production of bioethanol and biodiesel for blending with fossil counterparts. In each case, the environmental impact must be studied individually, i.e., at the local or community level, as it will depend, for example, on:

- Farming practices or the extraction of raw materials, the characteristics of these,
- type of conversion technology,
- distances that the raw materials and final product had to be moved,
- combustion of biofuels and
- indirect impacts or external influences associated to their production.

In the assessment of impact, a set of indicators can be used. Worldwide, various organizations have been created to promote the sustainable production of biofuel industries and several indicators have been proposed to address sustainable development goals as pursued by public policies. Environmental indicators of the biofuels industry can be classified into six groups: indicators of soil quality, water quality and quantity, air quality, climate modification, biodiversity, and plant productivity (McBride et al., 2011).

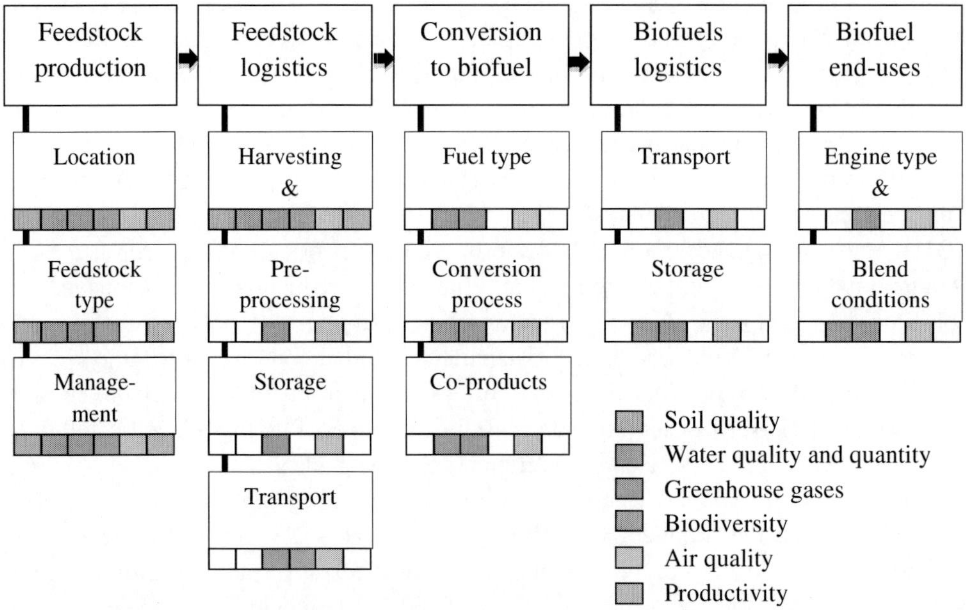

Adapted from Efroysom et al., 2012.

Figure 6. Stages of the production chain of biofuels and environmental indicator categories applicable to each stage.

Indicators of one or more groups are applicable to each stage of biofuel production, as shown in Figure 6. The design of a set of indicators to measure the environmental performance of a biofuels industry involves considering the context in which it is embedded, as certain potential impacts could be more important than others.

4.1. Features of the Best Biofuel in Environmental Terms

The promoters and producers of biofuels should be interested in knowing the best choice of raw materials and conversion methods to generate the greatest economic benefits, but with the lowest environmental and social damage.

To assess which is the best biofuel in environmental terms is a complex task considering all possible environmental issues involved at each stage of the production process of biodiesel or bioethanol. Currently a key indicator of environmental performance of biofuels is the carbon footprint, given the goal of reducing greenhouse gas concentrations that cause global warming. Other important indicators are those related to energy efficiency, because it would be illogical to employ more energy in the production of biofuel than the amount it could generate. These two indicators are considered the most important among 35 indicators of sustainable development in the context of bioenergy (Markevicius et al., 2010). In addition to these indicators, those related to the use and quality of soil, water, and biodiversity are less common in the literature. However, these are the ones that influence mostly at the local level when setting on the process of biodiesel or bioethanol production.

Considering the above mentioned the biofuel production process that is most energy efficient, from the stage of obtaining raw materials to the time of combustion, and that contributes further to climate change mitigation and to the conservation of natural resources, will be the best biofuel in environmental terms.

4.2. Methods for Assessing the Environmental Impact of Biofuels

There are several methods or tools applicable to the calculation of environmental indicators related to biofuel production According to reports from the FAO (2012).

Among the methods to evaluate agrobiodiversity are IBAT or tool for the integrated biodiversity assessment tool; RABA, which is a fast assessment method of biodiversity in the context of rewards for environmental services; and the RAP or rapid assessment program.

The state of the soil quality has been evaluated through the universal equation for soil loss RUSLE, the CQESTR model, which calculates the rate of decomposition of crop residues and organic amendments to the extent which they become organic matter or organic carbon of the soil; or the use of a manual for local level assessment of land degradation and sustainable land management, LADA.

Availability and quality of water in relation to bioenergetics production can be estimated through the AQUACROP and CROPWAT programs, which simulate the production performance based on the availability of water and water requirement of crops, respectively. Another tool is the SWAT model, which allows quantifying the impact of land management practices in large, complex watersheds, and to evaluate water quality problems, including problems of nonpoint source pollution.

GHG balance can be performed through various softwares, such as EXACT, GHGenius, global emissions model for integrated systems (GEMIS), resources and energy analysis programme, REAP, the GHG, regulated emissions and energy use in transport, GREET, and SIMAPRO, UMBERTO, etc. as life cycle analysis software.

The application of each of these tools involves gathering specific activity data and the participation of experts in the application of the models and software use.

4.3. The Life Cycle Analysis (LCA) Applied to Biofuels

Given the complexity of some models to calculate environmental indicators for each specific issue related to the impact of biofuel production, it may be convenient to use or specialize in one method such as the life cycle analysis (LCA), which allows to asses and group different indicators from different themes: soil water, air, biodiversity, and productivity.

4.3.1. Origins and Applications of LCA

The LCA originated in 1970, about the same time in the USA and Europe. The objects of this analysis were then used in household products such as containers, detergents, and diapers. At the beginning, LCA results were sometimes placed under doubt because the methodological bases were not defined and companies managed their own studies on their own methods placing the results in the direction most suited for themselves (Udu de Haes & Heijungs, 2007). But when done properly it can be a truly serious analytical tool.

Since 1989, the Society of Environmental Toxicology and Chemistry (SETAC) defines LCA as such, as it was formerly known under various names, and began to establish a methodological framework for its implementation.

In 1994, the International Organization for Standardization (ISO) began to apply its 14040 series to LCA. Between 1996 and 2006 standards 14040, 14041, 14042, 14043, 14044, 14047, 14048 and 14049 were published, dealing with general principles, the guidelines calculation, and interpretation of results, data documentation formats, and application examples of this impact assessment methodology. The standards are constantly reviewing.

In 2002, the United Nations Program for the Environment, to improve the use of the LCA approach by decision-makers in developing countries, began cooperation with SETAC to carry LCA into practice through programs focused on the development of best practices in the field of LCA, the development of simplified methods and public databases, and the support for the management of products or services during their life cycle (UNEP, 2010).

The LCA methodology allows the evaluation of environmental impacts over the life of a product or service. The LCA results are quantitative and allow comparisons of environmental performance of the products or services with similar functions. LCA is used to identify weaknesses in the design of a product, to make environmental product declaration, provide general information to consumers for marketing purposes, to establish product standards, taxes, possible subsidies or grant criteria for ecolabeling, to evaluate the level of a company's environmental performance and to develop product legislation with the corresponding plans of action to increase social awareness (Díaz et al., 2004).

A considerable amount of LCA studies has been conducted for the assessment of biofuels production worldwide, mostly in Europe and North America in the past years; although, in recent years, an increasing number of studies have been conducted in developing countries (Larson 2006, Cherubini & Hammer, 2011). Their purpose is the investigation and evaluation of the environmental impacts of biofuels production and to rank the best performing pathways. In terms of types of biofuels assessed, there is a similar number of studies evaluating 1st and 2nd generation biofuels, although the latter are mainly at a pre-commercial stage.

4.3.2. Overview of LCA Methodology

The LCA methodology comprises three phases: initial phase, inventory analysis and impact assessment phase. The initial phase comprises the goal and scope definition. During the inventory phase all data related to the system under study are collected. The life cycle inventory is a list of all the components that are included as part of the system under study in the analysis of life cycle. The impact assessment phase follows the inventory and calculates the potential environmental impacts associated with each input datum of materials and energy, emissions and waste outputs listed in the inventory.

According to ISO standard 14040 the general methodology of an LCA can be summarized in four steps: goal and scope definition, inventory analysis, life cycle impact assessment, and life cycle interpretation.

4.3.2.1. Goal and Scope Definition

During the goal and scope definition phase the LCA plan has to be defined as clearly and unambiguously as possible. The goal has to include the intended application, the reasons for carrying out the study, the intended audience, whether or not the results can be public and whether the result can be compared with other similar products or services. The scope includes the system boundary and the function of the product system, the impact categories to be assessed and the treatment of uncertainty.

System Boundary

The system boundary is defined by the spatial, temporal and production chain limits of the product system that is being analyzed (Davis et al., 2009). For example, a study may include all material inputs used in the process, the equipment, infrastructure and services, or only material inputs but not the others. Boundaries can be tied to a single unit process depending on the objective of the study, but the results would be substantially limited.

In biofuel's LCAs, the boundaries usually include upstream activities, such as biomass cultivation (which includes, for example, the production and use of fertilizers, machinery and equipment and other consumables), and transport, as well as the processes involved in converting the biomass to biofuel, the delivery of biofuel to vehicles, and consumption of the fuel.

When defining the system boundary it has to be taken into account that an LCA can be classified into two types: attributional and consequential. The attributional or accounting one assesses as the events occur within the product system and the impacts these generate internally on the overall result, while the consequential or change oriented attemps to evaluate the effects on external product systems due to changes and events occurred into the analyzed product system. In biofuels, this could be the indirect change of land issues.

Functional Unit

The functional unit is defined according to the primary function fulfilled by the system under study. Thus, it enables different systems to be treated as functionally equivalent and allows reference flow to be determined for each of them (Guineé et al., 2001). For instance, based on the functional unit of drying 260,000 pair of hands over 10 years, the environmental impact of using an electrical hand dryer against the use of paper towels can be compared.

Functional unit in LCA applied to biofuels is usually related to input units as quantity of biomass or energy used in the process, although more usually it refers to output units, such as biofuel or energy generated, units of agriculture land used and units of time, usually reported on a yearly basis. According to Larson (2006), few studies focus on the question of relative land-use efficiency for different biofuel pathways, which is somewhat surprising since land is the basic primary resource for biofuel production.

It has been reported that LCA results should be preferably shown using several functional units, which become indicators, and that the limiting factor of the system should be identified and used as the reference indicator of the assessment (Cherubini & Hammer, 2011).

4.3.2.2. Inventory Analysis

A life cycle inventory is a list of quantified process flows of all resources used and emissions that occur due to the use of materials and activities needed to deliver the product or service under study. In order to obtain the inventory, a mathematical model of the production system has to be built. This can be as simple as a spreadsheet computertool adding emission factors to the requirements for the product system, or, alternatively, LCA software can be used to link thousands of processes (Horne et al., 2009).

4.3.2.3. Impact Assessment

In the life cycle impact assessment phase, the set of results from the inventory analysis is interpreted in terms of impact indicators according to different impact categories. ISO 14040 (2000) defines an impact category as “class representing environmental issues of concern to which life cycle inventory analysis results may be assigned”.

To carry the impact assessment, each substance from the life cycle inventory has to be assigned to at least one impact category. This is named classification. Then, the quantity of each substance is multiplied by pre-existing factors to compute a quantified environmental impact. These factors are based on internationally accepted environmental models. For instance, the factors used to calculate the indicator of the climate change impact category are the global warming potentials of greenhouse gases. This step, where the impact of all the substances from each selected category is calculated and summed is known as characterization.

Classification and characterization are the only mandatory steps during the impact assessment phase according to ISO standards. However, other indicators can be estimated through normalization and weighting.

The goal of normalization of impact is a better understanding of the relative proportion or magnitude for each impact category of a product system under study (ISO 14042, 2000). It allows the impact category indicator results from characterization to be compared by a reference value. This means, the impact category is divided by the reference. A commonly used reference is the average yearly environmental load in a country or continent, divided by the number of inhabitants (Gooedkoop et al., 2008).

When weighting, the impact category indicator results are multiplied by weighting factors, and are added to create a total or single score. However, it should be taken into account that weighting indicators are subjective values, usually used at the discretion of the evaluator.

Allocation

When in a production process co-products or by-products are generated, it becomes necessary to allocate the environmental burdens among all the products and by-products. A preferred method of allocation is direct substitution where, for example, the heat produced from burning a residue from the process of study can replace heat that would otherwise have been supplied from an external system. If direct substitution cannot be used, simpler allocation methods can be applied, including allocation by economic value, calorific value or mass.

Processes for making biofuels typically involve generation of co-products. In the production of biodiesel, glycerin is co-produced; while in the wet milling process for making ethanol, from corn, there are multiple co-products including animal feed and corn oil. The choice of allocation applied in these cases significantly affects the biofuel LCA results.

4.3.3. Impact Assessment Methods and Impacts Analyzed

Impact assessment can be performed by various established methods. Each method entails a number of different categories of impact and characterization factors, normalization and weighting associated with a number of specific substances. The existing evaluation methods differ mainly in terms of the impact categories considered, the models used in the calculation of factors, the type of indicators (characterization, normalization and weighting) that can be calculated and the date of the last update.

Traditionally in LCA, the emissions and resource extractions are expressed as 10 or more different midpoint impact categories, like acidification, global warming, eutrophication, ozone layer depletion, ecotoxicity, resource extraction, etc. For example, EDIP 2003 is a Danish LCA methodology that is presented as an update of the EDIP 97 methodology. The EDIP 2003 methodology represents 19 different impact categories. Some of them are updated versions of EDIP 97, whereas others are modeled totally different. Another kind of methodology is the Eco-indicator which uses a damage-oriented approach and thus assess the seriousness of three damaging categories: damage to human health, damage to ecosystem quality, and damage to resources (Goetkoop et al., 2008b).

4.3.4. General Results from Biofuels LCA Studies

Although many LCAs applied to biofuels cover different types of impact categories, others are limited to GHG and/or energy balance. This approach is usually supported by the fact that mitigation of climate change and reduction of fossil fuel consumption are the main driving factors for worldwide bioenergy development. Lifecycle carbon reporting is required by policies for supporting biofuels, such as the EU's Renewable Energy Directive (RED), the Renewable Fuel Standard in the USA, and the UK's Renewable Transport Fuel Obligation (RTFO), to ensure that biofuels achieve greenhouse gas reductions relative to fossil fuels (Brander et al., 2009). Few studies include in their impact assessment the land use as impact category.

Several authors have concluded that concerning the human and ecotoxicity impact categories, most bioenergy systems lead to increased impacts when compared to fossil reference systems (Cherubini & Hammer, 2011). However, according to Davis (2008), the holistic LCA approach to biofuels is often not compared with an equally holistic view of the fossil fuels that they would replace. LCA research applied to biofuels leads to more negative results relative to fossil fuels because fossil fuel impacts´ have not been as thoroughly assessed.

Transportation biofuels produced from residue streams and second generation raw materials usually have larger GHG savings than first generation biofuels. This is also generally reported when by-products are used to generate energy that can further substitute the use of electric energy from the grid or fossil fuels conversion.

Comparison between biofuel LCA studies often becomes a difficult task because each author defines its own goal and specific boundaries for which particular assumptions are made. Furthermore, the components of each production system, the functional unit and the chosen allocation methods can vary according to the approach of each study.

Table 2. Functional units, allocation procedures and impact assessment methods used in different LCA of biodiesel production processes

Raw Material	Functional Unit	Allocation	Impact Assessment Method	Reference
Sunflower, rapeseed and soybean	1 kg of biodiesel	Not considered	Eco-Indicator 99	Sanz et al., 2011.
Waste vegetable oils	1000 kg of biodiesel	Mass	CML 2001	Morais et al., 2010
Used cooking oils	1000 kg of biodiesel	Not considered	CML 2 baseline 2000	Talens et al., 2010.
Soybean, Jatropha and Microalgae	1 MJ	Mass	CML 2001	Hou et al., 2011.
Palm	1 ha plantation	Mass	Ecosystem carbon payback time (ECTP)	Pereira de Souza et al., 2010.
Rape seed and palm	1 t refined vegetable oil	Avoided by system expansion	EDIP 97 plus land use impact indicators	Schmidth, 2010.
Rapeseed	1 t biodiesel	Economic	EDIP 2003	Stephenson et al., 2008
Microalgae, canola and ultra-low sulfur diesel	1 km (truck fuel used)	Not considered	IPCC	Campbell et al., 2011.

4.3.5. LCA of Biodiesel Production

In Table 2 the raw materials used, the functional unit, the allocation method and the impact assessment method used in some biodiesel LCA studies are shown.

The results obtained from an LCA study can be compared to another when system boundaries, applied allocation method, and functional unit are the same. Moreover, the results depend on the context of the production system under study. For example, the comparison of environmental impact from three different raw materials for biodiesel production in China would not be useful to decide which of these raw materials is better for biodiesel production in Mexico, since local conditions such as climatology and agricultural practices are not considered.

Taking into account the allocation method is necessary when analyzing results from LCA studies. For example, when assessing the impact of biodiesel production from used vegetable oil, compared to Talens et al., (2010) who did not applied any allocation method, Morais et al., (2010) estimated a much lesser energy consumption when mass allocation was considered.

Table 3 shows some biodiesel LCA results from different authors. It can be observed that most of them consider a cradle to door approach (production of raw materials and biofuel only) and evaluate the characterization of the global warming category. Normalization and weighting indicators are also evaluated in some cases, but are less common. Generated energy from the biofuel and energy consumption in the production process is also mentioned in some cases. From the table it can be observed that comparison between studies is difficult when functional units are different and if a common unit of measure is not provided.

When assessing the biodiesel production process using sunflower, rapeseed, and soybean as raw materials, Sanz el al. (2011) concluded that the process in which a greater effort should be made to reduce the environmental impact is in the production of seeds, through minimization in the use of fertilizers and simplification of labor.

According to Morais et al., (2010), who compared three different biodiesel production processes from waste vegetal oils, based on normalization results, marine aquatic ecotoxicity and depletion of abiotic resources are the most relevant potential environmental impact categories.

Talens et al., (2010) concluded that the stage with greater impact in the production of biodiesel from used cooking oil is the transesterification stage (68%); and that despite the consumption of further materials and energy, the environmental impact of the system is reduced if the by-products are recovered.

Hou (2011) compared the performance of soybean, jatropha, and microalgae-based biodiesel against fossil diesel, and found that biodiesel contributes to a reduction of the potential abiotic depletion and global warming significantly, but that it performs worse regarding other environmental impacts, including photochemical oxidation, eutrophication, acidification, and human- and eco-toxicity. Jatropha and microalgae resulted more competitive biodiesel feedstock compared to soybean in terms of all impacts due to the lower level of agricultural inputs per unit of oil produced.

Schmidth (2010) reported that palm oil tends to be environmentally preferable to rapeseed oil within all impact categories except global warming, biodiversity and ecotoxicity in which the difference is less pronounced and is highly dependent on the assumptions regarding system delimitation in the agricultural stage.

Table 3. Results from LCA applied to biodiesel production processes from different raw materials

Raw Material	Functional Unit	System Boundary	Output Energy	Input Energy	GWP Characterization	GWP Normalization	Total Weighting	Reference
Sunflower, rapeseed and soybean	1 kg of biodiesel	Cradle to door	na	na	na	0.0004	10.8	Sanz et al., 2011.
Waste vegetable oils	1000 kg of biodiesel	Cradle to door	na	3.636 MJ for the alkali-catalized process,	na	1.50E-10	na	Morais et al., 2010
Used cooking oils	1000 kg of biodiesel	Cradle to door	na	1,158 MJ	299.60 kg CO_2e	Na	na	Talens et al., 2010.
Soybean, Jatropha and Microalgae	1 MJ	Cradle to grave	1 MJ	na	0.035 kg CO_2e for soybean, 0.02 kg CO_2e for jatropha, and 0.018 kg CO_2e for microalgae	Na	na	Hou et al., 2011.
Palm	1 ha plantation	Cradle to grave	158 GJ	29.42 GJ	1436.51 kg CO_2e	Na	na	Pereira de Souza et al., 2010.
Rape seed and palm	1 t refined vegetable oil	Cradle to door	na	na	2740 kg CO_2e for rape seed and 3450 kg CO_2e for palm oil, considering the atributional scenario	na	na	Schmidth, 2010.
Rape seed	1 t biodiesel	Cradle to door	na	20,605 MJ for large scale scenario	2415 kg CO_2e	na	na	Stephenson et al., 2008
Microalgae, canola and ultra-low sulfur (ULS) diésel	1 km (truck fuel used)	Cradle to grave. Excludes production facility and its construction	0.89 MJ	na	-27.56 g CO_2e for biodiesel, 35.856 g CO_2efor canola, and 81.239 g CO_2e for ULS diesel.	na	na	Campbell et al., 2011.

GWP – global warming potential, na – not available information within the article.

4.3.5.1. LCA and Industrial Symbiosis of Biodiesel from Microalgae. Case Study in Instituto Politécnico Nacional (IPN), México

An LCA was applied to biodiesel production from *Neochloris oleabundans* in the Unidad Profesional Interdisciplinaria de Biotecnología (UPIBI-IPN) laboratories. The LCA analysis was run according to ISO 14040 series. Data from cultivation stage were experimental, while the algae oil processing information was collected from literature. The EDIP (2003) model was chosen as assessment method, and Simapro software (v. 7.3) was used to compute the input data.

The functional unit was 100 MJ generated by the biodiesel, considering a heating value of 38 MJ/kg. The selected system boundary approach was cradle to door. Production of machinery and installations was not considered. Limitations of the study include the fact that data from the cultivation downstream processes, oil extraction and transesterification were taken as average values from literature. Another important fact is that the production process from chitosan flocculant and some chemical compounds used in the culture medium had to be investigated, to compute their LCA, since they were not included in the Simapro data base. This carried higher level of uncertainty to the study.

The overall carbon footprint is 2.5 kg of CO_2 eq per MJ, while the ozone depletion is 2.28x10-3 kg CFC11 eq and resource consumption is 1 g. The cultivation and downstream processes of biomass production account for 99% of the environmental impact of biodiesel production microalgae in all impact categories except for the hazardous waste one, which is mostly affected by the oil extraction and transesterificaction stages. Most impact from these stages, which are the experimental ones, is due to high energy consumption (if compared to other studies) by the raceway pond and waste water treatment from the chitosan production process. This could be an indicator that energy has to be used more efficiently and that chitosan production process has to be reviewed.

In the case of energy, the most intensive stage is cultivation where energy consumption comes from the paddle wheel operation. The net energy ratio (NER), i.e the energy produced from the biofuel to the energy used for its production, is 0.12, meaning that the process consumes 88% more energy than the amount it generates.

Industrial Symbiosis, which is the sharing of services, utility, and by-product resources among industries in order to add value, reduce costs and improve the environment (Argawal 2008), was applied to the process as a way of reducing environmental impacts. Three Industrial Symbiosis scenarios were applied to residual biomass from the oil extraction stage: composting, animal feed production and energy production. Results show that energy generation reduces impacts in most of the categories except for the acidification and terrestrial and aquatic eutrophication. According to the database used, composting doesn't improve environmental behavior. Animal feed production decreases some impacts, such as global warming and ozone depletion, but acidification, eutrophication and chronic eco-toxicity are amplified.

When industrial symbiosis is applied to the transesterificaction process, by using the residues from this stage to produce pure glycerin and chemical fertilizer, the carbon footprint is reduced by 89.5%. All the others impacts are also minimized, while hazardous waste production is reduced in 99.9% since the glycerol and methanol mixture are separated and purified, and thus are not emitted as residues.

It is therefore expected that the environmental impact and the NER could be reduced by the application of industrial symbiosis and the use of renewable energy into the microalgae biodiesel production process.

Table 4. Carbon footprint of biodiesel from different raw materials

Raw Material	Carbon Footprint	Units	System Boundaries	Reference
Waste vegetable oil♣	-0.35	kg CO_2 e / t biodiesel	Collection of waste vegetable oils to biodiesel production.	Talens et al. 2010
Waste vegetable oil	299.60			Talens et al. 2010
Rapeseed♦	2415	kg CO_2 e / t biodiesel	Biomass production to biofuel transportation to filling station. Includes land use impacts.	Stephenson et al., 2008.
Rapeseed♦ ♦	2195			Stephenson et al., 2008.
Jatropha	123.73 ± 34.23	g CO_2 e / MJ	Biomass production to biofuel generation. Excludes land use change impacts	Achten et al., 2010
Jatropha♣♣	159.36 ± 34.41			Achten et al., 2010
Soybean	~ 380	g CO_2 e / MJ	Biomass production, biofuel generation, transportation stages and combustion in vehicles.Excludes land use change impacts	Hou et al., 2010
Jatropha	~ 200			Hou et al., 2010
Microalgae	~ 180			Hou et al., 2010
Palm oil	9.1	g CO_2 e / MJ		Pereira de Souza et al., 2010

♣ By-products are used for energy generation. ♣♣ Biogas is generated from meal by-product.
♦ Large scale production. ♦♦ Small scale production.

4.3.6. LCA of Bioethanol Production

Biofuels show much lower GHG emissions than fossil fuels. First and second generation bioethanol are, respectively, 60% and 90% less impacting than gasoline (Fazio 2011).

For first generation bioethanol, the lowest emissions and highest energy ratios are observed when ethanol is produced from direct sugarcane juice and when generating surplus electricity. In this case GHG emissions are 36.8 gCO_2e/GJ ethanol (Garcia 2011).

In most LCA studies of lignocellulosic ethanol, environmental benefits are reported, especially concerning GHG emissions and net energy output. But the production cost of ethanol is dependent on the feedstock, conversion technologies, and enzymes, generation of valuable coproducts, plant sizes; its economic viability remains doubtful at present.

In terms of LCA environmental impacts, lignocellulosic ethanol contributes significantly to a reduction of GHG emissions and when enzymatic hydrolysis is used. In average, 63% lower GHG emissions is shown in comparison to first generation biofuels (Fazio 2011). But the bioenergy system can release more GHG emissions than its fossil alternative when the energy used to feed the biomass conversion process comes from carbon-intensive fossil sources (Roy 2012). The use of E85 in fuel flexible vehicles instead of gasoline reduces 86–113% of GHG emissions when bioethanol comes from agro-residues (Davis et al. 2009). Also, abiotic resources and ozone layer depletion decrease when gasoline is replaced by stover ethanol fuels. Instead, acidification and eutrophication increase because phosphorus and nitrogen related environmental burdens are released from the soil during cultivation (Uihlein 2009).

4.4. Carbon Footprint of Biofuels

Accounting for carbon footprints consists in quantifying GHG in the whole life cycle of products in a consistent manner (Weidema et al., 2008). Although, years ago, transport biofuels were considered carbon neutral, many studies showed that land use change can make footprints highly carbon positive (Johnson, 2009). The result of the carbon footprint of a certain biofuel depends on the system boundary considered within the LCA study.

Table 5. Carbon footprint of bioethanol from different raw materials

Raw Material	Carbon Footprint (G CO_2e/ MJ)	Products Considered in the System	Allocation Method
Wheat	45.8	Ethanol and protein feed generated, excluding straw.	No allocation
Wheat	27.9		Energy
Wheat	37		Economic
Wheat	16.5		System expansion
Sugar beet	30.2	Ethanol and protein feed generated.	No allocation
Sugar beet	19.6		Energy
Sugar beet	25.4		Economic
Sugar beet	15.8		System expansion
Straw	2.1	Ethanol, electricity and biogas generated.	No allocation
Straw	1.8		Energy
Straw	1.8		Economic
Straw	-0.6		System expansion
Willow	21.2	Ethanol and pellets generated.	No allocation
Willow	11		Energy
Willow	16.7		Economic
Willow	18.8		System expansion
Poplar	9.2	Ethanol and pellets generated.	No allocation
Poplar	4.8		Energy
Poplar	7.3		Economic
Poplar	6.8		System expansion
Forest residues	7.1	Ethanol and pellets generated.	No allocation
Forest residues	3.7		Energy
Forest residues	5.6		Economic
Forest residues	4.7		System expansion

Table 4 shows the carbon footprint of biodiesel made from different raw materials. It can be observed that the carbon footprint of biodiesel from waste vegetable oils can be greatly reduced when by-products are used for energy generation. However, another study of biodiesel production from jatropha showed that producing biogas from by-products led to an increase in the carbon footprint (Achten et al. 2010) . The small scale production of biodiesel from rapeseed generates less GHG emissions if compared to the large scale production process; when comparing the biodiesel production process from soybean, jatropha, and microalgae, the latter one depicted the smallest carbon footprint.

Table 5 shows a comparison, according to Börjesson et al., (2012), of the carbon footprint of bioethanol made from various raw materials of first and second generation considering different allocation methods. It is observed that second generation materials (straw, willow, poplar and forest residues) have a lower carbon footprint than first generation ones (wheat and sugar beet). Considering the allocation method, system expansion greatly reduces the carbon footprint, specially when no allocation method is applied, while energy allocation for second generation raw materials produces the lowest carbon footprint. This could be due to high calorific value of the co-products.

CONCLUSION

As biomass is considered a potentially renewable source of energy, this natural resource must be used at a rate lower than its regeneration to avoid its depletion.

Life cycle analysis (LCA) is a valuable tool to calculate the potential environmental impacts associated with the production and consumption of biofuels.

Currently, the carbon footprint is the most studied environmental impact of biofuels despite that other impacts could be just as important. The carbon footprint is calculated using the methodology of life cycle assessment.

Limiting points to compare results between different biofuel LCA studies are the system boundaries, allocation method, functional unit and type of used indicators and the inclusion of certain key processes as land use change. That is, the LCA applied to biofuels depends on the context of the production process being studied and on all the considerations of the methodological phase. For this reason, the results among studies can seem contradictory.

In most of the LCA studies of lignocellulosic ethanol, environmental benefits are reported, especially concerning GHG emissions and net energy output. But most of these studies do not consider land use change. A worse environmental performance is expected if land use change is included in LCA studies.

Although there is extensive knowledge about how various assumptions and methodology choices influence the resulting GHG performance of various biofuel systems, quantification of land use change effects is complicated and most LCA studies on biofuels do not consider it. Land use change can have positive or negative effects on carbon capture and storage depending on the base line scenario. This is related to the fact that carbon sequestration by soils is affected by the type of agricultural management, plant species and climate zones.

Comparison of biodiesel against fossil diesel shows that biodiesel contributes to the reduction of some impact categories as abiotic depletion potential and global warming potential. Other environmental impacts, including photochemical oxidation, eutrophication, acidification, and human and eco-toxicity show a higher environmental impact.

When industrial symbiosis is applied to the biodiesel production process from microalgae, a reduction in environmental impact is obtained when energy is generated from biomass waste and when pure glycerin and fertilizer are obtained from glycerin waste. Moreover, significant environmental impact reduction could be achieved if Industrial Symbiosis is applied to cultivation stage.

REFERENCES

Achten W., Almeida J., Fobelets V., Bolle E., Mathijs E., Singh V., Tewari D., Verchot L., Muys B. (2010). Life cycle assessment of biodiesel as transportation fuel in rural India. *Applied Energy*. 87: 3652-3660.

ActionAid International USA (2012). Biofueling hunger: how US corn ethanol policy drives up food prices in Mexico. Report. May 2012. Washington: ActionAid. 14 pp.

Agarwal, A., Strachan, P. (2008). Is Industrial Symbiosis only a Concept for Developed Countries? *The Journal for Waste & Resource Management Professionals* 42

Borjesson, P., Ahlgren, S., Berndes, G. (2012). The climate benefit of Swedish ethanol: present and prospective performance. WIREs Energy Environ, 1: 81–97

Brander M., Tipper R., Hutchison R., Davis G. (2009). Consequential and Attributional Approachesto LCA: a Guide to Policy Makers with Specific Reference to Greenhouse Gas LCA of Biofuels. Technical paper TP-090403-A. UK. Ecometrica Press. 14 pp.

Campbell C. J. (2012). Recognition of peak oil. *WIREs Energy and Environment*. 1: 114 – 117.

Campbell P., Beer T., Batter D. (2011). Life cycle assessment of biodiesel production from microalgae in ponds. *Bioresource Technology*. 102: 50-56.

Caulyt F. União Europeia supera Brasil no consumo de biocombustíveis. Noticias DW 28 de febrero de 2013. Consultado 06 abril de 2013. <http://www.dw.de/uni%C3%A3o-europeia-supera-brasil-no-consumo-de-biocombust%C3%ADveis/a-16613830>

CEPAL (2008). Aporte de los biocombustibles a la sustentabilidad en el desarrollo en América Latina y el Caribe: elementos para la formulación de políticas públicas. CEPAL, GTZ. Santiago de Chile. Pp 13-19.

Cherubini F., Bird N., Cowie A., Jungmeier G., Schlamadinger B., Woess-Gallash S. (2009). Energy –and greenhouse gas- base LCA of biofuel and bioenergy systems: Key issues, ranges and recommendtions. *Resources, conservation and recycling*. 53: 434-447.

Cherubini F., Hammer A. (2011). Life cycle of bioenergy systems: state of the art and future challenges. *Bioresource Tecnology*. 102: 437-451.

Chisti Y. (2007). Biodiesel from microalgae. *Biotechnology Advances*. 25: 294-306.

Chisti Y. (2008). Biodiesel from microalgae beats bioethanol. *Trends in biotechnology*. 2:126-131.

CIFOR (2011). Technologies to produce liquid biofuels for transportation. Working paper 72. Center for International Forestry Research. 32 pp.

Collet P., Arnaud H., Lardon L., Ras M., Goy R., Steyer J. (2010). Life cycle assessment of microalgae culture coupled to biogas production. *Bioresource technology* 102: 207-214.

Costanza R., Daly H. E. (1992). Natural Capital and Sustainable Development. *Conservation Biology*, 6: 37–46.

Demirbas A., Demirbas M. (2010). Algae energy. Algae as new source of biodiesel. London: Springer.

Daly H., Cobb J. (1989). For the common good. Boston: Beacon Press.

Davis, S. C., Anderson-Teixeira, K. J. & DeLucia, E. H. (2009) Life-cycle analysis and the ecology of biofuels. *Trends Plant Sci.*, 14: 140-146.

Diaz A., Álvarez M., González P. (2004). Logística inversa y medio ambiente. Madrid: McGraw Hill. Pp 183-192.

Earley J., Mceown A. (2009). Red, White and Green: Transforming U.S. biofuels. *Worldwatch Report* 180. WWI. Washington. 47 pp.

Efroysom R., Dale V., Kline K., McBride A., Bielicki J., Smith R., Parish E., Shweizer P., Shaw D. (2012). Environmental indicators of biofuel sustainability: what about context. *Environmental management*. 51: 291-306.

Fazio S., Monti A. (2011) Life cycle assessment of different bioenergy production systems including perennial and annual crops. *biomass and bioenergy* 35: 4868-4878.

Fargione, J., Hill J., Tilman D., Polasky S. and Hawthorne P. (2008) Land clearing and the biofuel carbon debt. *Science* 319: 1235 – 1238.

Food and Agriculture Organization of the United Nations (FAO) (2008). The state of food and agriculture. Biofuels: prospects, risks and opportunities. Rome: FAO. 138 pp.

Food and Agriculture Organization of the United Nations (FAO) (2012). A compilation of tools and methodologies to assess the sustainability of modern bioenergy. Rome: FAO. 118 pp.

Food and Agriculture Organization of the United Nations (FAO) (2013). FAOSTAT Online Statistical Service. Rome: FAO. Available online at: http://faostat.fao.org/.

García C., Fuentes A., Hennecke A., Riegelhaupt E., Manzini F., Masera O. (2011). Life-cycle greenhouse gas emissions and energy balances of sugarcane ethanol production in Mexico. *Applied Energy* 88: 2088–2097.

Gerbens-Leenes P., Hoekstra A., Van der Meer T. (2008). The water footprint of bioenergy. Research report Series n.34. UNESCO-IHE, Delft. 108 pp.

Goedkoop M., Schryver A., Oele M. (2008a). Introduction to LCA with SimaPro. PRé Consultants.The Netherlands. 82 pp.

Goedkoop M., Oele M., Schryver A., Vieira M. (2008b). SimaPro Database Manual Methods Library. The Netherlands. PRé Consultants. 45 pp.

Guineé J., Gorreé M., Jeijungs R., Huppers G., Kleijn R., Koning A., Oers L., Wegener A., Suh S., Udo de Haes H., Bruijn H., Duin R., Huijbregts M. (2001). Life cycle assessment: an operational guide to the ISO standards. Part 2. Ministry of Housing, Spatial Planning and the Environment (VROM) and Centre of Environmental Science – Leiden University (CML). 101 pp.

Guyomard H., Forslund A. and Dronne Y. (2011). Biofuels and World Agricultural Markets: Outlook for 2020 and 2050. Economic Effects of Biofuel Production, Dr. Marco Aurelio Dos Santos Bernardes (Ed.), ISBN: 978-953-307-178-7, InTech, Available from: http://www.intechopen.com/books/economic-effects-ofbiofuel-production/biofuels-and-world-agricultural-markets-outlook-for-2020-and-2050

Horne R., Grant T., Verghese K. (2009). Life Cycle Assesment Principles, Prectice and Prospects. Collingwood: CSIRO Publishing. 192 pp.

Hou J., Zhang P., Yuan X., Zheng Y. (2011). Life cycle assessment of biodiesel from soybean, jatropha and microalgae in China conditions. *Renewable and Sustainable Energy Reviews*. 15: 5081-5091.

INEGI. Banco de Información económica [en línea]. México D.F. Instituto Nacional de Estadística y Geografía, [fecha de consulta: 18 Junio 2012]. Disponible en: http://www.inegi.org.mx/sistemas/bie/default.aspx

INRA French National Institute for Agricultural Research http://www7.international.inra.fr/research/directories/agriculture/biofuels_and_biomaterials. Consultado, 13 abril 2013

IEA International Energy Agency (2011). Statistics: total primary energy supply. Available from: http://www.iea.org/stats/pdf_graphs/29TPES.pdf

IPCC (2011). Fuentes de energías renovables y mitigación del cambio climático. Resumen para responsables de políticas y resumen técnico. Grupo Intergubernamentald e Expertos sobre Cambio Climático. 242 pp.

ISO (2000). ISO 14042:2000 Environmental management: Life cycle assessment - Life cycle impact assessment. International Standardization Organization. 16 pp.

Johnson E. (2009). Goodbye to carbon neutral: Getting biomass footprints right. Environmental impact assessment review. Volumen 29(3): 165-168.

Lal R. (2005). Crop residues and soil carbon. Communication to the CarbonMeeting Consultation. Disponible en línea: http://www.fao.org/ag/ca/Carbon%20Offset%20Consultation/CARBONMEETING/3FULLPAPERSBYCONSULTATIONSPEAKERS/PAPERLAL.pdf

Lardon L., Hélias A., Sialve B., Steyer J., Bernard O. (2009). Life cycle assessment of biodiesel production from microalgae. *Environmental science and Technology*. 43: 6475-6481.

Larson E. (2006). A review of life cycle analysis studies on liquid biofuel systems for the transport sector. *Energy for sustainable development*. Volumen 10. No. 2:109-126.

LPDB ley de promoción y desarrollo de los Bioenergéticos (2008). Diario Oficial de la Federación. México 1 de febrero de 2008.

Markevicius A., Katinas V., Perednis E., Tamasauskiene M. (2010). Trends and sustainability criteria of the production and use of liquid biofuels. *Renewable and sustainability energy reviews*. 14: 3226-3231.

Mata T., Martins A, Caetano N. (2010). Microalgae for biodiesel production and other applications: a review. *Renewable and sustainable energy reviews* 14: 217-232.

McBride A., Dale V., Baskaran L., Downing M., Eaton L., Efroysom L., Garten C., Kline K., Jager H., Mulholland P., Parish E., Schweizer P., Storey J. (2011). Indicators to support environmental sustainability of bioenergy systems. *Ecological indicators*. 11: 1277-1289.

Morais S., Mata T., Martins A., Pinto G., Costa C. (2010). Simulation and life cycle assesment of process design alternatives for biodiesel production from waste vegetable oils. *Journal of Cleaner Production*. 18: 1251-1259.

Ortega M., Cervantes G., Torres L., Fernández L. (2012). Análisis de ciclo de vida del biodiesel producido a partir de microalgas. Comunicación en el 1r Congreso Iberoamericano sobre Biorefinerías. San José del Cabo, BC , 24-26 octubre 2012.

Pereira de Souza S., Pacca S., Turra de Ávila M., Borges J. (2010). Greenhouse gas emissions and energy balance of palm oil biofuel. *Renewable Energy*. 35: 2552-2561.

Roy P., Tokuyasu K., Orikasa T., Nakamura N. and Shiina T. (2012). A Review of Life Cycle Assessment (LCA) of Bioethanol from Lignocellulosic Biomass. *Japan Agricultural Research Quarterly* 46 (1): 41 – 57.

Sander K., Murthy G. (2010). Life cycle analysis of algae biodiesel. *International Journal of Life Cycle Assess*. 15: 704-714.

Sanz J.F., Guimaraes A., Quirós S., Relea E., Hernández-Navarro S., Navas L., Martin-Gil J., Fresneda H. (2011). Life cycle assesment of the biofuel production process from sunflower oil, rapeseed oil and soybean oil. *Fuel processing technology*. 92: 190-199.

Schmidth J. (2010). Comparative life cycle assesment of rapeseed oil and palm oil. *International Journal of Life Cycle Assessment*. 15: 183-197.

SEMARNAT (2009). Cuarta Comunicación Nacional ante la Convención Marco de las Naciones Unidas sobre el Cambio Climático. Comisión Intersecretarial de Cambio Climático. México: SEMARNAT. 274 pp.

Singh A., Singh Nigam P., Murphy J. (2010). Renewable fuels from algae: An answer to debatable land base fuels. *Bioresource Technology* 102: 10-16.

Stephenson A., Dennis J., Scott S. (2008). Improving the sustainability of the prduction of biodiesel from oilseed rape in the UK. *Process Safety and Environmental Protection*. 86: 427-440.

Stephenson A., Kazamia E., Dennis J., Howe C., Scott S., Smith A. (2010). Life Cycle Assessment of Potential Algal Production in d United Kingdom: A Comparison of Raceways and Air-Lift Tubular Bioreactors. *Energy Fuels*. 24: 4062-4077.

Stoytcheva M., Montero G., (Eds) (2011). Biodiesel: Feedstocks and Processing technologies. Rijeka: InTech. 470 pp.

Talens L., Lombardi L., Villalba G., Gabarrelli X. (2010). Life cycle assesment and exergetic life cycle assement of the production of biodiesel from used cooking oil. *Energy*. 35: 889-893.

Timilsina Govinda R., Shrestha A. (2010). How much hope should we have for biofuels? *Energy* 36: 2055-2069.

Udu de Haes H. & Heijungs R. (2007). Life cycle assesment for energy analysis and management. *Applied energy*. 84: 817-827.

Uihlein, A. & Schebek, L. (2009) Environmental impacts of a lignocellulose feedstock biorefinery system: an assessment. *Biomass and Bioenergy*, 33: 793-802.

United Nations Environmental Programme. (2010). The lifecycle initiative. Available online: http://lcinitiative.unep.fr/ Consulted: June 2013.

Vieira Costa J. A., Greque de Morais M. (2010). The role of biochemical engeneering in the production of biofuels from microalgae. *Bioresource Technology* 102: 2-9.

Weidema B., Thrane M., Christensen P., Schmidth J., Lokke S. (2008). Carbon footprint. A catalyst for life cycle assessment? *Journal of industrial ecology*. Volumen 12(1): 1-6.

Xercavins J., Cayuela D., Cervantes G., Sabater A. (2005). Desarrollo Sostenible. Barcelona: Edicions UPC.

In: Energy and Environment Nowadays
Editors: Luis G. Torres and Erick R. Bandala
ISBN: 978-1-63117-398-1

Chapter 5

HUMIC ACIDS PRODUCTION FROM SEWAGE SLUDGE

Victor A. Ramìrez Coutiño, Francisco J. Rodríguez*[*]*,
Luis A. Godinez, Erika Bustos Bustos,
Adrián Rodríguez García and Juan Manriquez Rocha
Centro de Investigación y Desarrollo Tecnológico en Electroquímica SC,
Mexico

1. INTRODUCTION

The fast growth of waste sludge production arising from domestic and/or urban wastewater treatment processes is posing serious problems in terms of their storage and, above all, for their elimination, as they contain inorganic, organic matter, and the contaminants removed from the treated waters. Since 1991 construction of wastewater treatment plants in Mexico has grown rapidly. As of 2012 there were 2186 facilities (Conagua, 2012).

The main problem posed by the sludge is its high content of pathogenic microorganisms, which can cause health problems in humans and animals, including diarrhea and other gastroenteric effects. Diseases associated with microbial pathogens include meningitis, myocarditis, respiratory infections, certain type of diabetes, eye and skin infection, etc., (Romdhana, 2009). Despite these problems, large amounts of generated sludge are discharged into the sewage or without any previous treatment to dams, empty lots, or into the same sources, and, in the best conditions, they are disposed into lagoons or sanitary landfills.

This sludge possesses beneficial characteristics that can be exploited once the sludge has been treated properly. Their potential uses are as fertilizers, soil improvers, or as covers for sanitary landfills. Taking into account the aforementioned there are a double set of problems: on one side, discharge of sludge into inadequate sites that can generate severe contamination issues, and, on the other side, the beneficial effects of sludge are being wasted while they could be used advantageously in agriculture or soil improvement.

[*] Email: frodriguez@cideteq.mx

2. Sewage Sludge of Municipal Wastewater Treatment Plants

The wastewater from households that reaches treatment plants is generally constituted by 99.9% water and 0.1% of total solids, 50% of these solids are dissolved and 50% are suspended. Around 70% of total solids are proteins and urea, sugars, cellulose and starches, greases, oils, and soaps. The remainder 30% corresponds to inorganic compounds such as metals, sands, among others (Spellman, 1997).

During wastewater treatment, a solid, semisolid or liquid residue is generated rich in organic matter with a variable content of moisture called sewage sludge (Marambio & Ortega, 2003). This waste sludge is obtained by means of biological and/or physical-chemical treatments, containing a high percentage of water (50-80%) and organic matter (Metcalf & Eddy, 2003). Composition of sludge is variable depending on the amount of water and the treatment it has received.

Recently, studies have been performed reporting that stabilized waste sludge can be reused without any risk to health or the environment, because of their characteristics they are an important resource of organic matter and of fertilizing agents, which makes sludge a potential source of nutrients for agricultural use. Approximately 50% of its dry weight corresponds to organic matter, providing, also, variable amounts of nitrogen(1-7%), phosphorus (1-5%), potassium (0.3-3%), and micronutrients diversely available to plants (Grigatti et al., 2004; Metcalf & Eddy, 2003).

3. Normativity Applicable to Sewage Sludge

The sludge from water treatment plants is generally not environmentally fit to be applied directly on soil, agriculture, gardening, etc., because, aside from being unpleasant due to its odor, it contains toxic substances and pathogenic microorganisms (Spellman, 1997). Currently there are regulation that prohibit their storage and disposal that can only be achieved under certain restrictions. In the USA, the EPA (Environmental Protection Agency) under Federal Regulations Code 40CFR503, part D, indicates that sludge to be applied to the soil or to be disposed of superficially must comply with two basic requirements, low content of pathogens and low potential attraction of contamination vectors (rodents, birds, insects, and other insect-carrying organisms).

4. Treatment Options for Urban Sewage Sludge

Diverse types of treatment can be applied to sludge generated in wastewater treatment plants that are described in the following (McFraland, 2000; National Research Council, 2002; Metcalf & Eddy, 2003).

Anaerobic digestion: This process includes biological stabilization of the sludge inside closed reactors by which content of organic matter, mass, smells, and pathogens are reduced. During the process, methane is generated that can be used as energy source and temperatures of up to 55° C are generated. The produced biosolids are used mainly in agriculture and forestry.

Aerobic digestion: Oxygen or air is used in the aerobic digestion process to stabilize biologically the sludge in either open or closed recipients. During the process, the organic matter is converted to CO_2, water, nitrogen, pathogens, and smells are also reduced. Biosolids obtained by this process can be used in agriculture.

Alkaline stabilization: The principle of this process is the increase of the pH to reduce pathogens and attraction of vectors. This process yields products that can be used in agriculture or that can be disposed of in sanitary landfills.

Drying by heating: This treatment alternative involves the use of dryers to remove the water from the sludge. The use of these dryers destroys pathogens and removes water, thereby reducing the initial volume of the material. The biosolids obtained from this process are used mainly in agriculture.

Composting: By using this process, biological degradation of the organic matter is achieved. In general, during this process, temperatures above 50 °C are reached what is sufficient to destroy pathogens and to reduce smell. Composting produces biosolids that can be used mostly in agriculture, public parks, and soil remediation.

Among these options, composting is undoubtedly the most practical way to stabilize waste sludge, using scarce infrastructure and at reasonable costs. This process yields a product with a pH between 6.5 and 8, destroys almost 100% of pathogens and mineralizes nutrients making them available for vegetal growth and development (EPA, 1999). In addition, it can also be used beneficially as soil conditioner (Kuter et al., 1995).

5. Composting Process

Composting has been widely used since ancient times. Israelis, Greeks, and Romans composted or used directly the organic wastes for agricultural purposes. The first civilizations of South America, China, Japan, and India practiced an intensive agriculture using for it human and animal wastes or residues as fertilizers (Epstein, 1997). The concept of composting at a large scale in a methodical way has been attributed to Sir Albert Howard and its INDORE composting process, developed at the Institute of Plant Industry in Central India between 1924 and 1931 (Epstein, 1997). The basic concept of this process was to use animal, vegetal residues, and human feces, mixing them with an alkaline material to neutralize its acidity and to handle the mass by turning it to provide aeration and water.

The composting process has been defined by diverse authors. The most accepted definition of the composting process was established by the EPA (1989) that defines it as: "the process of biological decomposition and stabilization of organic substrates under such aerobic conditions that allow the development of thermophilic temperatures as the result of the biological activity and which allow the final product to be sufficiently stable for its storage and application without affecting the environment where it is found". This process is schematically shown in **Figure 1**.

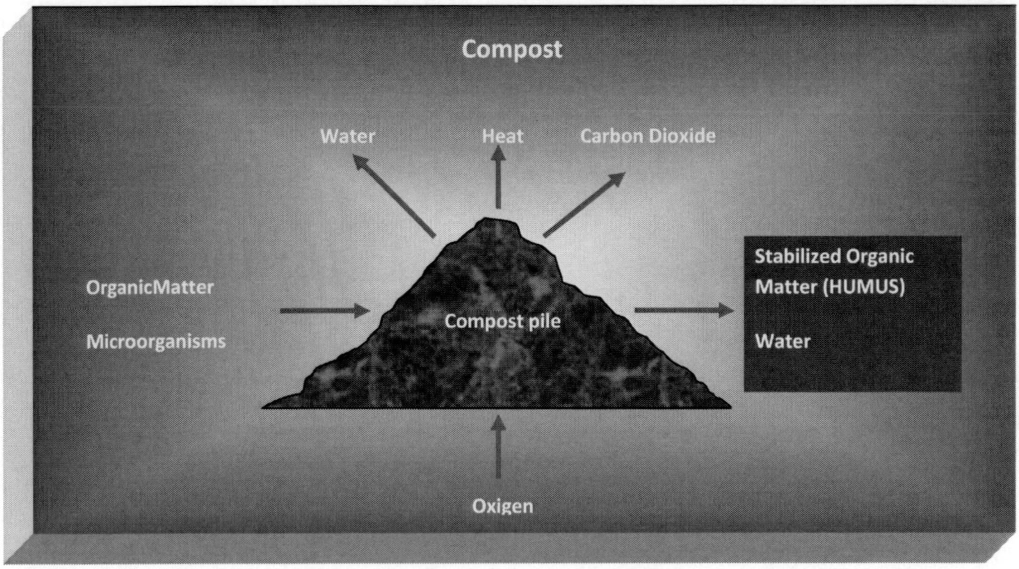

Figure 1.Composting process (Cooperband, 2002).

The general equation of organic matter decomposition during the composting process is as follows:

$$Organic\ matter + Microorganisms + O_2 \rightarrow Stable\ org.\ matter(Humus) +$$

$$Water + CO_2 + Heat + Microorganisms$$

6. Objectives of the Composting Process

Objectives of the composting process comprise a series of benefits, among them the most important ones are described in the following (Polprasert 1996):

Stabilization of organic residues. Biological reactions occurring during composting convert the putrescible forms of organic residues in stable complex material (humic substances) and inorganic forms. Hence, the composting products can be disposed of safely on the soil, using them as fertilizers or for soil amendment.

Reduction of pathogenic microorganisms. The heat produced biologically during composting can produce temperatures close to 60 ºC or even higher, which is sufficient to reduce some pathogenic microorganisms.

Returning nutrients to soil. Nutrients (N, P, K) present in organic wastes are usually found in complex organic forms, making it difficult to be taken up by crops. After composting, these nutrients are transformed into inorganic forms such as nitrates $(NO_3)^-$ and phosphates $(PO_4)^{-3}$, which are better used by crops. Application of compost as fertilizer on the soil reduces the loss of nutrients because they are in insoluble forms released slowly and, hence, are used more profitably than the nutrients that could come from unprocessed wastes.

Drying of sludge. Waste sludge contain approximately 80% to 95% of water, which elevates storage, transport and disposal costs. Drying the sludge by composting is an alternative in which the biologically produced heat can evaporate some of the moisture content of the sludge.

7. Fundamental Concepts of Humus

In general, organic matter is divided in humic and non-humic. Carbohydrates, amino acids, proteins, lipids, nucleic acids, and lignins constitute the organic matter that is not part of humus. The humic fraction is commonly known as humus, humic compounds, humic substances, or humic matter, and it is formed by the decomposition of plant and animal wastes. The process of humic substances formation is called humification.

Humic substances are defined as amorphous mixtures, extraordinarily complex and heterogeneous, produced during decaying or putrefaction of biomatter and formed in the environment by means of chemical reaction of species chosen randomly from a group of diverse molecules (Hayes & Malcolm, 2001). These humic substances exhibit macromolecular characteristics that have been found as resulting from the aggregation of relatively small primary molecular structures (100-2000 Da), which form clusters by hydrogen bonds, non-polar interactions, and polyvalent cation interactions. (Sutton & Sposito, 2005; Piccolo, 2002; Senesi et al., 2003). Depending on their origin, humic substances vary considerably in the concentration of functional groups, elemental composition, and molecular size distribution.

The most common and accepted way to classify humic substances is based on the different solubility according to the pH depicted by the fractions of which said substance are being formed. **Figure 2** depicts the extraction of humus and its classification according to the solubility exhibited to different media. Using the solubility criterion, humic substances in turn are divided in three fractions: humic and fulvic acids, and humins.

Fulvic acids: They constitute a series of solid or semisolid compounds, amorphous, yellowish in color and of colloidal nature. **Figure 3** shows the structure of fulvic acid.

Humic acids: They occur as amorphous solids, dark brown in color, generally insoluble in water and in most non-polar solvents, but easily dispersible in aqueous solutions of hydroxides and basic salts of alkali metals. From a structural point of view, its molecule seems to be constituted by a nucleus of aromatic nature more or less condensed and by a region with greater presence of aliphatic radicals, presenting as a whole the character of condensed heteropolymers. **Figure 4** shows the molecular structure of humic acid.

Humins: These are humic compounds non extractable with alkaline reagents, they comprise a group of relatively different substances, constituted fundamentaly by amino acids, carbohydrates and lipids, which are retained in the heavy fraction aggregates of the soil by bonds that cannot be disrupted through common mechanical agitation but with ultrasound. They predominate in those soils that have a vegetation not easily biodegraded.

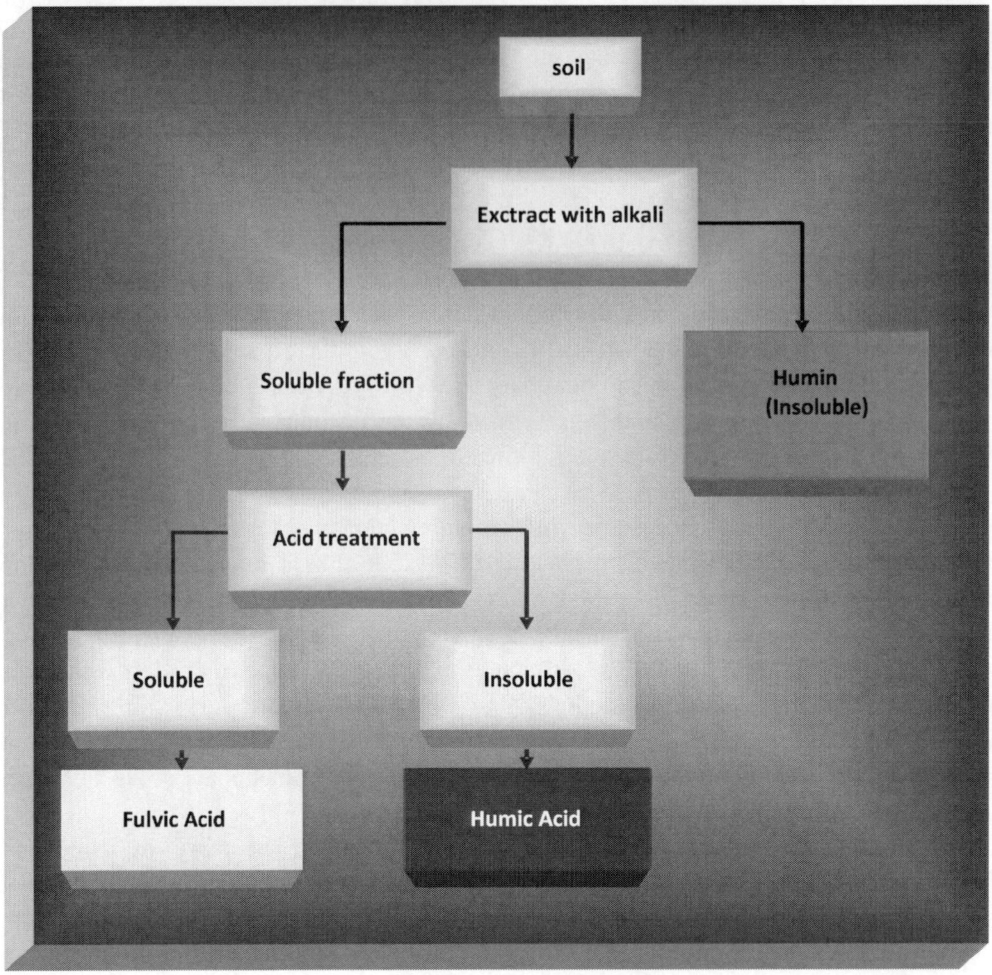

Figure 2. General method for the extraction and fractionation of humic matter.

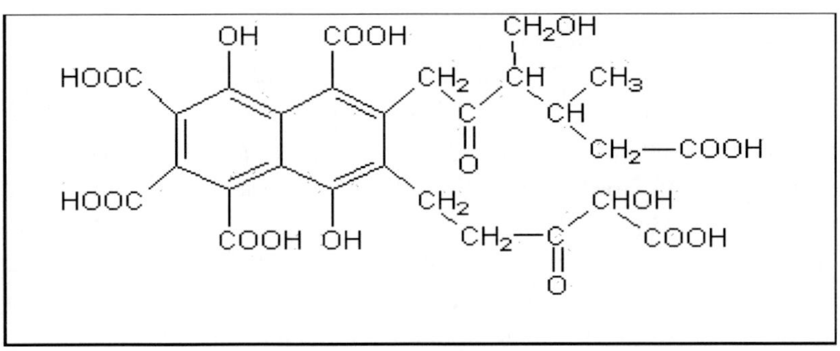

Figure 3. Molecular structure of fulvic acid.

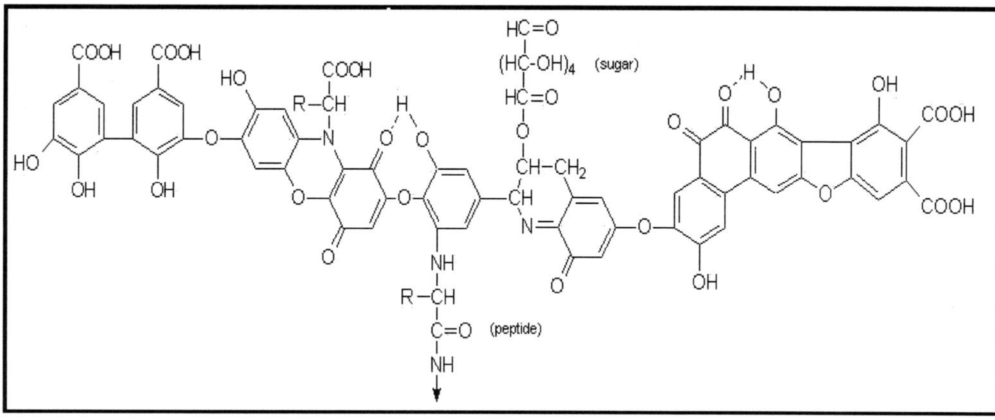

Figure 4. Molecular structure of humic acid (Stevenson, 1994).

In general, it can be said that humic acids are larger macromolecules than fulvic acids, they present a higher content of carbon and nitrogen, and fulvic acids have a higher percentage of oxygen in their structures than humic acids. This higher oxygen content in fulvic acids increases their acidity and, hence, they present a higher capacity to retain metals. But the higher molecular weight of humic acids leads to a series of properties related with the colloidal state for example higher water retention (Perminova et al.,2002).

Several theories have been generated on the synthesis of humic matter. The most accepted is the one formulated in the last 10 to 15 years, according to which several authors have argued that it can be formed from lignin (Wershaw, 1993; Piccolo, 2002; Sutton &Sposito, 2005; Kelleher & Simpson, 2006).

7.1. Uses of Humic Substances

Humic substances are used mainly as an organic fertilizers because their addition to soil can stimulate crops growth. However the interest in exploring its physicochemical properties is due to its potential applications in environmental remediation (Perminova et al., 2002):

Enhances bioremediation: Humic acids can serve as a extracellular electron shuttles and accelerate microbial redox reactions. Also, in some cases can transform the contaminants in less toxic forms, reduce bioability or sequestrate them in a separate phase.

Enhances Phytoremediation (binding agent): Reduces the toxicity of the contaminants and increase the tolerance of the plants to the contaminants and stimulates the development of roots. This results in a better crops growth in contaminated soils.

In situ flushing: Humic acids solutions can be used in queous solutions injected into a zone of contaminated soil or aquifer. The injected fluid increases the mobility of the contaminants. There are many advantages of humic acids, mainly its low cost and the fact that these molecules are biologically recalcitrant avoiding the loss in soil permeability caused by microbial growth.

In our research group, we have proved surfactant properties of humic acids obtained from sewage sludge compost proving that humic acids lowered surface tension of aqueous solution in similar way to synthetic surfactants (Ramírez et al., 2013b). On the other hand, we also use biosolids and compost biosolids, as a raw humic subtances material, to stabilize heavy metals in mining tails lowering the lixiviation of Pb, Cd and Zn (Ramos et al., 2012). At this time, redox properties of humic acids are in evaluation to understand about the effect in microbial systems used in remediation process.

7.2. Physico-chemical Characterization of Humic Acids

To characterize completely humic acids, these are subjected to chemical and spectroscopic analyses, such as determination of functional groups, infrared spectroscopy, and size exclusion chromatography, which will be explained in the following.

Functional Groups (Total Acidity, Carboxylic Groups, Phenolic OH Groups)

Contents in functional groups determine its relation within the soil matrix, as they are responsible for the interaction with diverse molecules of the environment. Among the functional groups found in humic acids, the most important ones are the carboxylic and OH phenolic groups, in which total acidity is to be determined firstly to be able to calculate the mentioned groups. Total acidity or capacity of exchanging humic compounds is attributed to the presence of protons that can become dissociated or of H^+ ions in aromatic, carboxylic aliphatic and phenolic hydroxyl groups.

The phenolic hydroxyl groups are OH bound to phenolic rings, in which estimation of the phenolic groups of each simple of humic acids is calculated by the difference between the total acidity and the carboxylic groups.

$$mEq\ OH\ phenolic = mEq\ total\ acidity - mEq\ COOH$$

Visible Absorption Spectroscopy

Visible absorption spectra of humic substance exhibit absorbances that decrease in intensity when increasing the wavelength of radiation (Stevenson, 1994), so that most of the energy absorbed by humic materials is found between 300 and 665 nm (He X et al., 1995; Gieguzy´nska et al., 1998), providing information on their molecular size and structure. The obtained spectra can be expressed as a ratio or coefficient of the absorbance in two arbitrarily selected wavelengths (for example absorbances at 465 and 665 nm), and is called color relation E_4/E_6.

$$E_4/E_6 = (\text{absorbance at 465 nm})/(\text{absorbance at 665 nm})$$

This color relation is used as an index for the proportion of light absorption in the visible range. A high proportion of light, 7-8 or higher, corresponds usually to fulvic acids or fractions of humic acids of relatively small molecular weights. A low color proportion, 3-4, corresponds to humic acids of high molecular weights.

Infrared Spectroscopy

Each molecule possesses a characteristic vibration and interaction with electromagnetic energy to absorb or radiate in the infrared spectrum region, yielding bands that associate to specific functional groups. Therefore, infrared spectroscopy has been used widely to identify functional groups contained in humic substances.

Molecular Size Exclusion Chromatography(SEC)

Molecular size exclusion chromatography has been proposed for purification processes or fractioning of humic compounds within different components or in fractions of different molecular sizes, and it has also been used to determine molecular weights (Stevenson, 1994; Tan H, 1998). This molecular fractionating method is based on a column of packed gel spheres through which a solution of humic acid is passed at a controlled flow to separate the components at different elution times.

8. Composting of Sludge from Water Treatment Plants

Our research group has performed studies on composting sludge from water treatment plants characterizing humic substances produced in this process (Ramírez et al., 2013a; Ramírez et al., 2013b). For this purpose, composting piles of 1m^3 in volume were constructed with sludge samples coming from a wastewater treatment plant, after having been subjected to aerobic stabilization and further dried in a bands filter. We tried small pieces of wood (SGW) and “tezontle” (extrussive, igneous volcanic rock; SGT) as a bulking agents to test their effect on the performance and compared with the sludge sample without treatment (SWT). Mixtures were prepared by adding grass as nitrogen source at a proportion of sludge-bulking agent-grass of 1-1/3-2. We followed the behavior of the mixtures for 13 weeks, performing mechanical turnings each week to aerate the mixture and measuring the generated temperature, the pH, and moisture of the mixture, characterizing the humic materials generated in the process.

Figure 5 shows the temperatures generated during the process. There was a gradual increase of temperature in the first days until reaching temperatures between 50 and 67 °C for the SGW compost between days 3 and 21, as well as temperatures between 35 and 58 °C for the SGT compost sample in the same period of time. The temperature generated in the process was sufficient to decrease the amount of total coliforms to a value below 1×10^3, that is the permissible limit established for a treated sludge of the best quality (sludge class A).

According to the established norms, the quality of the obtained sludge allows its use for forestry and agricultural purposes, as well as to amend the soil and for urban uses with public contact during its application.

The pH of the mixture during the process was kept at values between 7.6 and 8.9, humidity was maintained above 59% during the whole process so that the experimental conditions were at all times appropriate for sludge composting. Regarding removal of volatile solids, in the SGWcompost we achieved a 36% removal, whereas compost SGT reached a 26% removal, indicating that in both conditions the process performed adequately.

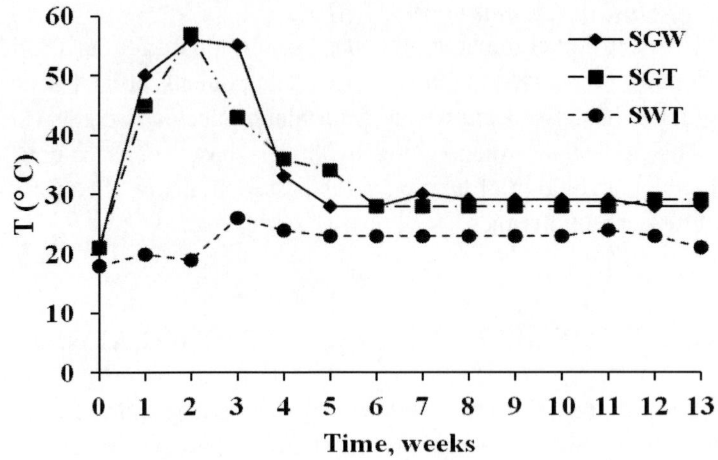

Figure 5. Temperatures obtained in the sludge composting system (Ramírez et al., 2013a).

9. CHARACTERIZATION OF HUMIC SUBSTANCES OBTAINED IN A COMPOSTING PROCESS

Characterization of the humic material generated in the composts was based on determinations of functional groups (total acidity, carboxylic and phenolic groups), E_4/E_6 absorbance relations, and infrared spectroscopy of purified humic acids. Additionally, we performed size exclusion chromatography with visible UV detection (SEC-UV/Visible) of humic substances in order to know the distribution of their molecular sizes along the process.

Acidity Determination

At the beginning of waste sludge composting, the sludge had a total acidity of 4.83 mEq/g, when incorporating the support material and the carbon source for compost formation, a drop in the pH occurred, obtaining a total acidity of 3.41 mEq/g for SGW and 3.28 meq/g for SGT (**Figure 6a**). Afterwards, as a general pattern, the compost piles originated an increase in carboxylic and OH phenolic groups, resulting consequently an increase in total acidity of the humic acids. At the end of composting, concentrations of 5.11 mEq/g for SGW and 4.82 mEq/g for SGT occurred for total acidity, which are similar to the range of humic acids in composts made with organic wastes as reported by Sánchez Monedero in 2002 (between 2.66 and 6.66 mEq/g). For the sludge used as control, the value remained along the whole process at around 4.8 mEq/g.

When the analyses of carboxylic and phenolic groups are made separately, the same growth tendency is shown (**Figure 6b** and **6c**). At the beginning of composting, carboxylic groups concentrations were 2.25-2.08 mEq/g and at the end were 2.91-2.81 mEq/g for SGW and SGT respectively; for phenolic groups, at the beginning concentrations of 1.26-1.18 mEq/g and at 91 days of 2.19-2.04 mEq/g were obtained for SGWand SGT respectively, as shown in figure 7. For the control sludge, no important changes occurred during the process.

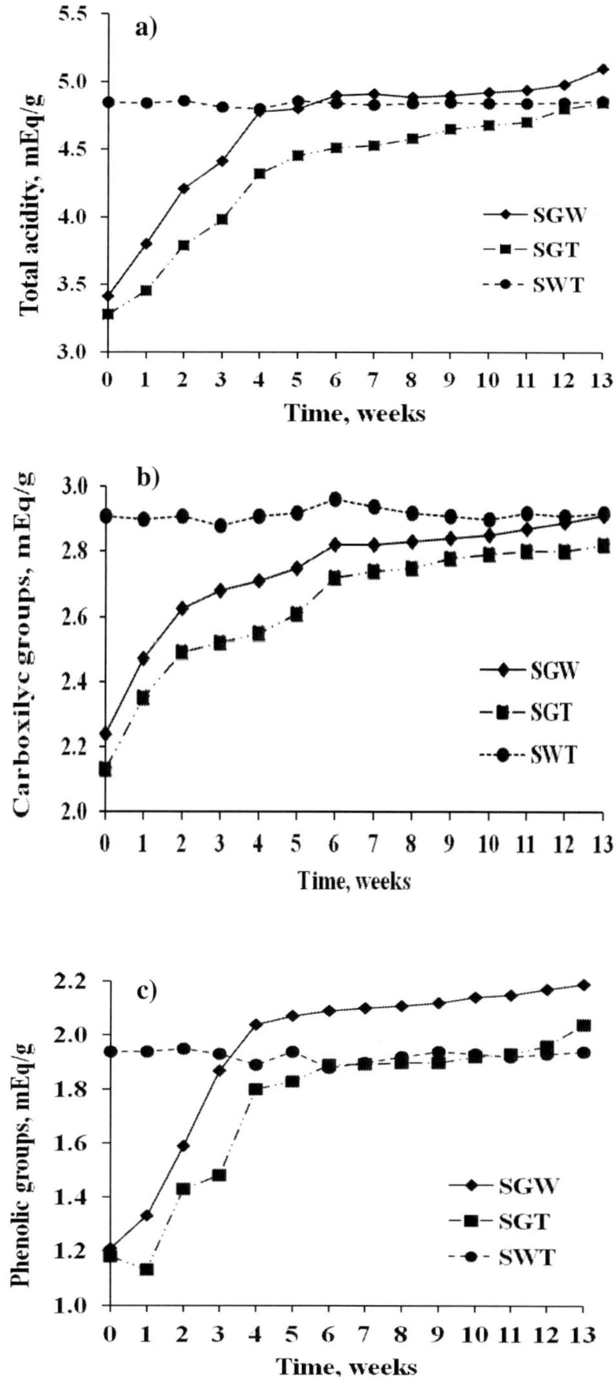

Figure 6. Results of a) total acidity, b) acidity per carboxylic and c) phenolic group in humic acids from composts and waste sludge (Ramírez et al. 2013a).

UV-VIS E_4/E_6spectroscopy

Results show that the E_4/E_6 ratio at the beginning of composting was lower for the waste sludge with a ratio of 5.2 in contrast to 5.79 and 5.68 for SGW and SGT, indicating an increase in molecular sizes in the waste sludge samples were higher than those of SGW and SGT due to the addition of grass (**Figure 7**).

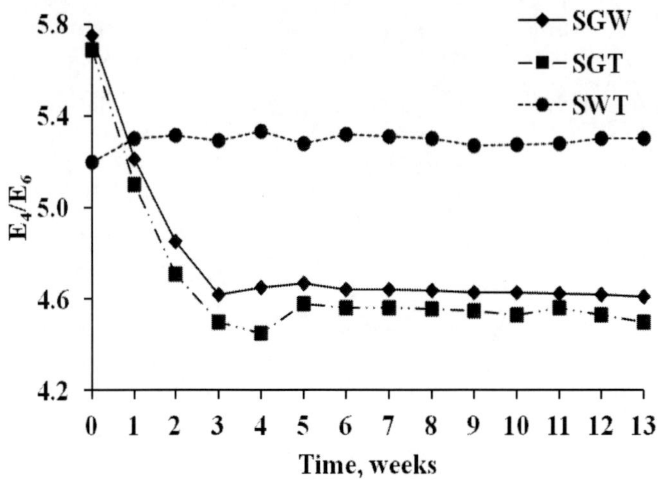

Figure 7. Results of E_4/E_6spectrophotometric ratios in humic acids in composts and waste sludge (Ramírez et al. 2013a).

E_4/E_6 ratios in the SGW and SGT compost piles decreased during the process, reaching a minimum on day 21 for the SGW pile (E_4/E_6 = 4.6) and for SGT on day 28 (E_4/E_6 = 4.4).This decrease in the E_4/E_6 ratio indicates that, during composting, compounds are being synthesized, increasing their molecular weight, which coincides with the humic formation models.This decrease in the E_4/E_6 ratio during composting agrees with the studies reported by Zbytniewski (2002) for composts with sewage sludge and woodshavings during the 53 days of the process.

The fast decay of the E_4/E_6 ratio during composting up to days 21 and 28 is due to the fact that degradation of organic matter occurs more rapidly during this period. The low E_4/E_6ratios at the end of the process reflect a high degree of aromatic condensation and indicate a highlevel of organic matter humification, and provide information on the increase in molecular size.

Infrared Spectroscopy

Figure 8 shows the infrared spectra of all studied humic acids at the beginning and end of composting, which are very similar. The main bands around 3300 cm^{-1} were due to H bonds to branched OH groups, the two bands found in the 3000-2850 cm^{-1} region were due to aliphatic branched C-H, the band in the1730-1720 cm^{-1} were dueto branched C=O groups of COOH, ketones or aldehydes; the strong band in the 1660-1630 region were due to C=O in aromatic structures, branched amide groups and quinones. We also found a band within the

1590-1517 cm^{-1} region related to the presence of peptide groups and NH deformation (amides), a peak in the 1460-1450 cm^{-1} region produced by C-H aliphatic groups, and finally a peak in the 1100-1050 cm^{-1} region produced by polysaccharide branched C-O.

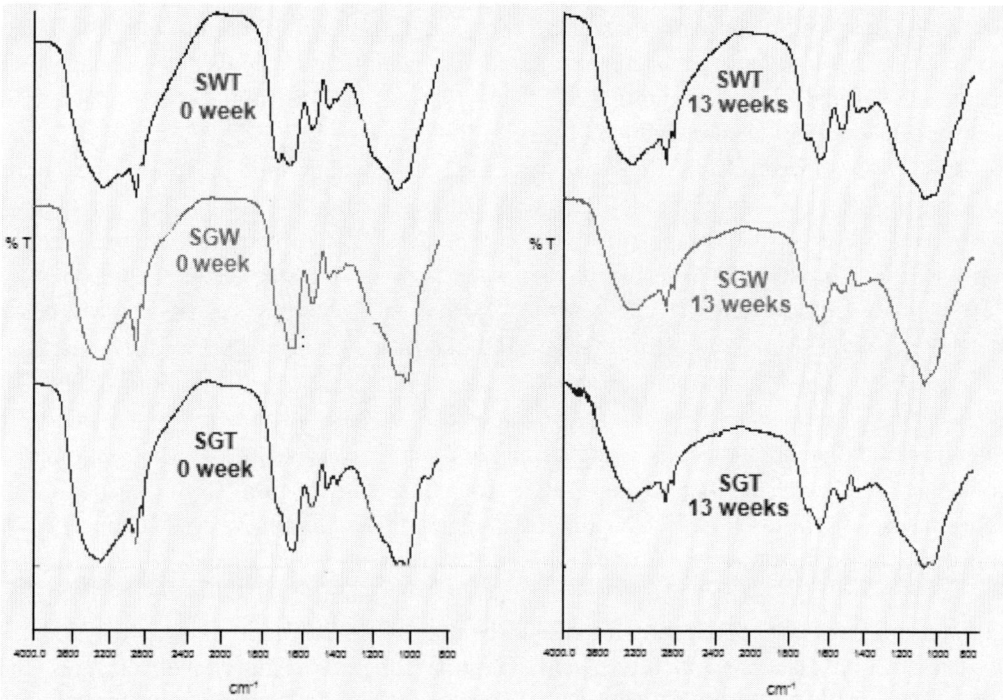

Figure 8. Infrared spectra of humic acid samples obtained from the SGW, SGT composts and sludge (SGT) from water treatment plant.

The most notable change induced by composting in the infrared spectra of the humic acids was the evident decrease in intensity of the bands around 3300 and 2850 cm^{-1}, indicating that there is decrease of aliphatic fractions in the humic structure. There were also other changes such as the decrease in the 1660-1630 cm^{-1} region originated by decomposition of quinones to possibly form the humic material; a slight decrease in the band intensity in the 1590-1517 cm^{-1} region that corresponds to peptide groups, which could indicate the rupture of peptides within amino acids. Finally, in the 1460-1450 cm^{-1} region, there is a decrease in bands intensity that could correspond to the continuous degradation of aliphatic fractions.

Similar infrared spectra were obtained in the studies performed by Sánchez Monedero (2003) with humic acids extracted from composts obtained from waste sludge at the beginning and end of composting (102 days). Stevenson (1994) performed infrared studies for humic acids from different soils, obtaining similar spectra to those produced by the composts studied in this project. The main bands in both studies were found at 3300 cm^{-1}, 2900 cm^{-1}, 1720 cm^{-1},1630 cm^{-1},1540 - 1500 cm^{-1}, and 1460 cm^{-1}.

Analysis of Chromatographic Separation by SEC-UV/Visible

SEC tests were performed to the humus samples extracted from the composts, without separating humic and fulvic acids in order to know the evolution both fractions. The obtained chromatograms exhibit similar signals in all cases, revealing two different fractions. The first appeared at 12.5 min and is related to humic acids, which have a higher molecular weight, whereas the signal identified close to 14.5 min corresponds to fulvic acids, which is the fraction with lower molecular weight (**Figure 9**).

SEC test allows identifying that, as the composting time elapses, the signals corresponding to both humic and fulvic acids also increase, indicating the generation of humic material along the composting process. For the SGW compost, the signal increased from 28 mAU to 48 mAU from the second week to week 13, whereas the signal of fulvic acids increased from 22 mAU to 30 mAU. For the SGT compost, in the same time, the increase was from 23 to 32 mAU for humic acids and from 16 to 21 mAU for fulvic acids.

Figure 9 shows that there is also a small displacement in elution times. This displacement is related to the increase in molecular weight, so that using the same SEC technique and known molecular weight markers, such as proteins, we can estimate the molecular weight of the humic and fulvic fractions. **Figure 10a** shows the calculated molecular weight for SGT and SGW composts obtained from the first peak that appears in the chromatogram of figure 9. It can be observed that there is an increment in both composts from week 1 to 3, reaching a value of 52 kDa for both composts. From here SGW remains without significant variation during the rest of the time while SGT increases up to 57 kDa.

The fraction associated to fulvic acids (**Figure 10b**) shows only small changes. SGT remains between 23 and 24 kDa, while SGW increases from 24 kDa to 28 kDa after 12 weeks of treatment. The increase in molecular weights of humic and fulvic fractions coincides with the decrease of E_4/E_6 ratios showed previously. Likewise, Sánchez Monedero et al., (2003) reported an increase in molecular weights of the humic substances generated in municipal sewage sludge composts, associated to the condensation reactions that are carried out during the degradation of organic matter.

10. Importance of Physicochemical Characterization of the Humic Substances Generated during Composting

Characterizing the humic substances generated during composting is important because the behavior of these materials is determined by their physicochemical characteristics. Their properties as complexing agents, their redox features, as well as their effects as surfactants can be determined by the size of the molecule, predominance of the aliphatic or the aromatic fraction and the content of acid groups. In studies performed by our group, we observed that the humic acids obtained from sludge composts can decrease surface tension (Ramírez et al., 2013b). **Figure 11** shows how the obtained humic acids can decrease surface tension from 72 mN/m to close to 50 mN/m when using humic acids obtained from different compost (HACOMP1, HACOMP2, HALCOMP) at a concentration of 2000 mg/L. In agreement with this latter study, it could be observed that the surfactant properties are favored when there is predominance of the aliphatic fraction in the molecules and E_4/E_6values are higher. Likewise, humic acids that show a higher effect on surface tension are those that have a lower content of COOH groups and a lower molecular weight.

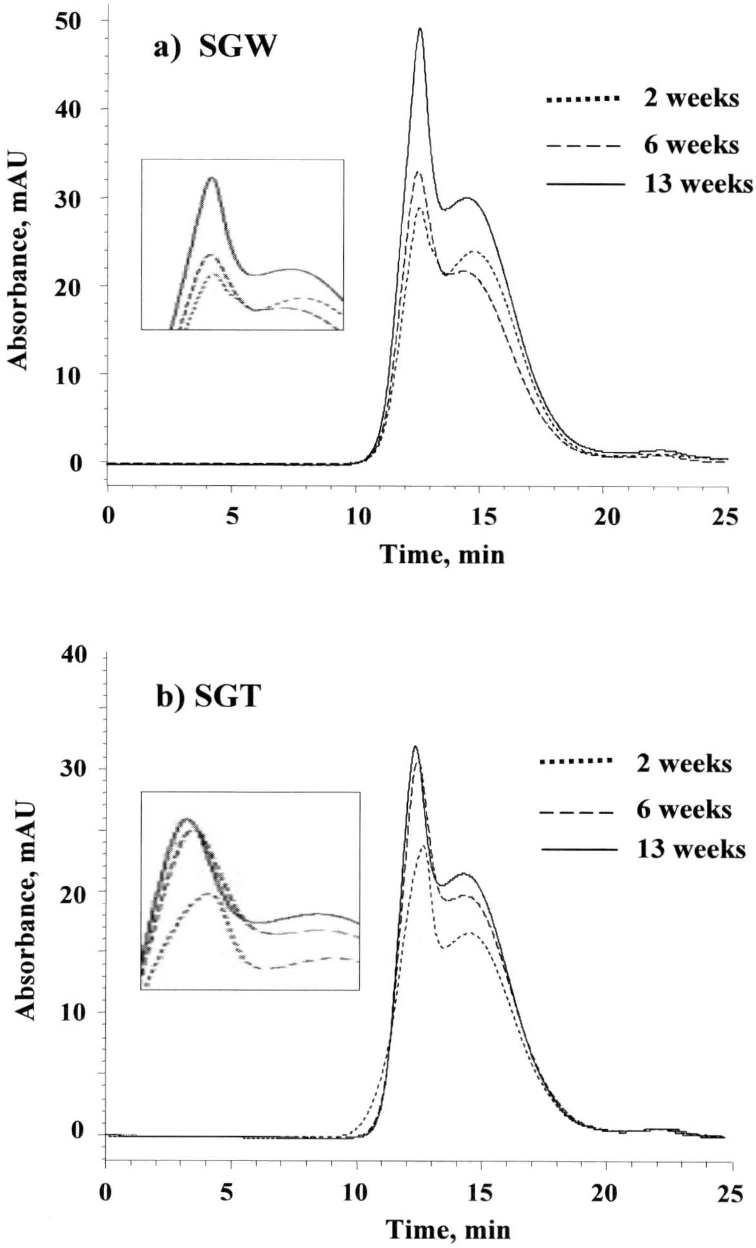

Figure 9. Size exclusion chromatographic separation of humic substances extracted from composts, detection at 500nm(Ramírez et al. 2013a).

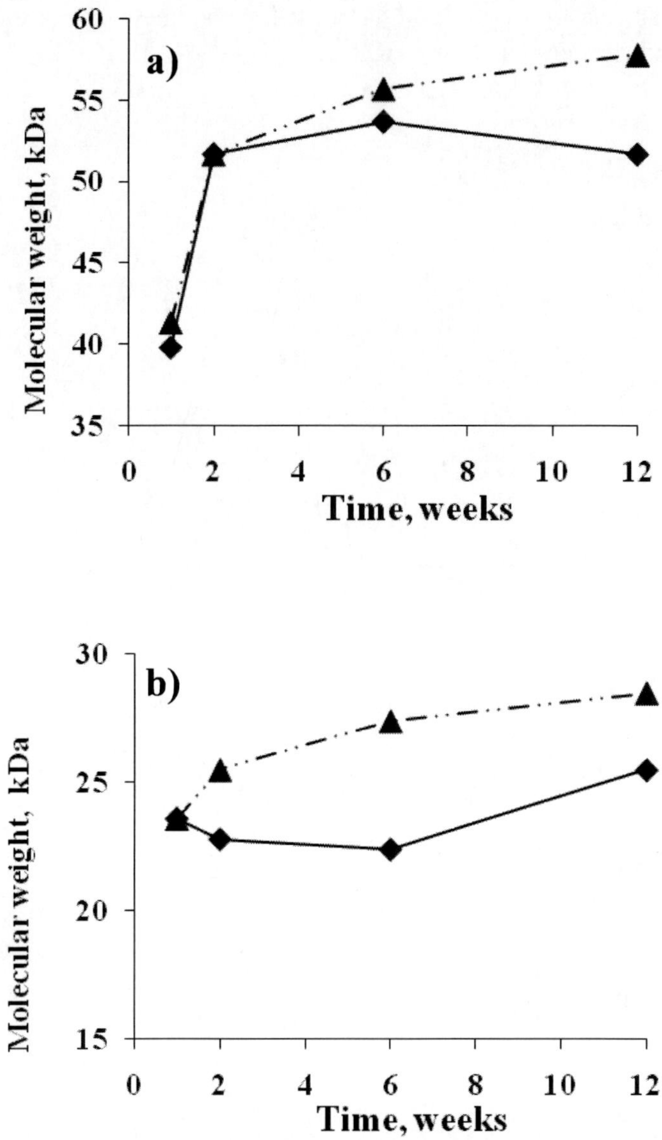

Figure 10.Changes of the molecular mass of humic extracts during the composting process: (—◆—) SGW, (—··▲··–) SGT. (a) humic substances with MM > 30 kDa; (b) humic substances with MM < 30 kDa. (Ramìrez et al., 2013a).

The characterization of humic substances during composting of sludge from treatment plants shows important diversity in terms of the concentration they can attain and their size. Likewise, the content of acid groups, carboxylic and phenolic acids groups, can vary during the composting process. The way each of these parameters affects the complexing, surfactant and redox properties will be the object of future studies.

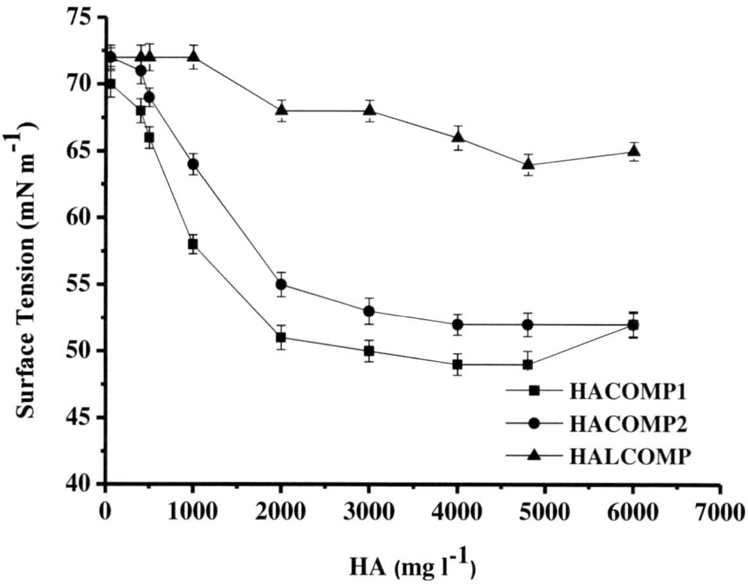

Figure 11. Surface tension at different concentrations of humic acid (Ramírez et al., 2013b).

REFERENCES

Cooperband Leslie, (2002). The art and science of composting, A resource for farmers and compost producers. University of Wisconsin-Madison, Center for integrated agricultural systems, , pp. 14.

Chen, Y., Imbar, Y., (1993). Chemical and spectroscopic analysis of organic matter transformation during composting in relation to compost maturity. In: Hoitink, H. A. J., Keener, H. M. (Eds.), Science and Engineering of Composting. The Ohio State University, pp. 551–600.

Conagua (2012). Atlas digital del agua Mexico, Epstein, Eliot (1997). The science of composting. Boca Raton London by CRC.

Gieguzy´nska E., Kócmit, A, Golebiewska D., (1998). Studies on humic acids in eroded soils of Western Pomerania. In Zaujec, A., Bielek, P., Gonet, S.S. Humic Substances in Ecosystems. Slovak Agricultural University, Nitra, pp.35-41.

Hayes, M. H. B. & Malcolm, R. L. (2001). Considerations of compositions and aspects of the structures of humic acids. In: Humic Substances and Chemical Contaminants (eds C. E. Clapp, M. H. B. Hayes, N. Senesi, P. R. Bloom & P. M. Jardine). pp. 3–40. Soil Science Society of America, Madison, WI.

He X, Logan TJ, Traina SJ (1995). Physical and chemical characteristics of selected U.S. municipal solid waste composts. *J. Environ. Qual.* 24:543–552.

Kelleher, B. P. & Simpson, A. J. (2006). Humic substances in soils: are they really chemically distinct? *Environmental Science and Technology*.40, 4605–4611.

Kluwer Academic Publishers. Printed in the Netherlands. Grigatti M, Ciavatta C, Gessa C. (2004). Evolution of organic matter from sewage sludge and garden trimming during composting. *Bioresour. Technol.* 91:163–169.

Kuter, G., Blackwood, K., L. F. Díaz, J., (1995). Biosolids Composting. Water Environment Federation, Washington, 187 p.

Marambio, C. y Ortega, R. (2003). Uso potencial de lodos derivados del tratamiento de aguas servidas en la producción de cultivos en Chile. Revista Agronomía y Forestal UC 20: 20-23.

Mc Fraland M. J, (2000). Biosolids Engineering. New York:McGraw-Hill.

Metcalf & Eddy, (2003). Wastewater Engineering, Treatment and Reuse. 4th edn. McGraw Hill.

National Research Council, (2002). Biosolids Applied to Land: Advancing Standards and Practices.

Perminova I. V., Hatfield K. and Hertkorn N. (2002). Use of humic substances to remediate polluted environments; from theory to practice. IV. *Earth and environmental science-* Vol. 52.

Piccolo, A. (2002). The supramolecular structure of humic substances: a novel understanding of humus chemistry and implications in soil science. *Advances in Agronomy*, 75, 57–134.

Polprasert, c. (1996). Organic waste recycling. Technology and management. Ed. Wiley & Sons, Inglaterra. 412 pp.

Ramírez, V., Wrobel, K. Kazimierz, W., Navarro, R., Godinez, L., Rodriguez, F., (2013a). Evaluation of composting process in sewage sludge from municipal waste treatment plant in the city of San Miguel de Allende, central Mexico. *Revista Internacional de Contaminación Ambiental.* In Press.

Ramírez, V., Torres, L., Godinez, L., Guerra, R., Rodriguez, F. (2013b). pH effect on surfactant properties and supramolecular structure of humic substances obtained from sewage sludge composting. *Revista Internacional de Contaminación Ambiental.* 29: 191-199.

Ramos, S., Avelar J., Medel A., YAmamoto L., Godinez L., Ramírez M., Guerra R., Rodriguez F. (2012). Movilidad de metales en jales procedentes del distrito minero de Guanajuato. *Revista Internacional de Contaminación Ambiental.* 28: 49-59.

Romdhana M., Lecomte D., Ladevie B., Sablayrolles C. (2009). Monitoring of pathogenic microorganism contamination during heat drying process of sewage sludge. *Process Safety and Environmental Protection.* 87: 377-386.

Sánchez-Monedero Miguel A., Cegarra Juan (2003). Chemical and structural evolution of humic acids during organic waste composting. *Biodegradation* 13: 361-371.

Senesi, N., D'Orazio, V., Ricca, G., (2003). Humic acids in the first generation of EUROSOILS. *Geoderma* 116, 325–344.

Spellman, Frank R., (1997). Wastewater Biosolids to Compost. 1ra. Edicion. Technomic, publishing Co. Ing.

Stevenson, F. J., (1994). Humus Chemistry: Genesis, Composition, Reactions. Wiley-Interscience, New York.

Sutton, R. & Sposito, G. (2005). Molecular structure in soil humic substances:The new view. *Environmental Science and Technology*, 39,9009–9015.

Tan Kim H. (1998). Principles of soil chemistry. Third edition.

United States Environmental Protection Agency (USEPA),(1989). Standards for the Use or Disposal of Sewage Sludge. Washington, DC.

United States Environmental Protection Agency (USEPA), (1999). Biosolids Generation, Use, and Disposal in the United States U.S. Municipal and Industrial Solid Waste. Division Office of Solid Waste EPA530-R-99–009.

Wershaw, R. L. (1993). Model for humus in soils and sediments. *Environmental Science and Technology*, 27, 814–816.

Zbytniewski, Kosobucki, P., Kowalkowski, T., Buszewski. B., (2002). The comparison study of compost to natural organic matter samples. Environ. *Sci. Pollut. Res*. 1, 68-74.

In: Energy and Environment Nowadays
Editors: Luis G. Torres and Erick R. Bandala
ISBN: 978-1-63117-398-1

Chapter 6

MATHEMATICAL MODELING AND SIMULATION OF HYDROGEN PRODUCTION BY DARK FERMENTATION USING AN ADM1-BASED MODEL

Guillermo E. Baquerizo Araya
Bioengineering Department, Unidad Profesional Interdisciplinaria de Biotecnología,
Instituto Politécnico Nacional, Mexico

INTRODUCTION TO THE ANAEROBIC DIGESTION TECHNOLOGY

General Overview

Anaerobic digestion is a well-known technology aimed to treat and revalue organic wastes. Biochemical reactions involved during anaerobic conversions have been used over many centuries, originally for food and beverage production (Batstone et al., 2002). Hence, anaerobic digestion is considered as one of the oldest biological technology utilized by mankind. Nevertheless the main dramatic advances and applications of this technology have been assessed during the last decades with the introduction of large scale configurations to treat a wide range of organic substrates. Advantages demonstrated by anaerobic processes over other biological technologies include the capacity to treat high organic loading rates together with the slow production of sludge. However one of the major driven-attribute to increase the application of anaerobic treatments deals with the energy production. The generation of biogas, mainly constituted by methane, has been seen as an alternative to replace fossil-based sources, contributing to reduce the greenhouse gas emissions (Speece, 1996).

Anaerobic digestion is currently considered a consolidated technology with more than 2,200 high-rate reactors implemented worldwide (Van Lier, 2008). In Europe the number of installed plants increased from 15 to 200, in the period of 1995 to 2010, which implies an installation capacity of about 6,000,000 tons per year (de Baere et al., 2010). This biological process is currently used for both sludge stabilization and energy recovery from the treatment of different organic wastes, including manure, municipal and industrial wastewaters, the organic fraction of municipal solid waste and sewage sludge (Appels et al., 2011; Rajeshwari et al., 2000; Speece, 1996;).

The main issues related to anaerobic digestion such as reactor design features, optimal operation conditions, utilization of substrate mixtures (co-digestion), microbial aspects, inhibition phenomena, among others have been broadly described in the literature and can be found elsewhere (Appels et al., 2008; Chen et al., 2008; Speece, 1996; Ward et al., 2008). Additionally Rittmann (2008) provided a detailed description of metabolic aspects in microorganisms involved in different bioenergy production processes, including the anaerobic digestion.

Biohydrogen Generation by Dark Fermentation

Although the anaerobic digestion of organic wastes aimed to produce methane is a well-established technology, the scientific community has been focused on developing other renewable and environmental friendly energy sources due to the expected energy crisis as a result of depleting foil fuel reserves coupled with the concerns for environmental pollution associated with greenhouse emissions. In this context, fermentation processes are receiving revived attention as a sustainable and economically technology to produces energy such as ethanol, butanol, and hydrogen from renewable and low-cost biomass and waste material (Gadhamshetty et al., 2010). Among biofuels, hydrogen has gained importance over competing technologies since it exhibits the highest energy density (122 kJ/g) of any currently available fuel (Park et al., 2005) and its combustion is environmentally clean since only water vapor and heat energy are produced (Nath and Das, 2011). Biological hydrogen can be metabolically produced by photosynthetic and fermentative microorganism (i.e., bacteria and algae). Also biohydrogen can be efficiently converted into electrical and thermal energy while supplied to a fuel cell (Hallenbeck and Ghosh, 2010).

Additionally, fermentative hydrogen production processes can overcome some of the typical limitations of methane production by anaerobic digestion, mainly in relation to the difficulty to achieve a stable operation regime due to the complex syntrophic relationships between the microbial populations present in anaerobic digesters. Indeed the application of two-stage anaerobic digestion process for sequential H_2 and CH_4 production has been proposed as a feasible and promising technology to improve energy yields as compared to the traditional one-stage CH_4 production process (Antonopoulou et al., 2008; Cooney et al., 2007, Liu et al., 2006; Pakarinen et al., 2009). The two stage system is based on the fact that both the growth rates and the optimal pH are different for acidogen and methanogen populations (Liu et al., 2004) and thus, faster growing of acidogens are developed in the first-stage hydrogenic reactor. A methanogenic bacterial population is mainly developed in the second-stage methanogenic reactor, in which the end-products from hydrogen generation are further converted to CH_4 and CO_2. In addition, the optimal temperature for hydrolysis/acidogenesis can differ from optimal temperature for methanogenesis (Ward et al. 2008).

A large number of microbial species, including types with significantly taxonomic and physiological differences can produce hydrogen by fermentation, including strict anaerobes such as clostridia (*Clostridium butyricum, Clostridium welchii, Clostridium pasteurianum, Clostridium beijerincki*), Rumen bacteria, hyperthermophilic archaebacteria, and facultative anaerobes like Enterobacter, *Escherichia coli*, Citrobacter, or even some aerobes like Alcaligenes and Bacillus (Nath and Das, 2011).

Anaerobic fermentations can generate hydrogen from a wide spectrum of potentially utilizable substrates and also from refuse and waste products (Karlsson et al., 2008; Kim and Lee, 2010; Nath and Das, 2011; Ntaikou et al., 2009a). In dark fermentation, organic matter is basically degraded to volatile fatty acids (VFAs) while hydrogen is produced as a secondary metabolite. This is possible because microorganisms use protons (H^+) for two equivalent electrons: $2H^+ + 2e^- \leftrightarrow H_2$. The reaction is performed by hydrogenase enzymes and strong reducing agents (ferrodoxin (Fd) and NADH) (Nath and Das, 2004). Fermentative microorganisms have two basic types of metabolism: facultative anaerobe (i.e., E*scherichia coli*, Enterobacter sp.) and strict anaerobe (i.e., Clostridium sp.). Many microorganisms have both types but one is more active than the other (Hallenbeck and Ghosh, 2010). The facultative metabolism of bacteria may theoretically generate 2 moles of H_2 per mole of glucose when butyric acid is mainly produced (Eq. (1)). On the other hand, strict anaerobe bacteria can theoretically reach a maximum yield of 4 moles of H_2 per mole of glucose when acetic acid is primarily produced (Eq. (2)) (Fang et al., 2002; Hussy et al., 2003; Van Ginkel et al., 2005).

$$C_6H_{12}O_6 \rightarrow CH_3(CH_2)_2COOH + 2H_2 + 2CO_2 \tag{1}$$

$$C_6H_{12}O_6 + 2H_2O \rightarrow 2CH_3COOH + 4H_2 + 2CO_2 \tag{2}$$

The hydrogen yields obtained in dark fermentation are determined by the metabolic pathways that control the biological process (Mizuno et al., 2000; Oh et al., 2011) and by the thermodynamic conditions that influence these metabolic pathways. Butyric and acetic fermentations are associated with specific bacteria and characteristic optimum conditions such as pH and temperature. Also hydrogen partial pressure is a thermodynamic variable with an important effect on biohydrogen production due to its influence on metabolic pathways.

In spite of over-reported drawback regarding the moderate hydrogen yield from dark fermentation, studies are constantly conducted to optimize and improve rates and energy yields (Gadhamshetty et al., 2010). As mentioned before, benefits of dark fermentation could be improved if its effluents are used as feeding for downstream process such as photo fermentation, microbial electric cells, methane fermentation or microbial cells (Kaparaju et al., 2009; Liu et al., 2005; Sun et al., 2009).

Modeling the Dark Fermentative Process

Modeling Methane Production by Anaerobic Digestion: from Simple Models to the ADM1

Anaerobic digestion can be seen as a chain of interconnected biochemical reactions and physicochemical processes where the organic matter (in the form of carbohydrates, proteins, lipids or more complex compounds normally called composites) is transformed into methane, carbon dioxide and anaerobic biomass, in an oxygen-free environment (Donoso-Bravo et al., 2011). Biochemical reactions are normally catalyzed by intra- or extra-cellular enzymes. Disintegration of composites to particulate compounds which are subsequently hydrolyzed by enzymes to soluble monomers has been considered as extracellular processes (Batstone et al., 2002). Metabolism of soluble materials by microorganisms is an intracellular processes resulting in biomass growth and decay. Physicochemical processes are related to ion association/dissociation, gas-liquid mass transfer rates, and precipitations of inorganic salts.

In the early 70's the necessity of both understanding the interactions of process compounds and predicting the behavior of the system under different scenarios to ensure efficient operation motivated the development of the first mathematical models for anaerobic digestion processes. Hill and Barth (1977) reported the first simulation of animal waste digestion assuming anaerobic digestion as a multistep process where one slower stage controls the global rate. The model focused on identifying and describing such limiting step. However, this limiting step may differ under different operating conditions. The hydrolysis of suspended solids, the conversion of fatty acids into biogas or the methanogenesis stage have been considered as limiting steps (Eastman and Ferguson, 1981). These mathematical approaches were relatively simple due to the limited available knowledge of the process. Consequently these models were unable to adequately describe the process performance under different conditions (i.e., different substrates, reactor configurations, stationary or transient operation, etc.).

A second generation of models was characterized by considering the volatile fatty acids as the key compounds. Therefore the acidogenesis and acetogenesis steps were modeled separately (Hill, 1982). More bacterial groups were included distinguishing acetoclastic and hydrogenotroph methanogens. The hydrogen partial pressure was considered as a key regulatory parameter influencing the redox potential (as $NADH/NAD^+$ ratio) determining the VFAs production and consumption (Costello et al., 1991; Ruzicka, 1996). Further experimental and microbiological studies together with the increase in computing capacity led to the development of much more detailed models. (Angelidaki et al., 1993, 1999; Batstone et al., 2000; Haag et al., 2003; Kalyuzhnyi and Davlyatshina, 1997; Kalyuzhnyi, 1997; Kalyuzhnyi and Fedorovich, 1998; Keshtkar et al., 2003; Siegrist et al., 1993; Tartakovsky et al., 2002; Vavilin et al., 1994, 1995). These models included additional processes and species, more detailed kinetics and inhibition phenomena.

Several reviews of anaerobic digestion modeling can be found in literature, describing different models in terms of complexity or kinetic approaches (Gavala et al., 2003; Husain, 1998). Recently Donoso-Bravo et al., (2011) presented a comprehensive overview of the anaerobic digestion models available in the literature. On the other hand, a reduced number of works have addressed with modeling issues related to the type of anaerobic reactor. A broad description of different models for UASB (Up-flow Anaerobic Sludge Blanket), AF (Anaerobic Filter) and EGSB (Expanded Granular Sludge Blanket) reactors was provided by Saravanan and Sreekrishnan (2006).

In 2002 the IWA Task Group for Mathematical Modeling of Anaerobic Digestion Processes developed the generic Anaerobic Digestion Model No.1 (ADM1) (Batstone et al., 2002) following the stoichiometric matrix representation developed in ASM models (Henze et al., 2000). The main goal of this first generic model was to provide a common basis for further model development and validation studies with comparable and compatible results.

Additional benefits including the application of ADM1 to full-scale plant design, process optimization and process control were intended.

The resulting structured model describes the dynamics of 27 species, includes 19 biochemical processes and 3 physicochemical processes, divided in the following steps: disintegration, hydrolysis, acidogenesis, acetogenesis and methanogenesis. A schematic overview of the ADM1 models is depicted in Figure 1. The list of components (i.e., soluble, particulate and gaseous compounds) considered in the ADM1 is provided in Table 1.

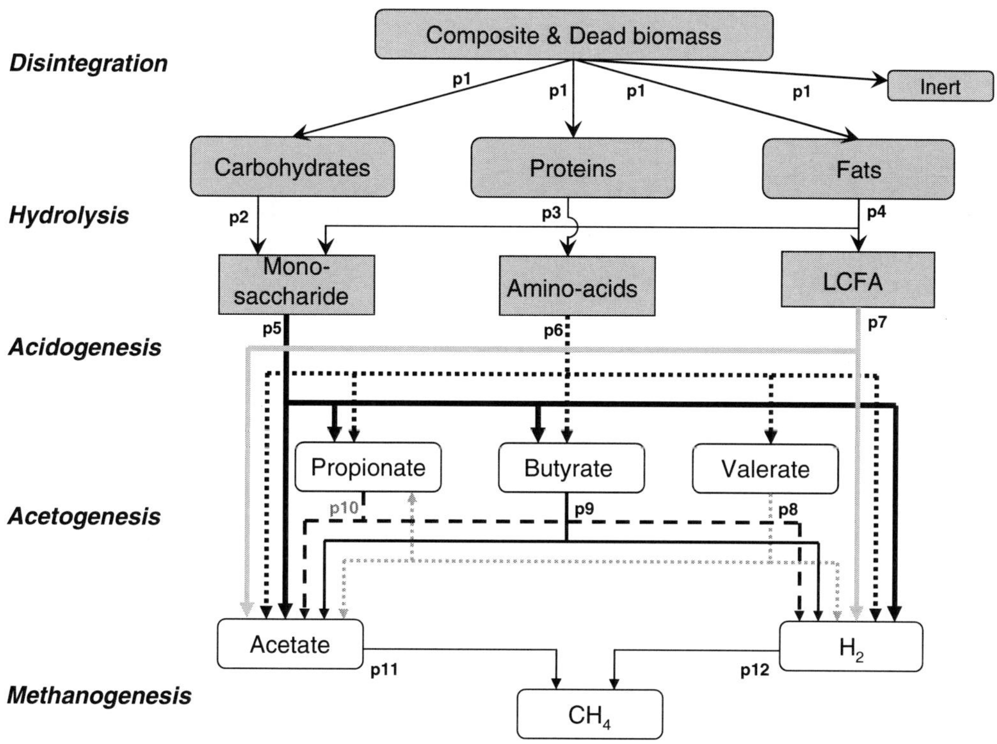

Figure 1. Scheme of different steps included in ADM1 (Batstone et al., 2002). The model is implemented including the following biochemical processes: disintegration (p1), carbohydrates hydrolysis (p2), proteins hydrolysis (p3), fats hydrolysis (p4), acidogenesis from sugars (p5), acidogenesis from amino acids (p6), acidogenesis from LCFA (p7), acetogenesis from butyrate and valerate (p8 & p9), acetogenesis from propionate (p10), aceticlastic methanogenesis (p11), hydrogenotrophic methanogenesis (p12). The remaining biochemical processes correspond to the decay of the 7 bacterial populations considered.

The ADM1 model neglects some processes and species, which are related to more specific applications, with the aim of avoiding extreme complexity. Despite that fact, the ADM1 model considers 41 stoichiometric parameters and 36 kinetic parameters for describing bioconversion processes. The large number of parameters and the difficulties associated for their estimation are the major drawbacks of ADM1, as well as, some structural weaknesses (Kleerebezem and Van Loosdrecht, 2006).

Table 1. List of compounds (soluble, particulate and gaseous) considered in the ADM1

Compound	Symbol	Units
Soluble		
Monosaccharides	S_{su}	kg COD m^{-3}
Amino Acids	S_{aa}	kg COD m^{-3}
Long Chain Fatty Acids (LCFA)	S_{fa}	kg COD m^{-3}
Total valerate	S_{va}	kg COD m^{-3}
Total butyrate	S_{bu}	kg COD m^{-3}
Total propionate	S_{pro}	kg COD m^{-3}
Total acetate	S_{ac}	kg COD m^{-3}
Dissolved Hydrogen	S_{h2}	kg COD m^{-3}
Dissolved Methane	S_{ch4}	kg COD m^{-3}
Inorganic carbon	S_{IC}	kmole C m^{-3}
Inorganic nitrogen	S_{IN}	kmole N m^{-3}
Soluble inerts	S_I	kg COD m^{-3}
Particulate (non-biomass)		
Composites	X_C	kg COD m^{-3}
Carbohydrates	X_{ch}	kg COD m^{-3}
Proteins	X_{pr}	kg COD m^{-3}
Lipids	X_{li}	kg COD m^{-3}
Particulate inerts	X_I	kg COD m^{-3}
Particulate (biomass)		
Sugar degraders	X_{su}	kg COD m^{-3}
Amino Acids degraders	X_{aa}	kg COD m^{-3}
LCFA degraders	X_{fa}	kg COD m^{-3}
Valerate and butyrate degraders	X_{c4}	kg COD m^{-3}
Propionate degraders	X_{pro}	kg COD m^{-3}
Acetate degraders	X_{ac}	kg COD m^{-3}
Hydrogen degraders	X_{h2}	kg COD m^{-3}
Gaseous		
Hydrogen	S_{gh2}	kg COD m^{-3}
Methane	S_{gch4}	kg COD m^{-3}
Carbon dioxide	S_{gco2}	kmole C m^{-3}

Later on, Rosen and Jeppsson (2006) improved the original ADM1 model in aspects related to carbon and nitrogen mass balances, inhibition expressions in kinetic equations, calculation of gas flow when system is overpressure, and the simulation of acid-base equilibrium using ordinary differential equations (ODE) or differential algebraic equations, among others. Authors noticed that in the original ADM1, the state variables for inorganic carbon and inorganic nitrogen acted as source or sink terms to close mass balances since the provided stoichiometric matrix was not properly defined to take this into account. Therefore, one of the major improvements proposed in their ADM1 implementation dealt with the correction of carbon and nitrogen mass balances by modifying the coefficients in the stoichiometric matrix. One example is referred to disintegration process, where a composite material (X_C) is transformed into several compounds (S_I, X_{ch}, X_{pr}, X_{li} and X_I). Assuming one

COD mass unit of X_C is completely disintegrated, original ADM1 gives the following mass balance in terms of COD:

$$f_{sI,xc}S_I + f_{xI,xc}X_I + f_{ch,xc}X_{ch} + f_{pr,xc}X_{pr} + f_{li,xc}X_{li} = 0.1S_I + 0.25X_I + 0.2X_{ch} + 0.2X_{pr} + 0.25X_{li} \quad (3)$$

where $f_{SI,xc}$, $f_{xI,xc}$, $f_{ch,xc}$, $f_{pr,xc}$, $f_{li,xc}$ are the fraction of soluble inerts, particulate inerts, carbohydrates, proteins and lipids in the composite, respectively, in kg COD (kg COD)$^{-1}$.

A COD balance is corroborated since the sum of all $f_{i,xc} = 1$. However, the nitrogen mass balance is not properly closed in the original ADM1. The proposed nitrogen content of X_C (N_{xc}) is set to 0.002 kmole (kg COD)$^{-1}$. Calculating the nitrogen content of the disintegration products (kmole N) using the parameters suggested by Batstone et al., (2002):

$$N_I 0.1S_I + N_I 0{,}25X_I + N_{ch}0.2X_{ch} + N_{aa}0.2X_{pr} + N_{li}0.25X_{li} = 0.0002 + 0.0005 + 0.0014 = 0.0021 \quad (4)$$

where N_I, N_{ch}, N_{aa}, N_{li} are the nitrogen content in inerts, carbohydrates, proteins and lipids, respectively, in kmole N (kg COD)$^{-1}$. Note that carbohydrates and lipids contain no nitrogen (i.e., N_{ch} and N_{li} are equal to zero). This result means that for every kg of COD that disintegrates, 0.0001 kmole of N is created (5% more than the originally proposed value). In their improved version of ADM1, Rosen and Jeppson (2006) suggested new values for $f_{xI,xc} =$ 0.2 and $f_{li,xc} = 0.3$. Also values for nitrogen contents were modified thus N_I was set to 0.06/14 in order to be consistent with ASM models and consequently N_{xc} was adjusted to a value of 0.0376/14 to maintain nitrogen balance.

Regarding carbon balances another example is given by the authors, related to the decay of biomass (processes 13 to 19) where only composite (X_C) is generated. In these processes the amount of composites is balanced in terms of COD. However the carbon content may vary from biomass to composites resulting from decay. It is suggested by Batstone et al., (2002) that the carbon content of bacteria (C_{bac}) is 0.03125 kmole C (kg COD)$^{-1}$ which results from assuming a typical bacterial composition (i.e $C_5H_7O_2N$). The latter value differs from the suggested value for carbon content of composite (C_{xc}). In such a case, a stoichiometric term ($C_{bac} - C_{xc}$) was added into the Petersen matrix for carbon balances.

Similarity, the nitrogen content of bacteria (N_{bac}) provided in the original ADM1 is 0.00625 kmole N (kg COD)$^{-1}$, which is three times higher than the suggested value for the composite (N_{xc}). Therefore the same term was added to the stoichiometric matrix ($N_{bac} - N_{xc}$). Similar modifications should be done for the 7, 8 and 9 processes (Figure 1) regarding carbon mass balance as well for disintegration and hydrolysis processes in both carbon and nitrogen balances. Finally Rosen and Jeppsson (2006) recommended including stoichiometric relationships for all 19 processes in relation with inorganic carbon and inorganic nitrogen in order to guarantee that the mass balances are closed and the conservation law fulfilled at all times for COD, carbon and nitrogen.

Modeling Hydrogen Production in Dark Fermentation

Mathematical models for hydrogen production by dark fermentation have been carried out to analyze and predict performance of the process. Several types of models including logistic and mechanistic-structured models have been applied to principally predict biohydrogen production and degradation of organics substrates even though formation of other end-products (VFAs) and microbial growth have also been included into modeling objectives. Normally carbohydrates, mainly glucose, are the preferred carbon sources for modeling dark fermentation which predominantly generates acetate and butyrate together with hydrogen.

Several simplified versions of ADM1 can be found in the literature to simulate biohydrogen generation but normally ignoring the stoichiometric structure in which the ADM1 is based. Thus CO_2 and compounds generated from biomass decay are constantly disregarded in these models causing that mass balances of COD, nitrogen and carbon could not be closed in the system. Moreover pH, which is an important parameter affecting metabolic activity, is normally excluded from the model framework and replaced by an experimentally measured pH profile (time-dependent) thus limiting the predictive ability of the model. Lin et al., (2007) used the kinetic structure of the ADM1 to evaluate model predictions of fermentative biohydrogen production including generation of alcohols from metabolism of glucose by selected clostridium species in batch cultures. A comparison of Gompertz equation and ADM1 model for describing batch hydrogen production by different consortia as inoculum was reported by Gadhamshetty et al., (2010). Using a refined version of ADM1 by varying half saturation constant as a function of reactor conditions, the authors satisfactorily predicted biohydrogen and VFAs formation together with COD consumption. The effect of carbohydrate-protein ratio on dynamic simulation of protons, biomass, fatty acids and hydrogen using ADM1 was demonstrated by Peiris et al., (2006). Penumathsa et al., (2008) presented a modified version of ADM1 using a variable stoichiometric approach derived from experimental information. The biomass and end-products yields (in this case, hydrogen, butyrate, propionate, acetate and lactate) from glucose were assumed to be dynamically depending on the total concentration of undissociated acids. Recently Ntaikou et al., (2009b and 2010) developed a kinetic ADM1-based model for the hydrogen production process by the bacterium *Ruminococcus albus* grown on glucose and sweet sorghum extract as the sole carbon sources. Authors included the generation of additional end-products that are not originally considered in the ADM1 such as ethanol and formate together with a substrate inhibition term.

A review of kinetic models used in fermentative hydrogen process describing hydrogen formation rate, substrate consumption, and biomass growth including inhibition phenomena was presented by Wang and Wan (2009). Recently Nath and Das (2011) presented a comprehensive work reporting different models used for also kinetically describing microbial growth, substrate utilization and product formation. In addition, the study dealt with experimental design methods to investigate the effects of various factors on fermentative hydrogen production. In both reviews, authors highlighted that biological production rate of hydrogen and molar yield are influenced by several variables including pH of mineral medium, temperature, substrate and products concentration, and the presence of inhibitors, among others. Some of these variables have been satisfactorily included in the kinetic expressions used for predicting biohydrogen generation. A brief overview of kinetic models reported for hydrogen production by dark fermentation is given in the following paragraphs.

Kinetic Models for Products Formation

Products generated from fermentative biohydrogen processes are mainly grouped into two categories: gaseous products (primarily H_2 and CO_2) and liquid products (VFAs and solvents). The modified Gompertz equation (Eq. (5)) has been widely used to describe the progress hydrogen formation and generation of some soluble metabolites during batch fermentative experiments. Several examples of studies using the modified Gompertz model are listed in Table 2.

$$H = H_{max} \, exp\left\{ - exp\left[\frac{R_{max} \, e}{H_{max}} (\lambda - t) + 1 \right] \right\} \tag{5}$$

where H represents the cumulative volume of hydrogen production (mL), H_{max} the maximum gas production potential (mL), R_{max} the maximum production rate (mL h^{-1}), λ the lag time (h), and t the cultivation time (h).

Despite the modified Gompertz model shows high correlation coefficients between experimental and predicted data, the three model parameters determined by curve fitting are restricted to specific experimental conditions and normally they cannot be used for predictive purposes. In this sense, the utilization the Gompertz equation is generally limited to biohydrogen simulation since its inability for predicting VFAs and substrate utilization.

Products formation during anaerobic hydrogen generation can also be modeled using the existing relationship between biomass and products by using the Luedeking-Piret model (Eq. (6)) (Mu et al., 2006; Obeid et al., 2009).

$$\frac{dP_i}{dt} = \alpha_i \frac{dX}{dt} + \beta_i X \tag{6}$$

where P_i corresponds to the concentration of product i, X the biomass concentration, α_i the growth-associated formation coefficient of product i, βi is non-growth-associated formation coefficient of product i, all in consistent units.

The classical Monod model (Eq. (7)) to simulate products generation has been proposed by several authors (Chittibabu et al., 2006; Lee et al., 2008; O-Thong et al., 2008; Sharma and Li, 2009; Whang et al., 2006). Indeed in the ADM1, kinetic expressions are obtained from Monod or Monod-based expressions. However the utilization of this approach is conditioned to the description of biomass growth on a particular substrate. Kinetic description for microbial growth and substrate utilization is provided in the following section.

$$\frac{dP}{dt} = \frac{1}{Y_{X/P}} \frac{dX}{dt} = \frac{u_{max}}{Y_{X/P}} \frac{S}{K_S + S} X \tag{7}$$

where μ_{max} is the maximum specific growth rate of the biomass (h^{-1}), $Y_{X/P}$ is the yield of product from biomass (mol $(mol)^{-1}$), S is the substrate concentration (mol L^{-1}), and K_S is the half saturation constant (mol L^{-1}).

Table 2. Studies using modified Gompertz model to simulate fermentative hydrogen generation

Microorganism	Substrate	Maximum H_2 production rate	Correlation coefficient	Reference
Enterobacter cloacae DM11	Glucose	14.1 mL h^{-1}	0.992	Nath et al., (2008)
Enterobacter cloacae IIT-BT 08	Glucose	72 mL (g $TVS)^{-1}$	0.99	Khanna et al., (2011)
Mixed anaerobic microflora	Cassava starch	477 mL h^{-1}	0.997	Lee et al., (2008)
Mixed microflora from digested sludge	Glucose	15.1 mL h^{-1}	0.99	Wang and Wan (2008)
Mixed microflora from anaerobic digester	Sucrose	10.6 mL h^{-1}	0.999	Chen et al., (2006)
Mixed microflora from anaerobic digester	Nonfat dry milk	22.0 mL h^{-1}	0.997	Chen et al., (2006)
Mixed microflora from anaerobic digester	Food waste	32.3 mL h^{-1}	0.998	Chen et al., (2006)
Anaerobic microflora from UASB	Lactose	6.8 mL h^{-1}	0.99	Davila-Vazquez et al., (2008)
Mixed anaerobic culture	Sucrose	1511 mL h^{-1}	-	Mu et al., (2006)
Mixed anaerobic culture	Sweet sorghum syrup	51.66 mL $L^{-1}h^{-1}$	-	Saraphiron and Reungsang (2010)
Mixed anaerobic culture	Starch	37.9 mL h^{-1}	0.999	Lin et al., (2008)
Seed sludge from anaerobic dgestor	Coffee drink manufacturing wastewater	12.03 mL h^{-1}	-	Jung et al., (2010)

Source: Nath and Das, 2011; Wang and Wan, 2009.

Kinetic Models for Microbial Growth and Substrate Utilization

In batch fermentations aimed to produce hydrogen, the dynamic variation of biomass concentration depends on the concentration of limiting substrate. Monod equation (Eq. (8)) is able to fit a wide range of data and is the most commonly applied unstructured, non-segregated model to describe microbial growth linked to substrate utilization (Shuler and Kargi, 2002).

$$\mu = u_{max} \frac{S}{K_S + S} X \tag{8}$$

where μ is the specific growth rate ((mol biomass) $L^{-1}h^{-1}$), and S is the limiting substrate concentration (mol L^{-1}).

However many authors have reported that the classical Monod model is unable to adequately predict experimental data in cases presenting inhibition phenomena due to high substrate concentration, pH, or presence of certain inorganic salts, among others. Growth kinetics of pure and mixed microorganism culture producing biohydrogen under high substrate concentration have been described by several inhibitions models (Kumar et al., 2000; Nath et al., 2008; Ntaikou et al., 2009b, 2010; Wang and Wan, 2008). Andrew-based equations (Eq. (9) and (10)) which incorporate a substrate inhibition factor into the classical Monod equation have been widely used. Nath et al., (2008) compared the classical Monod model and the Andrew model (Eq. (9)) for describing both the glucose degradation and the *Enterobacter cloacae* DM11 growth in batch experiments and concluded that the latter was the most suitable one. Similarly Kumar et al., (2000) compared the ability of classical Monod model and modified Andrew model (Eq. (10)) to simulate the evolution of glucose consumption and *Enterobacter cloacae* IIT-BT 08 growth in batch tests and again concluded that Andrew's equation was the most appropriate option.

$$\mu = u_{max} \frac{S}{K_S + S + S^2 / K_I} X \tag{9}$$

$$\mu = u_{max} \frac{S}{K_S + S - S^2 / K_I} X \tag{10}$$

where K_I is the substrate inhibition constant (mol L^{-1}).

Another alternative to describe the nature of substrate inhibition is the use of a generalized Monod type model (Eq. (11)), originally proposed by Han and Levenspiel (1998) which could probably better describe both substrate stimulation at low concentration and substrate inhibition at high concentration (Nath and Das, 2011; van Niel et al., 2003; Wang and Wan, 2008).

$$r = r_{max} \frac{S\left(1 - S / S_{max}\right)^n}{S + K_S\left(1 - S / S_{max}\right)^m} \tag{11}$$

where r is the biomass growth rate ((mol biomass) $L^{-1}h^{-1}$), r_{max} is the maximum growth rate of the biomass ((mol biomass) $L^{-1}h^{-1}$), S_{max} is the maximum substrate concentration above which fermentation stops (mol L^{-1}), m and n are the exponent constants determining the type of substrate inhibition such as non-competitive, competitive, uncompetitive and mixed inhibition.

The inhibition caused by low pH in microbial species producing hydrogen has been widely reported (Gadhamshetty et al., 2009; Van Ginkel et al., 2001). Low extracellular pH conditions caused by the presence/formation of acidic metabolites (i.e., VFAs) increase the energy requirements to transport the undissociated acids outside the cell wall which results in a lowering intracellular level of ATP; also the proton uptake decreases the availability of coenzyme A and phosphate pools causing a subsequent reduction in glucose flux through glycolysis (Jones and Woods, 1986). pH inhibition has been incorporated in Monod-type kinetic expressions by adding an inhibition term (I_{pH}) (Lin et al., 2007; Ntaikou et al., 2009b).

For the simulation of batch fermentative production of hydrogen from glucose and sweet sorghum using *Ruminococcus albus*, Ntaikou et al., (2009b) included the pH inhibition term to the Monod equation together with a biomass decay constant (Eq. (12)). Inhibition term (I_{pH}) has been also described using the Ratkowsky-based model (Wang and Wan, 2009), although the expression proposed by the ADM1 for I_{pH} (which is discussed in the next section) has been the most used one.

$$\frac{dX}{dt} = u_{max} \frac{S}{K_S + S} X \cdot I_{pH} - k_d X \tag{12}$$

where k_d is the biomass decay constant (h^{-1}).

Inhibition caused by other variables (i.e., secondary substrates, inorganic salts, etc) could be also incorporated into the kinetic expression for substrate consumption which in turn is linked to the expression for biomass growth. In any case, inhibitions terms should be defined on the basis of each studied case (Lin et al., 2007; Nath et al., 2008; Ntaikou et al., 2009b, 2010; Kumar et al., 2000; Gadhamshetty et al., 2010).

Development of an ADM1-Based Model to Simulate Fermentative Biohydrogen Production

As mentioned above, ADM1 is a mechanistic model characterized by a common nomenclature describing the stoichiometric and kinetics of biological reactions interacting with physicochemical phenomena (acid-base equilibrium, gas-liquid transfer) in anaerobic processes. Several authors have adapted the ADM1 framework to predict hydrogen and VFAs formation by dark fermentation excluding the final methanogenic step.

The development of a dynamic ADM1-based model to predict the evolution of the main process compounds during fermentative biohydrogen production is presented in this section. Mathematical equations are obtained from general mass balances taking into account gas-liquid transfer process of volatile compounds (i.e., H_2, CO_2, VFAs, N_2). The model includes biokinetic expressions considering inhibition terms as suggested in the original ADM1. Carbon and nitrogen balances are properly closed in the Peterson matrix following the analysis presented earlier. pH is also incorporated into the model framework as a state variable. Further improvements deal with the incorporation of both gas and dissolved nitrogen (N_2) as model variables, since this gas is normally used in batch fermentative tests to initially purge the headspace and then ensure anaerobic conditions. This inclusion allows simulating the overpressure evolution, which can be easily measured throughout biohydrogen batch production tests.

Model Kinetics

Model considers only carbohydrates as carbon source in which glucose was selected as a representative substrate. The biological reactions, in which the model is based, include microbial growth (kinetically expresses as substrate consumption) and biomass decay.

According to the ADM1, the main end-products from substrate consumption, in terms of COD, are butyrate, propionate, acetate, hydrogen, and biomass. Other compounds involved in biomass growth such as inorganic carbon (i.e., carbon dioxide, as end product) and inorganic nitrogen (as nitrogen source for biomass formation) are also included into the model. Regarding biomass decay some important considerations were done in order to describe this process. As discussed before, original ADM1 considers that biomass decay generates composite as a unique end-product which is disintegrated to form carbohydrates, proteins, lipids and inerts. Excluding inerts, the remaining compounds can be subsequently hydrolyzed to generate monosaccharides, amino acids and long chain fatty acids. In order to restrict model to only 2 biological reactions (i.e., biomass growth and decay), the biomass decay was assumed to only generate soluble and particulate inert compounds.

Total phosphate and ionic compounds from inorganic salts, which are normally present in mineral mediums, are considered into model compounds since they are required for modeling pH as a state variable (more details are given in the following section). A summary of model compounds considered in this work is shown in Table 3. The elemental composition of model compounds and thus unit conversion from COD to molar basis are also provided. The composition of particulate and soluble inerts was based on Rodriguez et al., (2004) while a typical composition was assumed for biomass (Batstone et al., 2002).

Table 3. Summary of compounds considered for developing the ADM1-based model

Compound	Symbol	Units	Elemental Composition	g COD mol^{-1}
Glucose	S_{su}	kg COD m^{-3}	$C_6H_{12}O_2$	192
Total butyrate	S_{bu}	kg COD m^{-3}	$(C_4H_8O_2 + C_4H_7O_2^-)$	160
Total propionate	S_{pro}	kg COD m^{-3}	$(C_3H_6O_2 + C_3H_5O_2^-)$	112
Total acetate	S_{ac}	kg COD m^{-3}	$(C_2H_4O_2 + C_2H_4O_2^-)$	64
Dissolved hydrogen	S_{h2}	kg COD m^{-3}	H_2	16
Dissolved nitrogen	S_{n2}	kmole N_2 m^{-3}	N_2	-
Inorganic carbon	S_{IC}	kmole C m^{-3}	$(CO_2 + HCO_3^- + CO_3^=)$	-
Inorganic nitrogen	S_{IN}	kmole N m^{-3}	$(NH_4^+ + NH_3)$	-
Inorganic phosphate [(a)]	S_{PHOS}	kmole P m^{-3}	different forms of PO_4^{-3}	-
Cations [(a)]	S_{cat}	kmole m^{-3}	-	-
Anions [(a)]	S_{an}	kmole m^{-3}	-	-
Soluble inerts	S_I	kg COD m^{-3}	$CH_{1.946}O_{0.6754}N_{0.1429}$	33.3
Particulate inerts	X_I	kg COD m^{-3}	$CH_{1.4}O_{0.4024}N_{0.1429}$	33.3
Monosaccharides degraders	X_{su}	kg COD m^{-3}	$C_5H_7NO_2$	160
Hydrogen	S_{gh2}	kg COD m^{-3}	H_2	16
Carbon dioxide	S_{gco2}	kmole C m^{-3}	CO_2	-
Propionic acid	S_{gHPro}	kg COD m^{-3}	$C_4H_8O_2$	160
Butyric acid	S_{gHBu}	kg COD m^{-3}	$C_3H_6O_2$	112
Acetic acid	S_{gHAc}	kg COD m^{-3}	$C_2H_4O_2$	64
Nitrogen	S_{gn2}	kmole N_2 m^{-3}	N_2	-

[(a)] Ionic compounds from mineral medium

Kinetics and stoichiometric parameters related to growth and biomass decay are shown in Tables 4 and 5. Nitrogen and carbon content in bacteria and inerts were calculated from the bacterial and inert composition, respectively (Table 3). Model kinetic expressions were developed on the basis of the two main biological reactions taking place in the process: uptake of monosaccharides and biomass decay. In Monod-based kinetics, substrate consumption is generally expressed as a function of biomass growth using a yield coefficient (Eq. (13)). However in ADM1 model framework, a generic term lumping biomass maximum specific growth rate and biomass yield coefficient is used to state the kinetic equation for maximum specific substrate rate. Thus the kinetic rate equation for substrate consumption is given by Eq. (14) where inhibition term for pH is added. Biomass decay rate is depicted in Eq (15). Notice that both biokinetic rates are given in units of kg COD $m^{-3}d^{-1}$.

$$r = \frac{u_{max}}{Y_{X/S}} \frac{S}{K_S + S} X \tag{13}$$

$$\rho_{substrate} = k_m \frac{S_{su}}{K_S + S_{su}} X_{su} \cdot I_{pH} \tag{14}$$

$$\rho_{decay} = k_d \cdot X_{su} \tag{15}$$

Table 4. Kinetic parameters used in the ADM1-based model for simulating fermentative hydrogen production from sugar-based substrate

Parameter	Value	Units	Parameter definition	Reference
k_m	30	d^{-1}	Specific Monod maximum monosaccharides uptake rate	Batstone et al., (2002)
K_S	0.5	kg COD m^{-3}	Monod half saturation constant for monosaccharides degraders	Batstone et al., (2002)
pH_{UL}	5.5	-	Upper limit for pH inhibition in monosaccharides uptake	Batstone et al., (2002)
pH_{LL}	3.0	-	Lower limit for pH inhibition in monosaccharides uptake	Adapted to this study
k_d	0.02	d^{-1}	First order decay rate for monosaccharides degraders	Batstone et al., (2002)

As already mentioned, pH has a significant effect on metabolic pathway of fermentative hydrogen bacteria, influencing end-products distribution. The ADM1 provides two different terms to model pH inhibition for intracellular processes, the first for systems that are strongly buffered by ammonia or other bases, and a second for low pH inhibitions likely to occur in carbohydrates systems (Batstone et al., 2002). The second empirical lower pH inhibition term was selected to be used in the model (Eq. (16)).

$$I_{pH} = \exp\left(-3\left(\frac{pH - pH_{UL}}{pH_{UL} - pH_{LL}} \right)^2 \right)\Bigg|_{pH<pH_{UL}} \tag{16}$$

$$I_{pH} = 1\Big|_{pH>pH_{UL}}$$

where the pH_{UL} denotes the upper limit at which the sugar degrader bacteria are not inhibited and pH_{LL} denotes the lower limit at which inhibition is complete.

Table 5. Stoichiometric parameters associated to biochemical reactions (biomass growth and decay)

Parameter	Value	Units	Parameter definition	Reference
$f_{si,bac}$	0.5	-	Fraction of soluble inerts on bacteria	Adapted to this study
$f_{xi,bac}$	0.5	-	Fraction of particulate inerts on composite	Adapted to this study
N_I	0.004287	kmole N (kg COD)$^{-1}$	Nitrogen content of inerts	Rosen and Jeppsson (2006)
N_{bac}	0.006250	kmole N (kg COD)$^{-1}$	Nitrogen content of bacteria	Batstone et al., (2002)
C_{su}	0.0313	kmole C (kg COD)$^{-1}$	Carbon content of monosaccharides	Batstone et al., (2002)
C_{bu}	0.0250	kmole C (kg COD)$^{-1}$	Carbon content of butyrate	Batstone et al., (2002)
C_{pro}	0.0268	kmole C (kg COD)$^{-1}$	Carbon content of propionate	Batstone et al., (2002)
C_{ac}	0.0313	kmole C (kg COD)$^{-1}$	Carbon content of acetate	Batstone et al., (2002)
C_{bac}	0.03125	kmole C (kg COD)$^{-1}$	Carbon content of bacteria	Batstone et al., (2002)
C_I	0.0300	kmole C (kg COD)$^{-1}$	Carbon content of inert	Rosen and Jeppsson (2006)
$f_{h2,su}$	0.1900	-	Fraction of hydrogen in monosaccharides	Batstone et al., (2002)
$f_{bu,su}$	0.1300	-	Fraction of butyrate in monosaccharides	Batstone et al., (2002)
$f_{pro,su}$	0.2700	-	Fraction of propionate in monosaccharides	Batstone et al., (2002)
$f_{ac,su}$	0.4100	-	Fraction of acetate in monosaccharides	Batstone et al., (2002)
Y_{su}	0.1000	kg COD (kg COD)$^{-1}$	Yield of monosaccharides degraders on substrate	Batstone et al., (2002)

In the ADM1 the values for pH_{UL} and pH_{LL} are recommended to be 5.5 and 4, respectively. However, some authors have reported that optimum pH range for hydrogen generation was dependent on the type of microorganisms. Furthermore some bacteria are capable to both uptake sugar and produce hydrogen under acidic conditions (Gadhamshetty et al., 2009, 2010; Nath et al., 2006). Thus, value of pH_{LL} was set to 3.0 in this study to cover a wider range of experimental conditions. As shown in Figure 2, the pH inhibition term increases with increasing pH values of the liquid phase.

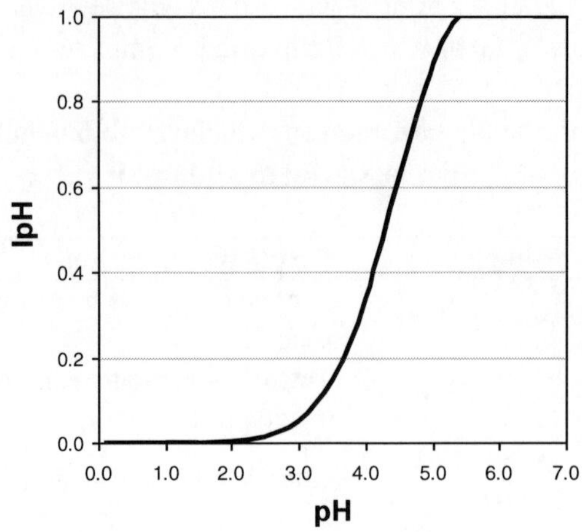

Figure 2. Profile of the pH inhibition term (pH_{UL} = 5.5 and pH_{LL} = 3).

Physicochemical Processes

In addition to the biochemical reactions described above, physicochemical processes described in the model are related to the non-equilibria liquid-gas transfer. As suggested in the original ADM1, hydrogen and carbon dioxide were included as gas components since they are the main biogas constituents. However other volatile end-products (i.e., VFAs) and nitrogen (N_2), which is normally used in batch fermentative tests to initially purge the headspace to ensure anaerobic conditions, were considered into the liquid-gas transfer processes. Other potentially important gases like hydrogen sulfide (H_2S) was not included because sulfate reduction was no considered as a biochemical process, and ammonia which will practically not be present in the liquid phase according to pHs normally reported in hydrogen fermentative processes (at pH of 7 and 25ºC less than 0.6% of inorganic nitrogen is present as ammonia). List of physicochemical parameters used to simulate the model are listed in Table 6.

Dynamic gas transfer equations were used to describe liquid-gas transfer following the well known two-film theory and using the Henry's law to describe the equilibrium relationship between gas and liquid phases. Equation for liquid-gas mass flux applied to hydrogen, carbon dioxide and nitrogen is stated as follows:

$$\rho_{T,i} = k_L a(S_i - \alpha_i \cdot K_{H,i} p_{gas,i}) \qquad (17)$$

where k_La is the overall mass transfer coefficient multiplied by the specific transfer area (d^{-1}), S_i is the concentration of compound i (H_2, CO_2, N_2) in the liquid phase (kg COD m^{-3} or kmole m^{-3}), $K_{H,i}$ is the Henry coefficient of component i (M bar^{-1}), α_i is the conversion factor of component i to transform molar units of $K_{H,i}$ to COD units (see Table 3) and $p_{gas,i}$ is the partial pressure of gaseous compound i (bars, see next section for calculation procedure).

Table 6. Physicochemical parameters used in the ADM1-based model for simulating fermentative hydrogen production

Parameter	Value	Units	Parameter definition f	Reference
R	0.083145	bar $M^{-1}K^{-1}$	Gas law constant	-
T_{base}	298.15	K	Reference temperature (standard)	-
T_{op}	308.15	K	Operation temperature	-
$K_{eq,w}$	$10^{-14} exp\left[\frac{55900}{R*100}\left(\frac{1}{T_{base}}-\frac{1}{T_{op}}\right)\right]$	M	Water dissociation constant ($H_2O \leftrightarrow H^+ + OH^-$)	Rosen and Jeppsson (2006)
$K_{a,bu}$	$10^{-4.82}$	M	Acid dissociation constant of butyrate ($HBu \leftrightarrow H^+ + Bu^-$)	Rosen and Jeppsson (2006)
$K_{a,pro}$	$10^{-4.88}$	M	Acid dissociation constant of propionate ($HPro \leftrightarrow H^+ + Pro^-$)	Rosen and Jeppsson (2006)
$K_{a,ac}$	$10^{-4.76}$	M	Acid dissociation constant of acetate ($HAc \leftrightarrow H^+ + Ac^-$)	Rosen and Jeppsson (2006)
$K_{a1,ic}$	$10^{-6.35} \exp\left[\frac{7646}{R*100}\left(\frac{1}{T_{base}}-\frac{1}{T_{op}}\right)\right]$	M	Acid dissociation constant of bicarbonate ($CO_2 \leftrightarrow H^+ + HCO_3^-$)	Rosen and Jeppsson (2006)
$K_{a2,ic}$	$10^{-29024 \cdot T_{op}^{-1}+6.498-0.02379T_{op}}$	M	Acid dissociation constant of carbonate ($HCO_3^- \leftrightarrow H^+ + CO_3^=$)	Loewenthal and Marais (1976)
$K_{a,in}$	$10^{-9.25} \exp\left[\frac{51965}{R*100}\left(\frac{1}{T_{base}}-\frac{1}{T_{op}}\right)\right]$	M	Acid dissociation constant of ammonia ($NH_4^+ \leftrightarrow H^+ + NH_3$)	Rosen and Jeppsson (2006)
$K_{a1,phos}$	$10^{-7993T_{op}^{-1}+4.5535-0.01349T_{op}}$	M	Acid dissociation constant of orthophosphate ($H_3PO_4 \leftrightarrow H^+ + H_2PO_4^-$)	Loewenthal et al., (1989)
$K_{a2,phos}$	$10^{-19795T_{op}^{-1}+5.3541-0.01948T_{op}}$	M	Acid dissociation constant of dihydrogen phosphate ($H_2PO_4^- \leftrightarrow H^+ + HPO_4^=$)	Loewenthal et al., (1989)
$K_{a3,phos}$	$10^{-12.023}$	M	Acid dissociation constant of hydrogen phosphate ($HPO_4^= \leftrightarrow H^+ + PO_4^{-3}$)	Loewenthal et al., (1989)
P_{atm}	1.013	bar	Atmospheric pressure	-
$p_{gas,h2o}$	$0.0313 exp\left[5290\left(\frac{1}{T_{base}}-\frac{1}{T_{op}}\right)\right]$	bar	Pressure of gaseous water	Rosen and Jeppsson (2006)
kp	5e-2	m3 d-1bar-1	Resistance coefficient for biogas release	Adapted from Rosen and Jeppsson (2006)
kLa	200	d-1	Mass transfer coefficient	Rosen and Jeppsson (2006)

Table 6. Continued

Parameter	Value	Units	Parameter definition f	Reference
$K_{H,h2}$	$7.08\cdot10^{-4}\,exp\left[\frac{-4157}{R\cdot100}\left(\frac{1}{T_{base}}-\frac{1}{T_{op}}\right)\right]$	M_{liq} bar^{-1}	Henry coefficient of hydrogen	Rosen and Jeppsson (2006)
$K_{H,co2}$	$0.035\,exp\left[\frac{-19410}{R\cdot100}\left(\frac{1}{T_{base}}-\frac{1}{T_{op}}\right)\right]$	M_{liq} bar^{-1}	Henry coefficient of carbon dioxide	Rosen and Jeppsson (2006)
$K_{H,HBu}$	4700	M_{liq} bar^{-1}	Henry coefficient of butyric acid	Khan et al., (1995)
$K_{H,HPro}$	5700	M_{liq} bar^{-1}	Henry coefficient of propionic acid	Khan et al., (1995)
$K_{H,HAc}$	$4100\,\exp\left[\frac{-52381}{R\cdot100}\left(\frac{1}{T_{base}}-\frac{1}{T_{op}}\right)\right]$	M_{liq} bar^{-1}	Henry coefficient of acetic acid	Johnson et al., (1996)
$K_{H,n2}$	$6.5\cdot10^{-4}\,\exp\left[\frac{-10809}{R\cdot100}\left(\frac{1}{T_{base}}-\frac{1}{T_{op}}\right)\right]$	M_{liq} bar^{-1}	Henry coefficient of nitrogen	Wilhelm et al., (1977)

Conversion factor is not applied to carbon dioxide (CO_2) and nitrogen (N_2) since concentrations of these components are not expressed in COD basis. The concentration of CO_2 in the liquid phase is calculated as a function of the total inorganic carbon concentration and proton concentration, using a dissociation factor (f_{co2}) (Eq. (18)).

$$S_{co2} = S_{IC}\cdot f_{co2} = S_{IC}\cdot\frac{\frac{1}{K_{a2,ic}}\left(\frac{[S_{H^+}]^2}{K_{a1,ic}K_{a2,ic}}+\frac{2[S_{H^+}]}{K_{a1,ic}}\right)}{\left(\frac{[S_{H^+}]^2}{K_{a1,ic}K_{a2,ic}}+\frac{[S_{H^+}]}{K_{a2,ic}}+1\right)^2} \tag{18}$$

where $K_{a1,ic}$ and $K_{a2,ic}$ are the dissociation constant of bicarbonate and carbonate, respectively (M), and S_{H+} is the proton concentration (M).

The liquid-gas mass transfer for VFAs is calculated using Eq. (19).

$$\rho_{T,i} = k_L a(S_i - \alpha_i\cdot K_{H,i}p_{gas,i}) \tag{19}$$

where S_i is the concentration of the undissociated volatile fatty acid i (S_{HBu}, S_{HPro}, S_{HAc}) in the liquid phase (kg COD m^{-3}), and α_i is again the conversion factor of VFA i to transform molar units of $K_{H,i}$ to COD units (see Table 3). The concentration of the undissociated fraction for each VFA is calculated as a function of the total concentration of VFA and proton concentration, using a dissociation factor (f). As an example, concentration of acetic acid is shown in Eq. (20). Concentrations of the butyric and propionic acid are calculated in the same way using the corresponding acid dissociation constant.

$$S_{HAc} = S_{ac} \cdot f_{HAc} = S_{ac} \cdot \frac{K_{a,ac}}{S_{H^+} + K_{a,ac}} \tag{20}$$

where $K_{a,ac}$ is the acid dissociation constant of acetate (M).

The temperature dependence of several physiochemical parameters (including several dissociation constant, water partial pressure, and Henry coefficients of N_2, CO_2, acetic acid and N_2, see Table 6) were considered for model development since the overall effect on the anaerobic system due to changes in physicochemical parameters with temperature is generally more important than that due to changes in biochemical parameters (Batstone et al., 2002). The van't Hoff equation was generally used to recalculate physicochemical parameters at temperatures differ from the standard temperature. Acid dissociation constants of the VFAs together with Henry coefficients of butyric and propionic acids were assumed to no vary within the normal operational temperature range (0 – 60 °C) and thus were assumed to be constant (Batstone et al., 2002).

It was assumed that k_La values for all gaseous compounds have a similar order of magnitude since their diffusivities are similar and liquid-gas transfer is modeled as a liquid film controlled process. Values for k_La may highly vary depending on mixing, temperature and liquids properties (Batstone et al., 2002). In this work a same value of k_La was considered for performing simulations.

Note that each $\rho_{T,i}$ (Eq. (17) and (19)) is also a kinetic rate equation (in kg COD $m^{-3}d^{-1}$ or kmole $m^{-3}d^{-1}$) in addition to those related to biochemical reactions (Eq. (14) and (15)). Once all the model kinetic rates are defined, the process stoichiometry matrix for the system (Peterson Matrix) can be stated (Table 7). For this work, eight different specific kinetic rates were defined: two with regard biochemical reactions and six for liquid-gas transfer processes. Notice that stoichiometric coefficients in Table 7 are defined for model compounds expresses in terms of COD, inorganic carbon (IC) and inorganic nitrogen (IN). Verification of closing mass balances in terms of COD, IN and IC can be easily performed using values provided in Table 5.

Mass Balances of Liquid and Gas Phase

The mathematical equations for mass balances were implemented for a continuous-flow stirred tank reactor (CSTR), although simulation of batch and semi-batch operation can be easily assessed. The modeled system considers a constant volume completely mixed (Figure 3).

Mass balance equations for liquid phase are stated as follows:

$$\frac{dC_i}{dt} = \frac{q_{in}}{V_{liq}}\left(C_{i,in} - C_i\right) + \sum_{j=1..8} \rho_j v_{i,j} \tag{21}$$

Table 7. Stoichiometric coefficients for the biochemical and physicochemical processes including in the ADM1-based model

Process (ρ_i)	Compounds										
	S_{su}	S_{bu}	S_{pro}	S_{ac}	S_{h2}	S_{n2}	S_{IC}	S_{IN}	S_I	X_{su}	X_I
1. Uptake of sugar Eq.(14)	-1	$(1-Y_{su})\cdot f_{bu.su}$	$(1-Y_{su})\cdot f_{pro.su}$	$(1-Y_{su})\cdot f_{ac.su}$	$(1-Y_{su})\cdot f_{h2.su}$		$C_{su} - (1-Y_{su})\{f_{bu,su}\cdot C_{bu}+f_{pro,bu}\cdot C_{bu} + f_{ac,su} C_{ac}\} - Y_{su}\cdot C_{bac}$	$-Y_{su}\cdot N_{bac}$		Y_{su}	
2. Biomass decay Eq (15)							$C_{bac} - f_{xi,bac}\cdot C_{xi} - f_{si,bac}\cdot C_{si}$	$N_{bac} - f_{xi,bac}\cdot N_i - f_{si,bac}\cdot N_i$	$f_{si,bac}$	-1	$f_{xi,bac}$
3. H_2 gas/liquid transfer Eq.(17)											
4. CO_2 gas/liquid transfer Eq.(17)							-1				
5. HBu gas/liquid transfer Eq.(19)		-1									
6. HPro gas/liquid transfer Eq.(19)			-1								
7. HAc gas/liquid transfer Eq.(19)				-1							
8. N_2 gas/liquid transfer Eq.(17)						-1					
	Glucose kg COD m^{-3}	Total butyrate kg COD m^{-3}	Total propionate kg COD m^{-3}	Total acetate kg COD m^{-3}	Dissolved hydrogen kg COD m^{-3}	Dissolved nitrogen kmole m^{-3}	Inorganic carbon kmole C m^{-3}	Inorganic nitrogen kmole N m^{-3}	Soluble inerts kg COD m^{-3}	Monosaccharides degraders kg COD m^{-3}	Particulate inerts kg COD m^{-3}

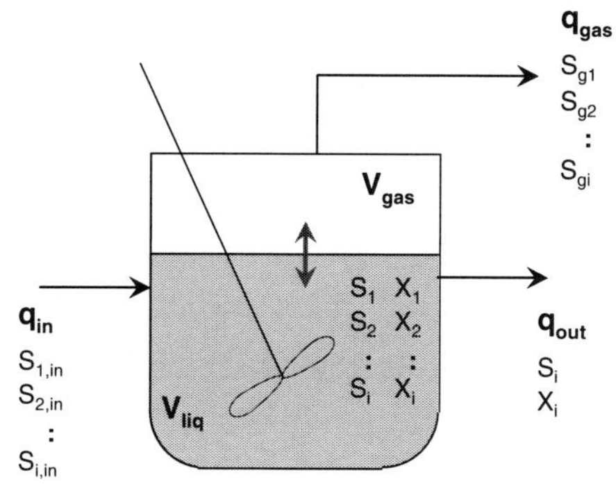

Figure 3. Scheme of a typical completely mixed bioreactor with constant volume ($q_{in} = q_{out}$).

where C_i is the bioreactor concentration of any soluble of particulate compound considered in the model (kg COD m^{-3} or kmole m^{-3}), $C_{i,in}$ is the inlet concentration of any compound (kg COD m^{-3} or kmole m^{-3}), q_{in} is the inlet liquid flow rate (m^3 d^{-1}), V_{liq} is the reactor liquid volume (m^3), ρ_j is the specific kinetic rate for process j (kg COD $m^{-3}d^{-1}$ or kmole $m^{-3}d^{-1}$) and $\upsilon_{i,j}$ is the stoichiometric coefficient for compound i associated to process j defined in Table 7.

It was assumed that the inlet liquid flow can only contain substrate and soluble compounds from mineral medium (inorganic carbon, inorganic nitrogen, inorganic phosphate, cations and anions). As an example, implementation of mass balances for biomass, substrate, acetate, hydrogen, and nitrogen are given in the following equations.

$$\frac{dX_{su}}{dt} = -\frac{q_{in}}{V_{liq}} X_{su} + Y_{su} \cdot k_m \frac{S_{su}}{K_S + S_{su}} X \cdot I_{pH} - k_d \cdot X \tag{22}$$

$$\frac{dS_{su}}{dt} = \frac{q_{in}}{V_{liq}} (S_{su,in} - S_{su}) - k_m \frac{S_{su}}{K_S + S_{su}} X \cdot I_{pH} \tag{23}$$

$$\frac{dS_{ac}}{dt} = -\frac{q_{in}}{V_{liq}} S_{ac} - (1 - Y_{su}) f_{ac,su} \cdot k_m \frac{S_{su}}{K_S + S_{su}} X \cdot I_{pH} - k_L a (S_{ac} \cdot f_{HAc} - 64 \cdot K_{H,ac} p_{gas,HAc}) \tag{24}$$

$$\frac{dS_{h2}}{dt} = -\frac{q_{in}}{V_{liq}} S_{h2} - (1 - Y_{su}) f_{h2,su} \cdot k_m \frac{S_{su}}{K_S + S_{su}} X \cdot I_{pH} - k_L a (S_{h2} - 16 \cdot K_{H,h2} p_{gas,h2}) \tag{25}$$

$$\frac{dS_{n2}}{dt} = -\frac{q_{in}}{V_{liq}} S_{n2} - k_L a (S_{n2} - K_{H,n2} p_{gas,n2}) \tag{26}$$

The mass balances for gaseous compounds are very similar to the liquid phase equations, except for the absence of the advectice inlet flow. The differential equations for the gas phase considering constant volumes for the liquid and gas phase can be written as follows:

$$\frac{dC_{gas,i}}{dt} = -\frac{q_{gas}}{V_{gas}} C_{gas,i} + \frac{V_{liq}}{V_{gas}} \sum_{j=3..8} \rho_j \nu_{i,j} \tag{27}$$

where $C_{gas,i}$ is the concentration of any gaseous compound considered in the model (kg COD m^{-3} or kmole m^{-3}), q_{gas} is the gas flow (m^3 d^{-1}), V_{gas} is the reactor gas volume (m^3), ρ_j is the specific kinetic rate for process j (see Table 7) related to the liquid-gas transfer (j = 3-8, in kg COD $m^{-3}d^{-1}$ or kmole $m^{-3}d^{-1}$) and $\upsilon_{i,j}$ is the stoichiometric coefficient for compound i associated to process j defined in Table 7.

For illustration purposes, mass balances for hydrogen, butyric acid and carbon dioxide are depicted in Eq. (28), (29) and (30).

$$\frac{dS_{gh2}}{dt} = -\frac{q_{gas}}{V_{gas}} S_{gh2} + \frac{V_{liq}}{V_{gas}} k_L a(S_{h2} - 16 \cdot K_{H,h2} \cdot p_{gas,h2}) \tag{28}$$

$$\frac{dS_{gHBu}}{dt} = -\frac{q_{gas}}{V_{gas}} S_{gHBu} + \frac{V_{liq}}{V_{gas}} k_L a(S_{bu} \cdot f_{HBu} - 160 \cdot K_{H,bu} \cdot p_{gas,HBu}) \tag{29}$$

$$\frac{dS_{gco2}}{dt} = -\frac{q_{gas}}{V_{gas}} S_{gco2} + \frac{V_{liq}}{V_{gas}} k_L a(S_{co2} - K_{H,co2} \cdot p_{gas,co2}) \tag{30}$$

The term V_{liq}/V_{gas} is required since the gas transfer kinetic rate is liquid volume-specific. The pressure of each gaseous compound ($p_{gas,i}$ in bars) can be calculated using the ideal gas law as given in Eq. (31).

$$p_{gas,i} = C_{gas,i} \cdot \frac{R \cdot T_{op}}{\alpha_i} \tag{31}$$

where R is the universal gas law constant (bar $M^{-1}K^{-1}$), and T_{op} is the reactor operation temperature (K).

The pressure in the reactor headspace is calculated from the contribution of partial pressures of each volatile compound considered in the model together with the partial pressure of water since reactor headspace is assumed to be water vapor saturated (Eq. (32)). The temperature dependence of water pressure is described using the van't Hoff-based equation (see Table 6 for physicochemical parameters). The gas flow is calculated on the basis of the difference between overpressure in the headspace and the atmospheric pressure as shown in Eq. (33).

$$P_{gas} = p_{gas,h2} + p_{gas,co2} + p_{gas,HPro} + p_{gas,HBu} + p_{gas,HAc} + p_{gas,n2} + p_{gas,h2o} \tag{32}$$

$$q_{gas} = k_p \cdot (P_{gas} - P_{atm}) \tag{33}$$

where k_p is the resistance parameter for biogas release ($m^3\ d^{-1}bar^{-1}$), generally adjusted as a function of the system operational overpressure, and P_{atm} is the external (atmospheric) pressure (bar).

The equations formulated for both the liquid and gas phase depend on the pH value which normally shows a dynamic behavior in unbuffered hydrogen fermentative experiments. Implementation of pH as an additional state variable is illustrated in the next section.

pH Implementation as State Variable

The strategy is based on previous works (Baquerizo et al., 2010; Campos and Flotats, 2003), where charge balance (Eq. (34)) is stated as function of concentration of both ionic compounds and total lumped compounds (i.e., total butyrate $S_{bu} = S_{Bu-} + S_{HBu}$; total propionate $S_{pro} = S_{Pro-} + S_{HPro}$, total acetate $S_{ac} = S_{Ac-} + S_{HAc}$; total inorganic carbon, $S_{IC} = S_{CO2} + S_{HCO3-} + S_{CO3=}$, total inorganic nitrogen, $S_{IN} = S_{NH3} + S_{NH4+}$; etc).

$$\varphi = S_{IN} f_{NH_4^+} + S_{H^+} + S_{cat} - S_{IC} f_{HCO_3^-} - 2S_{IC} f_{CO_3^=} - \frac{S_{bu}}{160} f_{Bu^-} - \frac{S_{pro}}{112} f_{Pro^-} - \ldots$$
$$\frac{S_{ac}}{64} f_{Ac^-} - 3S_{PHOS} f_{PO_4^{3-}} - 2S_{PHOS} f_{HPO_4^=} - S_{PHOS} f_{HPO_4} - S_{an} - \frac{K_{eq,w}}{S_{H^+}} = 0 \quad (34)$$

where f are the dissociation factors depending of pH and acid dissociation constants. Conversion factors to convert COD to molar units are used since the charge balance is stated in molar basis. For instance acetate dissociation factor is given in Eq. (35).

$$f_{Ac^-} = \frac{S_{H^+}}{S_{H^+} + K_{a,ac}} \quad (35)$$

Thus charge balance function (φ) can be seen as a function depending on compounds concentrations and pH (Eq. (36)). The latter equation is differentiated to obtain a differential equation for pH (Eq. (37) and (38)). Terms related to partial differentiation of φ with respect to total compounds are algebraic expressions depending of pH. Partial differentiations of total compounds with respect to time are obtained from mass balances (Eq. (21)) at each integration step.

$$\varphi = f(S_{bu}, S_{pro}, S_{ac}, S_{IN}, S_{IC}, S_{PHOS}, S_{cat}, S_{an}, pH) = 0 \quad (36)$$

$$\frac{\partial\varphi}{\partial t} = \frac{\partial\varphi}{\partial S_{IN}}\frac{\partial S_{IN}}{\partial t} + \frac{\partial\varphi}{\partial S_{IC}}\frac{\partial S_{IC}}{\partial t} + \frac{1}{160}\frac{\partial\varphi}{\partial S_{bu}}\frac{\partial S_{bu}}{\partial t} + \cdots + \frac{\partial\varphi}{\partial pH}\frac{\partial pH}{\partial t} = 0 \quad (37)$$

$$\frac{\partial pH}{\partial t} = \left(\frac{\partial\varphi}{\partial pH}\right)^{-1} \cdot \left(-\frac{\partial\varphi}{\partial S_{NI}}\frac{\partial S_{NI}}{\partial t} - \frac{\partial\varphi}{\partial S_{CI}}\frac{\partial S_{CI}}{\partial t} - \frac{1}{160}\frac{\partial\varphi}{\partial S_{bu}}\frac{\partial S_{bu}}{\partial t} - \ldots\right) \quad (38)$$

The resulting set of ordinary differential equations developed for modeling mass balance and pH were used to simulate the reactor performance under both continuous and batch operation mode. Model implementation was performed in MATLAB® in a home-made modeling environment. A low, variable order non-stiff integration method based on the numerical differentiation formulas was used to solve mathematical equations after testing different integration methods provided by MATLAB.

RESULTS

Simulation of Continuous Operation

A complete mixed bioreactor with a total volume of 1.2 L was selected as a case of study for continuous operation. A working volume of 1 L was chosen to maintain the V_{liq}/V_{gas} relationship proposed by Rosen and Jeppsson (2006) in their benchmarking study of the modified ADM1 for simulating continuous operation. Input concentrations for continuous reactor operation are shown in Table 8. As mentioned in the model development section, the inlet liquid flow was assumed to only contains soluble substrate (in this case glucose with a concentration of 10 kg COD m^{-3}) and soluble compounds from mineral medium. A mineral medium reported elsewhere (Baquerizo et al., 2013) was used for simulation purposes, containing 3 g/L $(NH_4)_2SO_4$, 0.6 g/L KH_2PO_4, 2.4 g/L K_2HPO_4, 1.5 g/L $MgSO_4 \cdot 7H_2O$, 0.15 g/L $CaSO_4$, 0.03 g/L $FeSO_4$ and 1.55 mL of HCl (38%) per liter of mineral medium (meaning 0.7 g of HCl per liter of mineral medium).

Initial conditions were defined assuming that bioreactor was originally filled with mineral medium containing inoculum, followed by a headspace purge using nitrogen gas (meaning that the headspace was initially composed only for N_2 at atmospheric pressure). Therefore, values for initial conditions of biomass, pH, gaseous N_2 and compounds present in the mineral medium were calculated while the remaining state variables were set to zero (Table 8).

According to the mineral medium composition, a pH value of 6.9 was calculated as initial pH condition. Inorganic carbon in the inlet flow was calculated considering carbonates equilibriums and assuming that liquid phase is saturated of CO_2. An operating temperature of 35°C and hydraulic retention time (HRT) of 20 days (meaning q_{in} equal to 5e-5 m^3 d^{-1}) were fixed also following Rosen and Jeppsson (2006) simulating conditions.

The model was simulated for 10 times the HRT (200 days) in order to provide the steady state data (Table 8). Almost complete glucose consumption was verified for the operating conditions together with the formation of the expected end-products (hydrogen, VFAs and inerts). Inorganic compounds from mineral medium (inorganic phosphate, cations and ions) which play an important role for pH calculation remained constant since they are treated as inert compounds with no consumption or reaction terms.

A slight decrease of biomass was observed at the selected HRT. Additionally a small decrease in the steady state value of inorganic nitrogen compared to that in the inlet flow was obtained since the nitrogen is utilized for biomass formation. The obtained biogas was composed by hydrogen (60.26 %), carbon dioxide (34.25%), water vapor (5.48%) and VFAs (less than 0.01 %). The limited presence of VFAs in the biogas is explained by their high

Henry coefficient values. The headspace initially composed of gaseous nitrogen was completely purged as a constant biogas production was established. The steady-state value for pH dropped to 3.37 showing the relatively low buffering capacity of the selected mineral medium. At that pH value, a strong inhibition caused by pH is attained (i.e., $I_{pH} = 0.11$) diminishing the substrate consumption rate. Despite this fact, almost complete glucose consumption was confirmed, suggesting that the selected HRT was overestimated.

Table 8. Initial, input and output values for steady-state simulations of fermentative hydrogen production in a CRST

Compound	Symbol	Input feeding	Initial state	Steady-state	Units
Glucose	S_{su}	10	0.0	0.128132	kg COD m^{-3}
Total butyrate	S_{bu}	0.0	0.0	1.154914	kg COD m^{-3}
Total propionate	S_{pro}	0.0	0.0	2.398683	kg COD m^{-3}
Total acetate	S_{ac}	0.0	0.0	3.642333	kg COD m^{-3}
Dissolved hydrogen	S_{h2}	0.0	0.0	0.007003	kg COD m^{-3}
Dissolved nitrogen	S_{n2}	0.0	0.0	1.2413e-28	kmole N_2 m^{-3}
Inorganic carbon	S_{IC}	9.4742e-6	0.0	0.012227	kmole C m^{-3}
Inorganic nitrogen	S_{IN}	0.045428	0.045428	0.040264	kmole N m^{-3}
Inorganic phosphate	S_{PHOS}	0.018189	0.018189	0.018189	kmole P m^{-3}
Cations	S_{cat}	0.039357	0.039357	0.039357	kmole m^{-3}
Anions	San	0.043014	0.043014	0.043014	kmole m^{-3}
Soluble inerts	S_I	0.0	0.0	0.141027	kg COD m^{-3}
Particulate inerts	X_I	0.0	0.0	0.141027	kg COD m^{-3}
Biomass	X_{su}	0.0	1.0	0.705133	kg COD m^{-3}
Hydrogen	S_{gh2}	0.0	0.0	0.382946	kg COD m^{-3}
Carbon dioxide	S_{gco2}	-	0.0	0.013603	kmole C m^{-3}
Propionic acid	S_{gHPro}	-	0.0	1.6138e-5	kg COD m^{-3}
Butyric acid	S_{gHBu}	-	0.0	9.3993e-6	kg COD m^{-3}
Acetic acid	S_{gHAc}	-	0.0	4.9947e-5	kg COD m^{-3}
Nitrogen	S_{gn2}	-	0.037365	8.6623e-27	kmole N_2 m^{-3}
pH	pH	6.9015	6.9015	3.374666	-

The effects of the HRT on the hydrogen production rate, substrate consumption and biomass concentration at different inlet glucose concentrations were investigated and the results are shown in Figure 4. Since a low steady-state value of pH was detected at inlet glucose concentration of 10 kg COD m^{-3} which clearly affects the hydrogen generation rate, additional simulations with a fixed pH value of 5.5 was also performed (emulating a pH-controlled bioreactor). For the selected range of HRT the same behavior for the substrate utilization and the hydrogen generation rate is observed which is characterized for decreasing rates as the HRT increases. At relative low values of HRT (i.e., HRT=10 days) the adverse effects of endogenous metabolism are minimized. In addition, and since reactor volume is kept constant, lower values of the HRT resulted in higher values of inlet liquid flow which implies larger values of both substrate inlet loads and biomass concentrations (Figure 4a, 4b and 4d). For glucose feeding of 10 kg COD m^{-3} slight variations were observed when pH was controlled. However the substrate consumption and the hydrogen production rates were dramatically improved when pH was controlled for inlet glucose concentration of 50 kg COD m^{-3}. Under this condition, the VFAs generation increased and thus pH inhibition phenomenon was accentuated causing low hydrogen production rates as observed in Figure 4b.

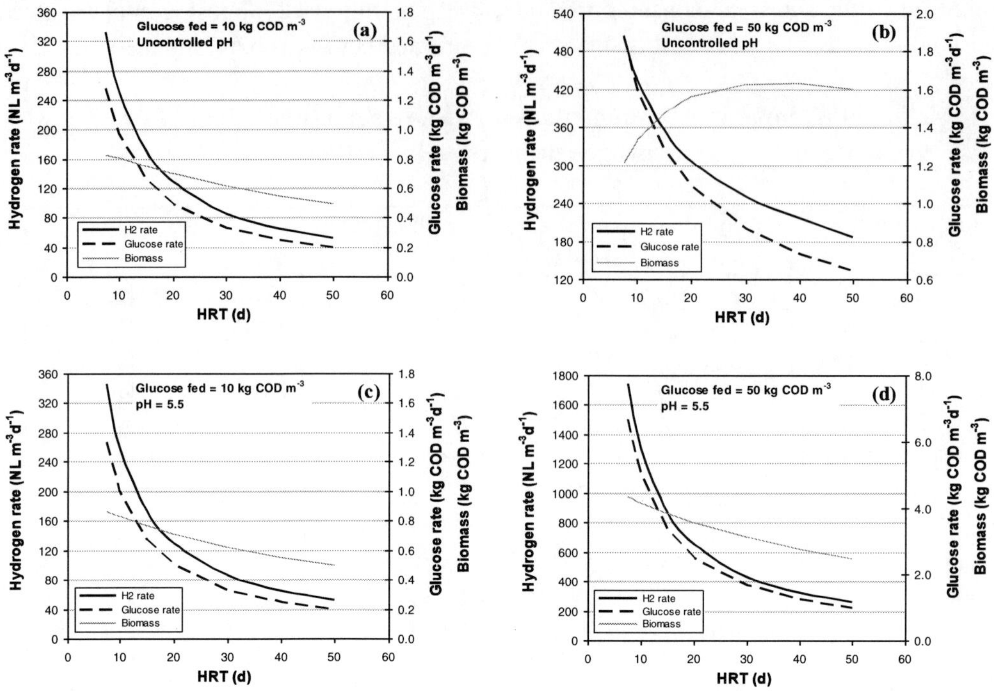

Figure 4. Numerical predictions of the HRT effects on the substrate consumption rate, hydrogen generation rate and biomass concentration at different reactor operating conditions: (a) glucose feeding of 10 kg COD m^{-3} without pH control; (b) glucose feeding of 50 kg COD m^{-3} without pH control; (c) glucose feeding of 10 kg COD m^{-3} with pH control; (d) glucose feeding of 50 kg COD m^{-3} with pH control.

It is worth noting that the HRT corresponds to the inverse of the dilution rate (D) under continuous operating conditions. Microbial populations with specific growth rates larger than the dilution rate are present in the reactor. In this study u_{max} was set to 0.125 h^{-1} meaning that the optimal HRT should be greater than 0.33 days in order to avoid substantial washout of hydrogen-producing bacteria from the CSTR and thus a rapid drop in hydrogen production rate. In Figure 4, higher hydrogen and glucose consumption rates are obtained as HRT decreases, until reaching a critical HRT in which washout phenomenon takes place (not shown). Pakarinen (2011) reported that short HRTs of 0.02 – 0.5 days can be set in continuous hydrogen fermentative processes fed with liquid substrates (i.e., sucrose or glucose) while a HRT higher than 0.06 days was suggested by Whang et al., (2006) to prevent microbial washout.

Simulation of Batch Operation

The developed model was used to simulate fermentative hydrogen production in a batch operation mode. The model was easily adapted to simulate batch conditions by setting the inlet flow rate equal to zero. The main goal of this section was to reproduce glucose-based tests widely reported in literature. Thus serum bottles of 150 mL with a working volume of 50 mL were selected as bioreactor model. Same mineral medium described above was used for simulating batch experiments with an initial glucose concentration of 10 kg COD m^{-3}. Initial conditions of state variables are detailed in Table 9. A lower concentration of initial biomass compared to that provided for continuous operation was chosen in order to study the progressive biohydrogen formation. Batch simulations were also performed at an operating temperature of 35°C.

Table 9. Initial and final values of the state variables for dynamic simulation of a dark fermentative hydrogen batch test

Compound	Symbol	Initial state	Final state [(a)]	Units
Glucose	S_{su}	10.0	8.1490e-7	kg COD m^{-3}
Total butyrate	S_{bu}	0.0	1.169981	kg COD m^{-3}
Total propionate	S_{pro}	0.0	2.429967	kg COD m^{-3}
Total acetate	S_{ac}	0.0	3.689898	kg COD m^{-3}
Dissolved hydrogen	S_{h2}	0.0	0.014573	kg COD m^{-3}
Dissolved nitrogen	S_{n2}	0.0	5.2929e-4	kmole N_2 m^{-3}
Inorganic carbon	S_{IC}	0.0	0.022361	kmole C m^{-3}
Inorganic nitrogen	S_{IN}	0.045428	0.039473	kmole N m^{-3}
Inorganic phosphate	S_{PHOS}	0.018189	0.018189	kmole P m^{-3}
Cations	S_{cat}	0.039357	0.039357	kmole m^{-3}
Anions	San	0.043014	0.043014	kmole m^{-3}
Soluble inerts	S_I	0.0	0.041362	kg COD m^{-3}
Particulate inerts	X_I	0.0	0.041362	kg COD m^{-3}
Biomass	X_{su}	0.1	1.017277	kg COD m^{-3}
Hydrogen	S_{gh2}	0.0	0.847714	kg COD m^{-3}
Carbon dioxide	S_{gco2}	0.0	0.024914	kmole C m^{-3}
Propionic acid	S_{gHPro}	0.0	1.6435e-5	kg COD m^{-3}
Butyric acid	S_{gHBu}	0.0	9.5763e-6	kg COD m^{-3}
Acetic acid	S_{gHAc}	0.0	5.0947e-5	kg COD m^{-3}
Nitrogen	S_{gn2}	0.037365	0.037100	kmole N_2 m^{-3}
pH	pH	6.9015	3.292097	-

[(a)] Values obtained after simulating 5 experimental days

The state variables values obtained after simulating a period of 5 days are given in Table 9 while the dynamic profiles of the main model variables are depicted in Figure 5. Similarly to the simulations of the continuous operation, concentration of inorganic compounds present in mineral medium (inorganic phosphate, cations and ions) remained constant. An expected decreasing of inorganic nitrogen which is used for biomass growth was confirmed. Nitrogen (N_2) equilibrium in the gas and liquid phases was also verified since a slight amount of gaseous nitrogen was dissolved into the liquid phase (it was assumed that nitrogen was initially present only in the gas phase).

Concurrent glucose consumption and end-products generation is observed as the model bases do not consider lag phase for microorganism metabolism. Glucose depletion was confirmed after 3 experimental days when main end-products (biomass, VFAs and biohydrogen) reached their maximum concentration values (Figure 5a and 5b). After the complete glucose utilization, dynamic evolution of biomass was governed by the decay process rate (Figure 5a). An additional simulation of 100 experimental days was run to examine the endogenous process involved in the biomass decay (results not shown) and thus biomass disappearance was confirmed at the end of this simulation period. The extremely low rate for biomass decay was attributed to the value chosen for k_d (from original ADM1) and explains why many authors have ignored this process when simulating biomass in ADM-based models.

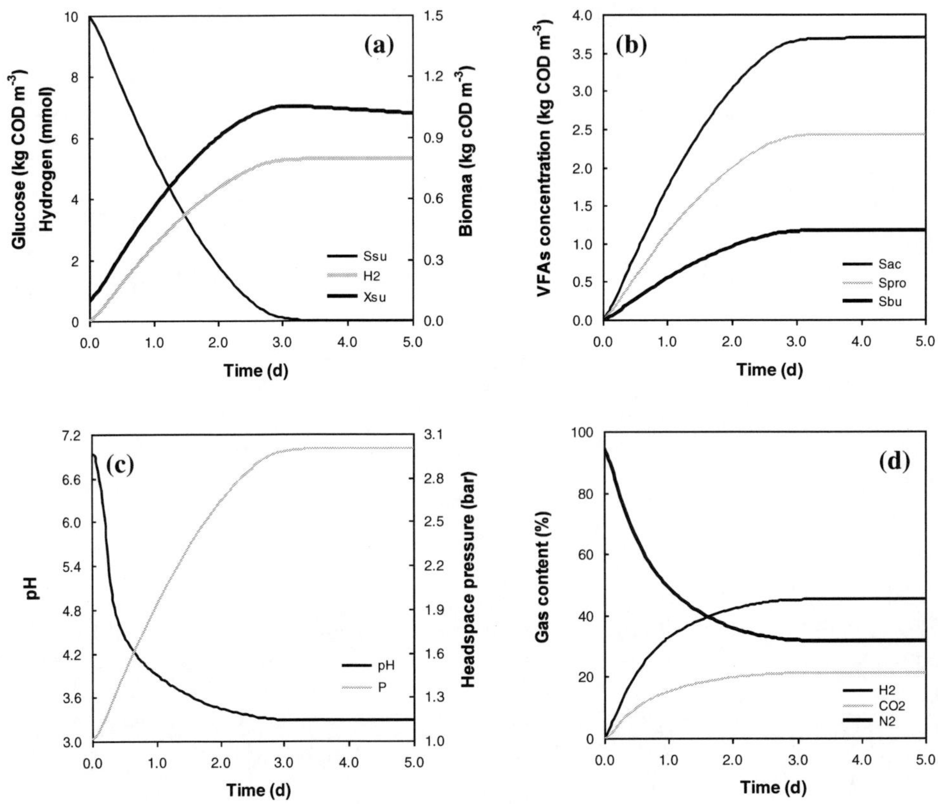

Figure 5. Dynamic profiles of the main model variables for the simulation of a batch glucose dark fermentation: (a) Glucose, biomass and accumulated hydrogen; (b) VFAs; (c) pH and headspace pressure; (d) Gaseous hydrogen, carbon dioxide and nitrogen.

Regarding pH, a decrease was observed as expected along the operational period since the volatile fatty acids are produced as a result of the metabolic activity (Figure 5c). The total pressure was also simulated and the resulting values are shown in Figure 5c. The model predicts an overpressure of 3.002 bars since gaseous H_2 and CO_2 are produced and no biogas release was included for simulating batch tests. Notice that hydrogen percentage in the final biogas reached a value of only 45.22 % since gaseous nitrogen was always present in the headspace, decreasing from 94.5% at the beginning of the experiment to 31.66% at the end of the simulation period (Figure 5d). Final percentages of CO_2 and VFAs were 21.26% and 0.001% respectively, confirming that VFAs were practically absent in the gas phase (see also previous section). The pressure of the headspace increased slightly (up 3.03 bars) after complete biomass disappearance (over 100 experimental days) since CO_2 is produced as a result of the endogenous respiration. It should be highlighted that disregarding nitrogen presence in the headspace (or other gas used for attaining anaerobic conditions at the beginning of batch experimental tests) may lead to erroneous estimation of H_2 percentage in the biogas and consequently the H_2 yield.

Distribution of the end products is determined by the stoichiometric coefficients provided in Table 7 which can be used to state the stoichiometrically balanced equation for the substrate uptake (in molar basis):

$$\begin{aligned} C_6H_{12}O_6 + 012\,N_{inorg} \rightarrow\ & 0.12\,C_5H_7NO_2 + 1.107\,CH_3COOH + 0.4166\,CH_2CH_2COOH + ... \\ & 0.1404\,CH_3(CH_2)_2COOH + 2.052\,H_2 + 1.3747\,C_{inorg} \end{aligned} \tag{39}$$

where N_{inorg} corresponds to the inorganic nitrogen acting as a nitrogen source for biomass growth (i.e., NH_4^+) and C_{inorg} corresponds to the inorganic carbon resulting from the microorganism metabolic activity (i.e., CO_2).

Table 10. Maximum biomass specific growth and hydrogen yield in dark fermentative hydrogen production models based on glucose utilization

Microorganism	μ_{max} (h^{-1})	Y_{H2} mol H_2 (mol glucose)$^{-1}$	Reference
Recombinant Escherichia coli BL-21	0.40	3.12	Chittibabu et al., (2006)
Enterobacter cloacae IIT-BT 08	0.57	2.20	Kumar et al., (2000)
Enterobacter cloacae DM11	0.398	3.31	Nath et al., (2008)
Ruminococcus albus	0.765	2.01	Ntaikou et al., (2009b)
Mixed microflora from organic farm soil	0.0013	2.10	Sharma and Li (2009)
Clostridium acetobutylicum M121	0.154	1.80	Lin et al., (2007)
Clostridium butyricum ATCC19398	0.405	2.29	Lin et al., (2007)
Clostridium tyrobutyricum FYa102	0.213	1.47	Lin et al., (2007)
Clostridium beijerinckii L9	0.236	2.81	Lin et al., (2007)
Mixed anaerobic culture	0.125	2.05	This study

This equation is balanced in terms of COD, carbon and nitrogen. Eq. (39) can be completely balanced (in terms of oxygen, hydrogen and charge), by adding other molecules into the equation (i.e., water, carbonates, proton). However, for the purpose of this chapter, Eq (39) provides valuable information regarding the process yields. Theoretical equations (Eq. (1) and (2)) predict a maximum hydrogen yield of 4 moles of H_2 per mole of consumed glucose when an acetic fermentation occurs. A molecular yield of 2.052 moles of H_2 per glucose is provided by the ADM1 which can be modified as a function of each particular case. In Table 10 several maximum specific biomass growth rates and hydrogen molar yields reported in the literature for models aimed to simulate glucose dark fermentation are presented.

A sensitivity analysis was performed for the kinetic and stoichiometric parameters due to the high variability of these parameters found in the literature (Table 10). The carbon content in both the substrate and the end-products; and also the nitrogen content in bacteria were not considered since they are assumed as true constants. The hydrogen fraction in monosaccharides ($f_{h2,su}$) were also included into the sensitivity study, which implied to proportionally adjust the fractions of remainder end-products when $f_{h2,su}$ was increased or decreased, in order to maintain the mass balances (i.e., $f_{h2,su} + f_{bu,su} + f_{pro,su} + f_{ac,su} = 1$). Simulations were performed for a batch test characterized for an initial glucose concentration of 10 kg COD m^{-3}. A simulation period of 3 days was selected since at that time glucose is not completely depleted. The cumulative biohydrogen volume and biomass concentrations were chosen as model variables to perform the analysis. Sensitivity was assessed by increasing and decreasing 10% the values of the parameters in Tables 4 and 5 (the parameter default value), and comparing the relative change of the state variables to a relative change of the value of the parameter according to the Eq. (40).

$$sensitivity = \frac{\Delta V / V_d}{\left| \Delta P / P_d \right|} \tag{40}$$

where ΔV means the difference between the simulated variable under the new conditions and the value of the variable in the default conditions (V_d). Similarly, ΔP means the difference between the value of the parameter at the ±10% change and the value of the default parameter (P_d).

Table 11 shows that kinetic parameters have a variable impact on model predictions. Practically no influence was found for the half saturation constant and decay constant since low default values were used for these parameters together with the relatively high concentration of initial glucose. A larger influence of the specific maximum monosaccharides uptake rate was verified, especially when this value was decreased, suggesting that this kinetic parameter should be carefully estimated for the consortia or the strain used in experimental assays. The similar influence found for k_m on both cumulative hydrogen volume and biomass formation is explained for the characteristic coupling of differential equations, where cell growth and product formation dynamics are tied to substrate consumption (Table 11). This is consistent with other studies reported previously in the literature (Gadhamshetty et al., 2010, Sharma and Li, 2009).

The strong influence of the pH limits (which define the value of the pH inhibition term) in the sensitivity analysis must be emphasized. This was expected because the pH inhibition has a notable influence on biomass growth kinetic and thus in biohydrogen generation rate. The hydrogen evolution was more sensitive to pH_{LL} than to pH_{UL} (Table 11). This dramatic influence of low pH values on biohydrogen produced is in agreement with literature studies (Gadhamshetty et al., 2010; Van Ginkel et al., 2001). As an example, when pH_{LL} was varied from 3.0 to 3.3, the calculated I_{pH} decreased from 0.34 to 0.25 which means a diminishing near to 27%. Therefore, the inclusion of pH as a model variable into the model framework is strongly suggested to warrant consistent and reliable predictions.

Table 11. Sensitivity results for cumulative biohydrogen volume (H_2^{vol}) and biomass concentration (X_{su}) in a batch experiment for the selected parameters of the model

Parameter	Δ (%)	Sensitivity H_2^{vol}	Sensitivity X_{su}
k_m	+10	0.08	0.05
	-10	-0.26	-0.22
K_S	+10	-0.03	-0.03
	-10	0.03	0.03
pH_{UL}	+10	-0.68	-0.59
	-10	0.09	0.04
pH_{LL}	+10	-1.49	-1.35
	-10	0.10	0.05
k_d	+10	0.00	-0.04
	-10	0.00	0.04
$f_{h2,su}$	+10 (a)	1.02	0.01
	-10 (b)	-1.02	-0.02
Y_{su}	+10	-0.10	0.91
	-10	0.07	-0.93

(a) $f_{bu,su} = 0.127$; $f_{pro,su} = 0.264$; $f_{ac,su} = 0.400$
(b) $f_{bu,su} = 0.133$; $f_{pro,su} = 0.276$; $f_{ac,su} = 0.420$

Regarding the stoichiometric parameters, high influence was found for both fraction of hydrogen in monosaccharides ($f_{h2,su}$) and yield of monosaccharides degraders on substrate (Y_{su}). The effect of Y_{su} variations are mainly related to predictions of biomass concentrations. The large influence of $f_{h2,su}$ on cumulative biohydrogen production is expected since this f-factors define the COD distribution of end-products. The sensitivity effect of f-factors will achieve a major influence in variable stoichiometry ADM1 framework as was suggested by Rodriguez et al., (2006).

Conclusion

In this chapter a mathematical model was developed to simulate hydrogen production by dark fermentation. Model was based on the acidogenesis process depicted in the ADM1 considering only one biomass whereas the processes of hydrolysis, acetogenesis from volatile fatty acids and methanogenesis were excluded. Mass balances for liquid and gas phases describing kinetic of microbial growth (biomass), products formation (acetate, propionate, butyrate, hydrogen and carbon dioxide), substrate utilization (glucose in this case) and biogas formation were performed taking into account fundamental biological reactions and physicochemical processes. Biomass growth (expressed as substrate consumption rate) and biomass decay were considered as biochemical reactions while liquid-gas transfer rate for the all volatile compounds (H_2, CO_2, butyric acid, propionic acid, acetic acid, and N_2) were included as psychochemical processes.

In contrast to the most reported ADM1-based models aimed to predict biohydrogen production in dark fermentation, the model developed here correctly closed mass balances for COD, carbon and nitrogen by including inert compounds from biomass decay and CO_2 as state variables. Gaseous and dissolved nitrogen were also included as state variables since this inert gas is usually used to initially achieve anaerobic conditions in the bioreactor headspace. An additional improvement dealt with the incorporation of pH as a state variable into the model framework since the right prediction of this variable has a strong influence on the pH inhibition term which in turn defines the hydrogen production rate.

The model was successfully used to simulate glucose consumption under both continuous and batch operation modes. The influence of the HRT was studied for continuous operation obtaining similar results than those reported in the literature. The evolution profile of the headspace overpressure was satisfactorily simulated in batch experimental tests where the contribution of the nitrogen partial pressure was taken into account. In this sense the overpressure measurement, which can be simply performed in batch experiments, may contribute to the right estimation of hydrogen percentage in the biogas and thus the proper hydrogen yield for the selected substrate. The extremely low presence of VFAs in the biogas explained by the high values of their Henry coefficients suggests that liquid-gas transfer of these compounds can be neglected in the model basis.

The model presented here can be simply adapted to different operating modes (continuous, semi-batch or batch), operating conditions (different temperatures) and substrates. The generation of additional end-products as a result of the selected substrate can be easily added into the model framework by modifying the stoichiometric parameters related to the end-product fraction in substrate (experimental determination). Likewise this model can be used for assessing different kinetic or stoichiometric constants in experiments conducted with pure strains or consortia growing on different substrates. Additionally, proper definition of the kinetic equations allows using the model for an efficient reactor design and scaling-up.

REFERENCES

[1] Angelidaki I., Ellegaard L., Ahring B.K. (1993). A mathematical model for dynamic simulation of anaerobic digestion of complex substrates: focusing on ammonia inhibition. *Biotechnology and Bioengineering* 42 (2): 159-166.

[2] Angelidaki I., Ellegaard L., Ahring B.K. (1999). A comprehensive model of anaerobic bioconversion of complex substrates to biogas. *Biotechnology and Bioengineering* 63 (3): 363-372.

[3] Antonopoulou G., Gavala H.N., Skiadas I.V., Angelopoulos K., Lyberatos G. (2008). Biofuels generation from sweet sorghum: Fermentative hydrogen production and anaerobic digestion of the remaining biomass. *Bioresource Technology* 99 (1): 110-119.

[4] Appels L., Baeyens J., Degreve J., Dewil R. (2008). Principles and potential of the anaerobic digestion of waste-activated sludge. *Progress in Energy and Combustion Science* 34 (6): 755-781.

[5] Appels L., Lauwers J., Degrve J., Helsen L., Lievens B., Willems K., Van Impe J., Dewil R. (2011) Anaerobic digestion in global bio-energy production: Potential and research challenges. *Renewable & Sustainable Energy Reviews* 15 (9): 4295-4301.
[6] Baquerizo G., Magrí A., Illa J., Revah S. (2010). Dynamic simulation of pH in biological systems: application to anaerobic digestion (In Spanish). In Proceedings of XXXI AMIDIQ National Meeting. Huatulco, Mexico.
[7] Baquerizo G., Garcia-Peña I. (2013). Modelling of pH and biological hydrogen production during glucose dark fermentation. In Proceedings of the 13th World Congress on Anaerobic Digestion: Recovering (bio) Resources for the World. Santiago de Compostela, Spain.
[8] Batstone D., Keller J., Newell R., Newland M. (2000). Modelling anaerobic degradation of complex wastewater. I: model development. *Bioresource Technology* 75(1): 67-74.
[9] Batstone D.J., Keller J., Angelidaki I., Kalyuzhnyi S.V., Pavlostathis S.G., Rozzi A., Sanders W.T., Siegrist H., Vavilin V.A. (2002). Anaerobic Digestion Model No. 1. (ADM1). IWA Scientific and Technical Report No. 13, IWA Publishing, London.
[10] Campos E., Flotats X. (2003). Dynamic simulation of pH in anaerobic processes. *Applied Biochemistry and Biotechnology* 109(1-3): 63-76.
[11] Chen W.H., Chen S.Y., Khanal S.K., Sung S. (2006). Kinetic study of biological hydrogen production by anaerobic fermentation. *International Journal of Hydrogen Energy* 31 (15): 2170-2178.
[12] Chen Y., Cheng J.J., Creamer K.S. (2008). Inhibition of anaerobic digestion process: a review. *Bioresource Technology* 99 (10): 4044-4064.
[13] Chittibabu G., Nath K., Das D. (2006). Feasibility studies on the fermentative hydrogen production by recombinant *Escherichia coli* BL-21. *Process Biochemistry* 41 (3): 682-688.
[14] Cooney M., Maynard N., Cannizzaro C., Benemann J. (2007). Two-phase anaerobic digestion for production of hydrogen-methane mixtures. *Bioresource Technology* 98(14): 2641–2651.
[15] Costello D.J., Greenfield P.F., Lee P.L. (1991). Dynamic modeling of a single-stage high-rate anaerobic reactor .1. Model derivation. *Water Research* 25 (7): 847-858.
[16] Davila-Vazquez G., Alatriste-Mondragon F., Leon-Rodrıguez A., Razo-Flores E. (2008). Fermentative hydrogen production in batch experiments using lactose, cheese whey and glucose: Influence of initial substrate concentration and pH. *International Journal of Hydrogen Energy* 33 (19): 4989-4997.
[17] de Baere L. A., Mattheeuws B., Velghe F. (2010). State of the art of anaerobic digestion in Europe. In Proceedings of 12th World Congress on Anaerobic Digestion (AD12). Guadalajara, Mexico.
[18] Donoso-Bravo A., Mailier J., Martin C., Rodríguez J., Aceves-Lara C.A., Wouwer, A.V. (2011). Model selection, identification and validation in anaerobic digestion: A review. *Water Research* 45 (17): 5347-5364.
[19] Eastman J.A., Ferguson J.F. (1981). Solubilization of particulate organic-carbon during the acid phase of anaerobic digestion. *Journal of Water Pollution Control Federation* 53 (3): 352-366.
[20] Fang H.H.P., Zhang T., Liu H. (2002). Microbial diversity of a mesophilic hydrogen-producing sludge. *Applied Microbiology and Biotechnology* 58 (1): 112-118.

[21] Gadhamshetty V., Johnson D.C., Nirmalakhandan N., Smith G.B., Deng S. (2009). Dark and acidic conditions for fermentative hydrogen production. *International Journal of Hydrogen Energy* 34 (2): 821-826.

[22] Gadhamshetty V., Arudchelvam Y., Nirmalakhandan N., Johnson D.C. (2010). Modeling dark fermentation for biohydrogen production: ADM1-based model vs. Gompertz model. *International Journal of Hydrogen Energy* 35 (2): 479-490.

[23] Gavala H.N., Angelidaki I., Ahring B.K. (2003). Kinetics and modeling of anaerobic digestion process. *Advances in Biochemical Engineering/Biotechnology* 81: 57-93.

[24] Haag J.E., Wouwer A.V., Queinnec I. (2003). Macroscopic modelling and identification of an anaerobic waste treatment process. *Chemical Engineering Science* 58 (19): 4307-4316.

[25] Hallenbeck P., Ghosh D. (2010). Improvements in fermentative biological hydrogen production through metabolic engineering. *Journal of Environmental Management* 95 (suppl.): S360-S364.

[26] Han K., Levenspiel O. (1988). Extended monod kinetics for substrate, product, and cell inhibition. *Biotechnology and Bioengineering* 32 (4): 430-437.

[27] Henze M., Gujer W., Mino T., van Loosedrecht M. (2000). Activated Sludge Models ASM1, ASM2, ASM2d and ASM3. IWA Scientific and Technical Report No. 9. IWA Publishing, London.

[28] Hill D.T., Barth C. (1977). A dynamical model for simulation of animal waste digestion. *Journal of Water Pollution Control Federation* 49(10): 2129-2143

[29] Hill D.T. (1982). A comprehensive dynamic model for animal waste methanogenesis. *Transactions of the ASAE* 25 (5): 1374-1380.

[30] Husain A. (1998). Mathematical models of the kinetics of anaerobic digestion - a selected review. *Biomass & Bioenergy* 14 (5-6): 561-571.

[31] Hussy I., Hawkes F.R., Dinsdale R., Hawkes D.L. (2003). Continuous fermentation hydrogen production from a wheat starch coproduct by mixed microflora. *Biotechnology and Bioengineering* 84 (6): 619-626.

[32] JianLong W., Wei W. (2008). The effect of substrate concentration on biohydrogen production by using kinetic models. *Sci. China Ser. B: Chem.* 51, 1110–1117.

[33] Johnson B.J., Betterton E.A., Craig D. (1996). Henry's law coefficients of formic and acetic acids. *Journal of Atmospheric Chemistry* 24 (2): 113-119.

[34] Jones D.T., Woods D.R. (1986). Acetone-butanol fermentation revisited. *Microbiological Reviews* 50 (4): 484-524.

[35] Jung K.-W., Kim D.-H., Shin H.-S. (2010). Continuous fermentative hydrogen production from coffee drink manufacturing wastewater by applying UASB reactor. *International Journal of Hydrogen Energy* 35 (24): 13370-13378.

[36] Kalyuzhnyi S.V., Davlyatshina M.A. (1997). Batch anaerobic digestion of glucose and its mathematical modeling. 1. Kinetic investigations. *Bioresource Technology* 59 (1): 73-80.

[37] Kalyuzhnyi S.V. (1997). Batch anaerobic digestion of glucose and its mathematical modeling .2. Description, verification and application of model. *Bioresource Technology* 59 (2-3): 249-258.

[38] Kalyuzhnyi S.V., Fedorovich V.V. (1998). Mathematical modeling of competition between sulphate reduction and methanogenesis in anaerobic reactors. *Bioresource Technology* 65 (3): 227-242.

[39] Kaparaju P., Serrano M., Thomsen A.B., Kongjan P., Angelidaki I. (2009). Bioethanol, biohydrogen and biogas production from wheat straw in a biorefinery concept. *Bioresource Technology* 100 (9): 2562-2568.

[40] Karlsson A., Vallin L., Ejlersson J. (2008). Effects of temperature, hydraulic retention time and hydrogen extraction rate on hydrogen production from the fermentation of food industry residues and manure. *International Journal of Hydrogen Energy* 33 (3): 953-962.

[41] Khan I., Brimblecombe P., Clegg S.L. (1995). Solubilities of pyruvic acid and the lower (C1-C6) carboxylic acids. Experimental determination of equilibrium vapour pressures above pure aqueous and salt solutions. *Journal of Atmospheric Chemistry* 22 (3): 285-302.

[42] Khanna N., Kotay S.M., Gilbert J.J., Das D. (2011). Improvement of biohydrogen production by *Enterobacter cloacae* IIT-BT 08 under regulated pH. *Journal of Biotechnology* 152 (1-2): 9-15.

[43] Keshtkar A., Meyssami B., Abolhamd G., Ghaforian H., Khalagi Asadi M. (2003). Mathematical modeling of non-ideal mixing continuous flow reactors for anaerobic digestion of cattle manure. *Bioresource Technology* 87 (1): 113-124.

[44] Kim M.-S., Lee D.-Y. (2010). Fermentative hydrogen production from tofu-processing waste and anaerobic digester sludge using microbial consortium. *Bioresource Technology* 101 (1 suppl.): S48-S52.

[45] Kleerebezem R., van Loosdrecht M.C.M. (2006) Critical analysis of some concepts proposed in ADM1. *Water Science and Technology* 54 (4): 51-57.

[46] Kumar N., Monga P.S., Biswas A.K., Das D. (2000). Modeling and simulation of clean fuel production by *Enterobacter cloacae* IIT-BT 08. *International Journal of Hydrogen Energy* 25 (10): 945-952.

[47] Lee K.-S., Hsu Y.-F., Lo Y.-C., Lin P.-J., Lin C.-Y., Chang J.-S. (2008). Exploring optimal environmental factors for fermentative hydrogen production from starch using mixed anaerobic microflora. *International Journal of Hydrogen Energy* 33 (5): 1565-1572.

[48] Lin C.-Y., Chang C.-C., Hung C.-H. (2008). Fermentative hydrogen production from starch using natural mixed cultures. *International Journal of Hydrogen Energy* 33 (10): 2445-2453.

[49] Lin P.Y., Whang L.M., Wu Y.R., Ren W.J., Hsiao C.J., Li S.L. (2007). Biological hydrogen production of the genus *Clostridium*: Metabolic study and mathematical model simulation. *International Journal of Hydrogen Energy* 32 (12): 1728-1735.

[50] Liu D., Liu D., Zeng R.J., Angelidaki I. (2006). Hydrogen and methane production from household solid waste in the two-stage fermentation process. *Water Research* 40 (11): 2230-2236.

[51] Liu H., Cheng S., Logan B.E. (2005). Production of electricity from acetate or butyrate using a single-chamber microbial fuel cell. *Environmental Science & Technology* 39 (2): 658-662.

[52] Liu J., Olsson G., Mattiasson B. (2004). On-line monitoring of a two-stage anaerobic digestion process using a BOD analyzer. *Journal of biotechnology* 109 (3): 263-275.

[53] Loewenthal R.E., Marais G.v.R. (1976) Carbonate Chemistry of Aquatic Systems: Theory and Application. Ann Arbor Science Publishers, Ann Arbor, MI.

[54] Loewenthal R.E., Ekama G.A., Marais G.v.R. (1989). Mixed weak acid/base systems. I. Mixture characterisation. *Water* SA 15(1): 3-24.

[55] Mizuno O., Dinsdale R., Hawkes F.R., Hawkes D.L., Noike T. (2000). Enhancement of hydrogen production from glucose by nitrogen gas sparging. *Bioresource Technology* 73 (1): 59-65.

[56] Mu Y., Wang G., Yu H.-Q. (2006). Kinetic modeling of batch hydrogen production process by mixed anaerobic cultures. *Bioresource Technology* 97 (11): 1302-1307.

[57] Nath K., Das D. (2004). Improvement of fermentative hydrogen production: various approaches. *Applied Microbiology and Biotechnology* 65 (5): 520-529.

[58] Nath K., Kumar A., Das D. (2006). Effect of some environmental parameters on fermentative hydrogen production by *Enterobacter cloacae* DM11. *Canadian Journal of Microbiology* 52 (6): 525-532.

[59] Nath K., Muthukumar M., Kumar A., Das D. (2008). Kinetics of two-stage fermentation process for the production of hydrogen. *International Journal of Hydrogen Energy* 33 (4): 1195-1203.

[60] Nath K., Das D. (2011). Modeling and optimization of fermentative hydrogen production. *Bioresource Technology* 102 (18): 8569-8581.

[61] Ntaikou I., Koutros E., Kornaros M. (2009a). Valorisation of waste paper using the fibrolytic/hydrogen producing bacterium *Ruminococcus albus*. *Bioresource Technology* 100 (23): 5928-5933.

[62] Ntaikou I., Gavala H.N., Lyberatos G. (2009b). Modeling of fermentative hydrogen production from the bacterium *Ruminococcus albus*: Definition of metabolism and kinetics during growth on glucose. *International Journal of Hydrogen Energy* 34 (9): 3697-3709.

[63] Ntaikou I., Gavala H.N., Lyberatos G. (2010). Application of a modified Anaerobic Digestion Model 1 version for fermentative hydrogen production from sweet sorghum extract by *Ruminococcus albus*. *International Journal of Hydrogen Energy* 35 (8): 3423-3432.

[64] Obeid J., Magnin J.P., Flaus J.M., Adrot O., Willison J.C., Zlatev R. (2009). Modeling of hydrogen production in batch cultures of the photosynthetic bacterium *Rhodobacter capsulatus*. *International Journal of Hydrogen Energy* 34 (1): 180-185.

[65] Oh S.-E., Zuo Y., Zhang H., Guiltinan M.J., Logan B.E., Regan J.M. (2009). Hydrogen production by *Clostridium acetobutylicum* ATCC 824 and megaplasmid-deficient mutant M5 evaluated using a large headspace volume technique. *International Journal of Hydrogen Energy* 34 (23): 9347-9353.

[66] O-Thong S., Prasertsan P., Karakashev D., Angelidaki I. (2008). Thermophilic fermentative hydrogen production by the newly isolated *Thermoanaerobacterium thermosaccharolyticum* PSU-2. *International Journal of Hydrogen Energy* 33 (4): 1204-1214.

[67] Pakarinen O.M., Tähti H.P., Rintala J.A. (2009). One-stage H_2 and CH_4 and two-stage $H_2 + CH_4$ production from grass silage and from solid and liquid fractions of NaOH pre-treated grass silage. *Biomass & Bioenergy* 33 (10): 1419-1427.

[68] Pakarinen O.M. (2011). Methane and Hydrogen Production from Crop Biomass through Anaerobic Digestion. PhD Thesis. University of Jyväskylä, Jyväskylä , Finland.

[69] Park W., Hyun S.H., Oh S.E., Logan B.E., Kim I.S. (2005). Removal of headspace CO_2 increases biological hydrogen production. *Environmental Science & Technology* 39(12): 4416–4420

[70] Penumathsa B.K.V., Premier G.C., Kyazze G., Dinsdale R., Guwy A.J., Esteves S., Rodríguez J. (2008). ADM1 can be applied to continuous bio-hydrogen production using a variable stoichiometry approach. *Water Research* 42 (16): 4379-4385.

[71] Peiris B.R.H., Rathnasiri P.G., Johansen J.E., Kuhn A., Bakke R. (2006). ADM1 simulations of hydrogen production. *Water Science and Technology* 53 (8): 129-137.

[72] Rajeshwari K.V., Balakrishnan M., Kansal A., Lata K., Kishore V.V.N. (2000). State-of-the-art of anaerobic digestion technology for industrial wastewater treatment. *Renewable & sustainable energy reviews* 4 (2): 135-156.

[73] Rittmann B.E. (2008). Opportunities for renewable bioenergy using microorganisms. *Biotechnology and Bioengineering* 100 (2): 203-212.

[74] Rodriguez J., Ruiz G., Roca E.; Lema J. (2004). Modification of the IWA-ADM1 for the application to anaerobic treatment of ethanolic wastewater from wine factories. In Proceedings of 10th IWA World Congress Anaerobic Digestion. Montreal, Canada.

[75] Rosen C., Jeppsson U. (2006). Aspects on ADM1 Implementation within the BSM2 Framework. IWA BSM

[76] TG Technical Report. Department of Industrial Electrical Engineering and Automation, Lund University, Lund, Sweden.

[77] Ruzicka M. (1996). The effect of hydrogen on acidogenic glucose cleavage. *Water Research* 30 (10): 2447-2451.

[78] Saraphirom P., Reungsang A. (2010). Optimization of biohydrogen production from sweet sorghum syrup using statistical methods. *International Journal of Hydrogen Energy* 35 (24): 13435-13444.

[79] Saravanan V., Sreekrishnan T.R. (2006). Modelling anaerobic biofilm reactors - a review. *Journal of Environmental Management* 81 (1): 1-18.

[80] Sharma Y., Li B. (2009). Optimizing hydrogen production from organic wastewater treatment in batch reactors through experimental and kinetic analysis. *International Journal of Hydrogen Energy* 34 (15): 6171-6180.

[81] Shuler L.M., Kargi, F. (2002). Bioprocess engineering. Basic concepts. 2nd ed. Prentice-Hall International Series in the Physical and Chemical Engineering Sciences.

[82] Siegrist H., Renggli D., Gujer W. (1993). Mathematical-modeling of anaerobic mesophilic sewage-sludge treatment. *Water Science and Technology* 27 (2): 25-36.

[83] Speece R.E. (1996). Anaerobic Biotechnology for Industrial Wastewaters. Archae Press. Nashville, TN, U.S.A.

[84] Sun M., Sheng G.-P., Zhang L., Xia C.-R., Mu Z.-X., Liu X.-W., Wang H.-L., Yu H.-Q., Qi R., Yu T., Yang M. (2008). An MEC-MFC-coupled system for biohydrogen production from acetate. *Environmental Science & Technology* 42 (21): 8095-8100

[85] Tartakovsky B., Morel E., Steyer J.P., Guiot S.R. (2002). Application of a variable structure model in observation and control of an anaerobic digestor. *Biotechnology Progress* 18 (4): 898-903.

[86] Van Ginkel S., Sung S., Lay J.-J. (2001). Biohydrogen production as a function of pH and substrate concentration. *Environmental Science & Technology* 35 (24): 4726–4730.

[87] Van Ginkel S.W., Oh S.-E., Logan B.E. (2005). Biohydrogen gas production from food processing and domestic wastewaters. *International Journal of Hydrogen Energy* 30 (15): 1535-1542.

[88] Van Lier J.B. (2008). High-rate anaerobic wastewater treatment: diversifying from end-of-the-pipe treatment to resource oriented conversion techniques. *Water Science and Technology* 57 (8): 1137-1148.

[89] van Niel E.W.J., Claassen P.A.M., Stams A.J.M. (2003). Substrate and product inhibition of hydrogen production by the extreme thermophile, *Caldicellulosiruptor saccharolyticus*. *Biotechnology and Bioengineering* 81 (3): 255-262.

[90] Vavilin V.A., Vasiliev V.B., Ponomarev A.V., Rytow S.V. (1994). Simulation model 'methane' as a tool for effective biogas production during anaerobic conversion of complex organic matter. *Bioresource Technology* 48 (1): 1-8.

[91] Vavilin V.A., Rytow S.V., Lokshina L.Y. (1995). Modelling hydrogen partial pressure change as a result of competition between the butyric and propionic groups of acidogenic bacteria. *Bioresource Technology* 54 (2): 171-177.

[92] Wang J., Wan W. (2008). The effect of substrate concentration on biohydrogen production by using kinetic models. *Science in China, Series B: Chemistry* 51 (11): 1110-1117.

[93] Wang J., Wan W. (2009). Kinetic models for fermentative hydrogen production: A review. *International Journal of Hydrogen Energy* 34 (8): 3313-3323.

[94] Ward A.J., Hobbs P.J., Holliman P.J., Jones D.L. (2008). Optimisation of the anaerobic digestion of agricultural resources. *Bioresource Technology* 99 (17): 7928-7940.

[95] Whang L.-M., Hsiao C.-J., Cheng S.-S. (2006). A dual-substrate steady-state model for biological hydrogen production in an anaerobic hydrogen fermentation process. *Biotechnology and Bioengineering* 95 (3): 492-500.

[96] Wilhelm E., Battino R., Wilcock R.J. (1977). Low-pressure solubility of gases in liquid water. *Chemical Reviews* 77 (2): 219-262.

In: Energy and Environment Nowadays
Editors: Luis G. Torres and Erick R. Bandala
ISBN: 978-1-63117-398-1

Chapter 7

METHANE PRODUCTION FROM TEQUILA VINASSES

H.O. Méndez-Acosta and V. González-Álvarez
University of Guadalajara, CUCEI, Department of Chemical Engineering, Mexico

1. INTRODUCTION

Tequila is the traditional Mexican alcoholic beverage with the "*appellation d'origine controlee* (aco)" (*i.e.*, geographic appellation) which has a significant impact not only in both social and cultural areas in Mexico but also in international markets with an increasing production that reached nearly 253.2 millions of liters in 2011 (CNIT, 2013; CRT, 2013), 50% of which were exported worldwide. In fact, the production of tequila has become one of the main economical activities in Mexico and particularly, in the western state of Jalisco. Figure 1 depicts the annual production of tequila in the last decades according to the database provided by the Regulation Council of Tequila (CRT).

Such an increasing demand of tequila and the regulations on product quality and production of the beverage have led the tequila industry to introduce major changes in the traditional elaboration process of tequila: like replacing the traditional old brick or stone furnaces by modern stainless steel ovens to cook the core of the *Agave tequilana Weber var. Azul* and steel roll mills to expel the solution of the hydrolyzed sugars which are then fermented in large stainless steel reservoirs for the classic tequila beverage. A number of alternatives have been also implemented to optimize the cooking and fermentation steps of the tequila process: better temperature control schemes in both processing steps, the use of various combinations of yeast strains grown under tight and well controlled conditions (Pinal et al., 1997), etc. Other studies have been conducted to optimize the effect of the distillation steps and the construction materials of the alembics on the organoleptic properties of the spirit (Prado-Ramirez et al., 2005). However, it is important to remark that such an increasing demand has brought about not only economical benefits and revenues to the well rooted and established tequila industry, but also the undesired generation of even higher solid and liquid wastes (it is well known that in order to produce a liter of tequila, 6 to 7 kg of agave are needed and 10 to 14 liters of tequila vinasses are generated). These liquid effluents are highly recalcitrant due to its low pH, high temperature and high organic matter content that unfortunately are discharged, untreated or partially treated, into the ultimate discharge point (usually surrounding water systems and crop fields) causing severe environmental problems (Valenzuela-Zapata, 2003).

To meet much more stringent limitations on discharging into natural receivers and to respond to the increasing legislative pressure to reduce the impact such effluents have on the environment, the Tequila industry has adopted policies, practices and strategies to achieve sustainability that called for a balance between the economical and environmental domains. These have included particular programs for the solid effluents (namely, agave leaves and bagasse) which did not render a solution to the solid wastes problems, such as pollution prevention, production of added value products (animal feedstuff, fiberboard, pulp and paper), and waste reduction (Idarraga et al., 1997; Iñiguez et al., 2001a,b). On the other hand, like all industrial and domestic wastewaters, tequila vinasses must comply with certain pretreatment conditions prior to their release into soil or crop fields, urban sewers and water streams or reservoirs. These conditions include cooling of vinasses, partial separation of solids, filtration, neutralization and nonconventional treatments, such as dilution and recycling (Mendez-Acosta, et al., 2010). However, the disposal in landfills has failed to achieve acceptable levels of chemical oxygen demand (COD) apart from been a major source of foul odors, pests (flies and mosquitoes), loss of soil fertility, leaching and groundwater contamination, while the discharge of vinasses into the sewer has caused significant problems to local water authorities since the additional treatment has not met their own effluent quality limitations. In general, the regulation bodies dictate that treated wastewaters should have a pH of 7 and the chemical oxygen demand (COD) should not exceed 50 mg/L, before discharging into the environment (NOM-001-SEMARNAT-1996, NOM-002-SEMARNAT-1997).

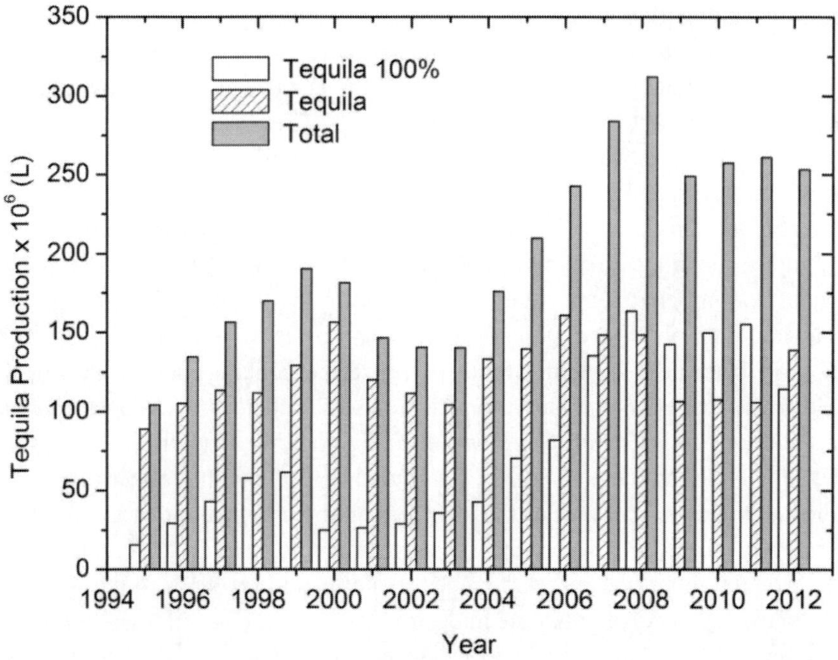

Figure 1. Annual tequila production during the last years expressed as 40% Alc. Vol. (CRT, 2013).

In recent years, the Mexican government has approved changes in the federal legislation to allocate funds from the oil and natural gas profits to conduct research projects on renewable energy sources, energy efficiency, clean technologies and diversification of primary sources of energy (Estrada-Gasca and Islas-Samperio, 2010). In 2009, in fact, the federal congress promulgated the law for renewable energy and funding of the energy transition, to regulate the renewable energy sources and the use and exploitation of clean technologies to generate electricity with end uses different to those of public service, as well as to establish the national strategy and the instruments to finance the energy transition. Moreover, Mexico has just began the exploitation of the so called first generation biofuels within the framework of the Bioenergy Promotion and Development Law by which biomass can be transformed into biofuels by means of combustion, digestion, decomposition, hydrolysis or fermentation. In the particular case of the tequila industry, the anaerobic digestion of more than 2500 million liters of vinasses represents a great bioenergy potential for the sustainable development of this industry. Nevertheless, economical and financial, and particularly technical aspects (related to the understanding and optimal operation of the anaerobic digestion systems) have hampered the implementation of appropriate treatment units in tequila factories for the biodegradation of their vinasses.

In this chapter, we present the state of the art on the treatment of tequila vinasses with emphasis on the anaerobic digestion to deal with the environmental problems caused by such vinasses and its potential in the production of biogas (composed mainly by carbon dioxide and methane, although hydrogen and hydrogen sulfide could be also present in lower concentrations). First, the tequila production process is reviewed and the environmental issues are then presented and discussed. Thereafter, the actual state of the art of tequila vinasses pretreatment and disposal is presented and typical conditioning and treatment facilities are also listed. Then, the fundamentals of anaerobic digestion and usual bioreactor configurations are reviewed. This chapter ends with current research development and advances on the application of anaerobic digestion to deal with the pollution problems caused by tequila vinasses and its potential for biogas production.

2. TEQUILA PRODUCTION

Tequila is traditionally associated with Mexico, particularly with Jalisco, a state located in the western region of the country.

There are two main types of tequila: tequila 100% obtained exclusively from sugars of the Agave tequilana Weber blue variety and tequila produced using 49% sugars from a source other than the agave (usually sugar cane or corn syrup) (CRT, 2013). Both, the production process of tequila and the agave plantation must be conducted in a land, determined by the declaration of protection of the geographic appellation of tequila, conformed by 181 counties in 5 states of the Mexican Republic: Jalisco (125 counties), Nayarit (8), Guanajuato (7), Michoacán (30) and Tamaulipas (11) (CNIT, 2012). Figure 2 depicts a flow chart of the whole tequila process.

The 241 registered companies in the Regulation Council of Tequila (CRT), which verifies and certifies the compliance of the tequila norms and its commercialization, produce different kinds of tequila. These companies are classified according to their production as big, medium, small and micro (see Table 1).

The big companies comprise the 7% of all tequila companies but they produce more than 80% of the beverage. Tequila products made by these companies differ mainly in proportions of agave used, production processes, microorganisms in the fermentation, distillation equipment and the maturation and aging times (Cedeño-Cruz and Álvarez-Jacobs, 2004).

Cultivation and harvest of agave. The *Agave tequilana Weber var. Azul* or blue agave is the only species out of hundreds of *Agavaceae* with the appropriate characteristics for tequila production. These include a high inulin concentration, low fiber content and the chemical compounds present in the plant that contribute to the final taste and flavor of tequila to give the beverage its particular character. The average maturity time for agave is 8 to 10 years. Agave is harvested and transported from the fields to the factories. Then, Agave heads are cut to sizes that facilitate uniform cooking and handling (Valenzuela-Zapata, 2003, Cedeño-Cruz and Álvarez-Jacobs, 2004).

The cooking step: hydrolysis of inulin. Cooking the agave hydrolyze inulin and other components of the plant. During this step some of the sugars are caramelized; and some of the compounds that contribute significantly to the aroma and flavor in wort formulation are due to its high content of fermentable sugars (>10% w/v). This step produces a syrup with a high sugar concentration (>10% by weight) that is later used to formulate the initial wort (Cedeño-Cruz and Álvarez-Jacobs, 2004).

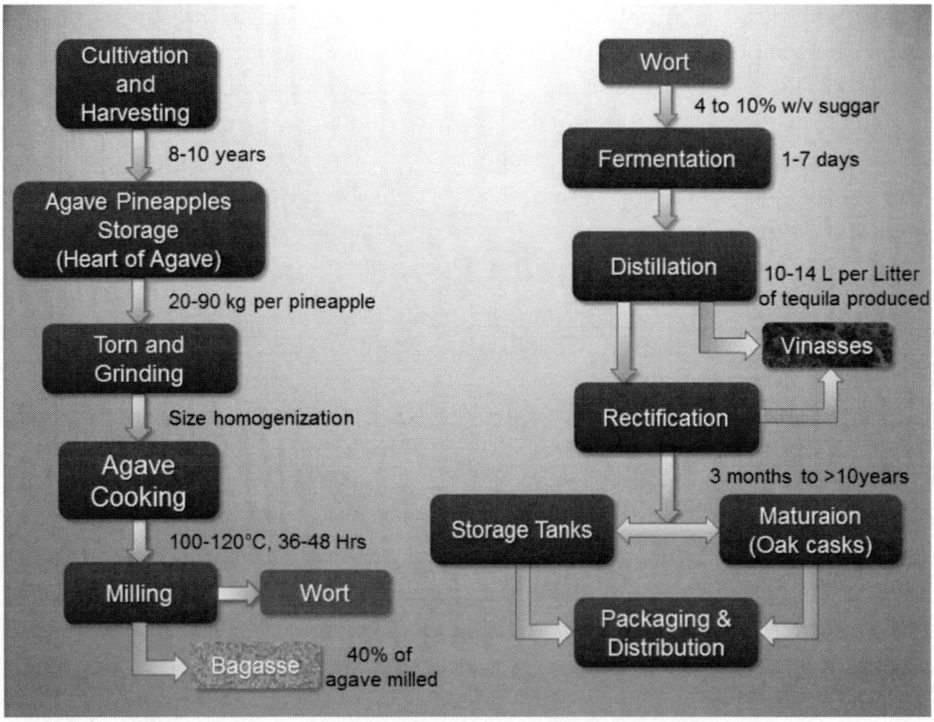

Figure 2. Flow chart of the tequila production process (figure constructed from information in Cedeno and Alvarez-Jacobs, 2004).

Table 1. Classification of tequila companies according to its production volume (CNIT, 2012)

Company Size	Annual tequila production (measured as 55% Alcohol)
Big	> 4,000,000 L
Medium	> 1,000,000 L
Small	> 100,000 L
Micro	< 100,000 L

Extraction of agave juice: milling. The mills used for agave are similar to those used in the sugarcane industry but are smaller in size. This system is still employed in most distilleries. The milling step generates a waste product called bagasse, which represents about 40% of the total weight of the milled agave on a wet weight basis (Cedeño-Cruz and Álvarez-Jacobs, 2004).

Fermentation: *wort formulation*. To produce 100% agave tequila, only agave may be used and the initial sugar concentration ranges from 4 to 10% w/v, depending on the amount of water added in milling. When other sugars are employed, they are previously dissolved and mixed with agave juice to obtain an initial sugar concentration of 8-16%, depending on sugar tolerance of the yeast strain. A few distilleries base wort formulation on the composition of raw materials and nutritional requirements for yeast growth and fermentation. Because the pH of the agave juice is around 4.5, there is no need for adjustment and the same wort composition is used for both inoculum growth and fermentation. Once a wort is formulated with the required nutrients and temperature is around 30°C, it may be inoculated with 5 to 10% (volume) of a previously grown *S. cerevisiae* culture (Cedeño-Cruz and Álvarez-Jacobs, 2004).

Distillation. As other distilled beverages, tequila is obtained after two consecutive differential distillations that are carried out in pot stills. In the first distillation (or stripping), the fermented mash is split into three different products: a light product which may be recycled to the fermented mash and then redistilled in the next batch, a tail product (the actual tequila wastewater or vinasses) which has to be disposed, and a slop cut product (better known as *ordinario*) whose ethanol content is 20-30% in volume. Then, the slop cut is fed into a second pot still for rectification where three additional products are collected: a second head product (which is discarded), a second tail product (which may be recycled to stripping) and a distillate (the actual tequila beverage), better known as the heart, whose ethanol content is around 55 % in volume (Prado-Ramirez et al., 2005).

2.1. Environmental Issues

As mentioned earlier, the production of tequila generates large quantities of solid and liquid wastes. The liquid wastes or vinasses result from the stripping step of the double distillation of the fermented agave juice. They contained fusel oil (obtained from the rectification step), solids, dead yeasts, non fermented sugars, minerals and other compounds such as aldehydes and ketones. They are dark brown in color, because they also contain phenolics (tannic and humic acids), melanoidins that are low and high molecular weight polymers formed as one of the final products of Maillard reaction (López-López et al., 2010 and references therein). Figure 3 shows actual photos of vinasses at the usual generation and discarding points.

Vinasses contain a high level of organic matter, usually measured as the chemical and biochemical oxygen demands (COD and BOD, respectively), with a high solids content, a relatively low pH and high content of nutrients such as nitrogen, phosphorous, potassium, calcium, etc. (Cedeno and Alavarez-Jacobs, 2005). Table 2 shows the main physico-chemical characteristics of the tequila vinasses reported in the current literature. As seen, the composition of the tequila vinasses varies from company to company and depends on the kind of agave, the efficiency of both cooking and fermentation steps as well as the distillation process and of course, on the classification of the beverage. Although, these are not classified as hazardous wastewaters, they are included in the category of complex and difficult to treat and their discharge without treatment into water systems or agricultural fields may cause severe environmental damages (Valenzuela-Zapata, 2003).

3. State of the Art in the Treatment of Tequila Vinasses

In the last decades, the tequila industry has tried to implement policies and practices to deal with the environmental problems caused by the tequila solid wastes and vinasses. In most tequila factories, agave leaves and bagasse are converted into compost and only 60% of these factories have reported partial treatment of vinasses prior to their discharge into water systems (rivers, water streams, lakes, reservoirs) and municipal sewer systems or directly onto crop fields. The most common pretreatment practices are cooling, solid separation, filtration, neutralization, dilution, recycling and biological oxidation (Linerio-Gil and Guzmán-Carrillo, 2004; López-López et al., 2010). These practices, however, have not proved to be a solution of the environmental problems; on the contrary, because of the vinasses low pH, high temperature and high organic load, their discharge onto landfills and the subsequent leaching have had severe environmental consequences on the soil fertility and the groundwater system. Furthermore, the efforts to adapt and develop biological and physicochemical processes for the treatment of tequila vinasses such as evaporation, ozonification and reverse osmosis, apart from being very expensive, have failed to comply with the regulations NOM-001-SEMARNAT-1996 and NOM-002-SEMARNAT-1997.

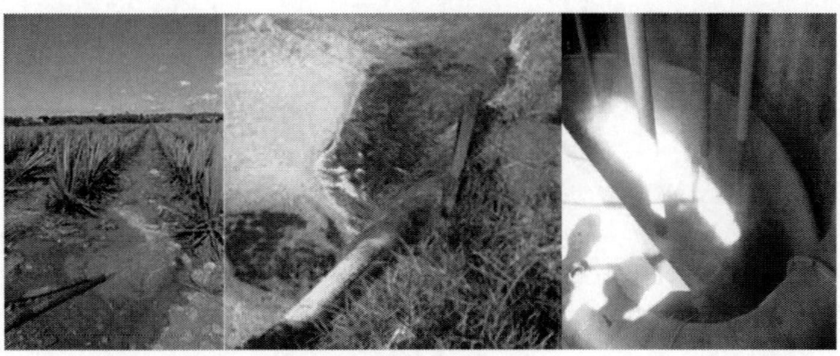

Figure 3. Generation and discarding points of tequila vinasses.

Table 2. Physico-chemical composition of the tequila vinasses reported by different authors

Parameters	García-Dueñas 1991	Cedeño and Álvarez 2004	Linerio and Guzmán 2004	Iñiguez et al. 2005	García, J 2007	Fricke 2008	López et al. 2010
BOD (g/L)	20-30	25-60	40	-	-	28.36	35-60
COD (g/L)	40-80	-	50-80	52.6	77.1	43.44	60-100
Total solids (g/L)	44-106	-	36	-	-	-	25-50
Settleable solids (mL/L)	-	-	-	333	-	0.119	-
Total suspended solids (g/L)	-	-	-	3.4	-	1.518	2.0-8.0
pH	3.0-4.0	< 3.9	< 4.0	3.35	-	3.38	3.4-4.5
Temperature (°C)	90-95	-	90	80	-	-	-
Total nitrogen (g/L)	0.6-10	-	-	-	25.5	0.220	0.02-0.05
Total phosphorus (g/L)	1.4-3.2	-	-	-	0.153	0.019	-
Calcium (g/L)	0.5-0.8	-	0.26-0.35	-	0.862	0.281	0.2-1.1
Potassium (g/L)	6.4-7.0	-	0.29-0.34	-	0.191	-	0.15-0.65
Magnesium (g/L)	2.0-4.0	-	0.22-0.23	-	0.211	0.092	0.1-0.3
Zinc (mg/L)	-	-	0.2-0.5	-	-	0.425	< 1
Hierro (mg/L)	-	-	35.2-78.1	-	15.8	-	<45
Sulphate (mg/L)	-	-	0.78-0.87	-	-	271	-

Nowadays, only the large and some medium size tequila factories (particularly those owned by foreign corporations) are committed to build the wastewater treatment facilities to deal with the vinasses pollution problem (the micro, small and medium-size factories are doing their best to solve the aforementioned problem without success due to their limited financial and human resources). The vinasses treatment facilities include a pretreatment for elimination of the settleable solids (SetS) and neutralization of pH, then an anaerobic biological process able to tolerate high organic loads and finally an aerobic process. This system succeeds in removing more than 90% of organic material as BOD; however, it does not reduce more than 90% of the COD and the treated effluent presents an elevated residual color (López-López et al., 2010)

From the above, it is clear that reducing pollution generated by a widespread network of tequila factories is a major issue. The treatment of vinasses and alcoholic wastewater is not only difficult but has a deep environmental impact. Because of the high organic carbon concentration (for example a middle size tequila factory produces a pollution equivalent to 125,000 inhabitants) and of the slow degradability of these vinasses, anaerobic digestion is well suited. However, this requires an understanding of all the technical, economical and financial aspects that limit its development. Within this situation, anaerobic digestion has grown in Mexico although not at the required rate and bigger investments are being made on conventional aerobic and physicochemical technologies. Presently, there are only two tequila distilleries that have implemented anaerobic digestion to treat their vinasses. Proper integration of the anaerobic digestion processes for water recycling and energy recovery has not been achieved because the treatment plants have been designed to treat diluted vinasses that resulted in bigger facilities and lower methane production. It seems that there is a significant need to understand the anaerobic treatment of tequila vinasses and demonstrate the economical and ecological sustainability of the well-rooted tequila industry.

4. Anaerobic Digestion Fundamentals

The anaerobic digestion process (AD) is in many ways ideal for wastewater treatment. It has several significant advantages over other available methods and is almost certainly assured of increased usage in the future. However, in spite of the present significance and large future potential of this process, it has not generally enjoyed the favorable reputation it truly deserves. The primary obstacle has been a lack of fundamental understanding of the process, required both to explain and deal with the occasional upsets that may occur, and to successfully extend this process to the treatment of a wide variety of industrial wastes. An increasing realization of the potentials of anaerobic treatment is evident from the reporting each year of larger numbers of research investigations on this process. Already, significant advances have been made extending the process so it can be used successfully on many more organic wastes (McCarty, 1964; Rittmann and McCarty, 2001;Moletta, 2005).

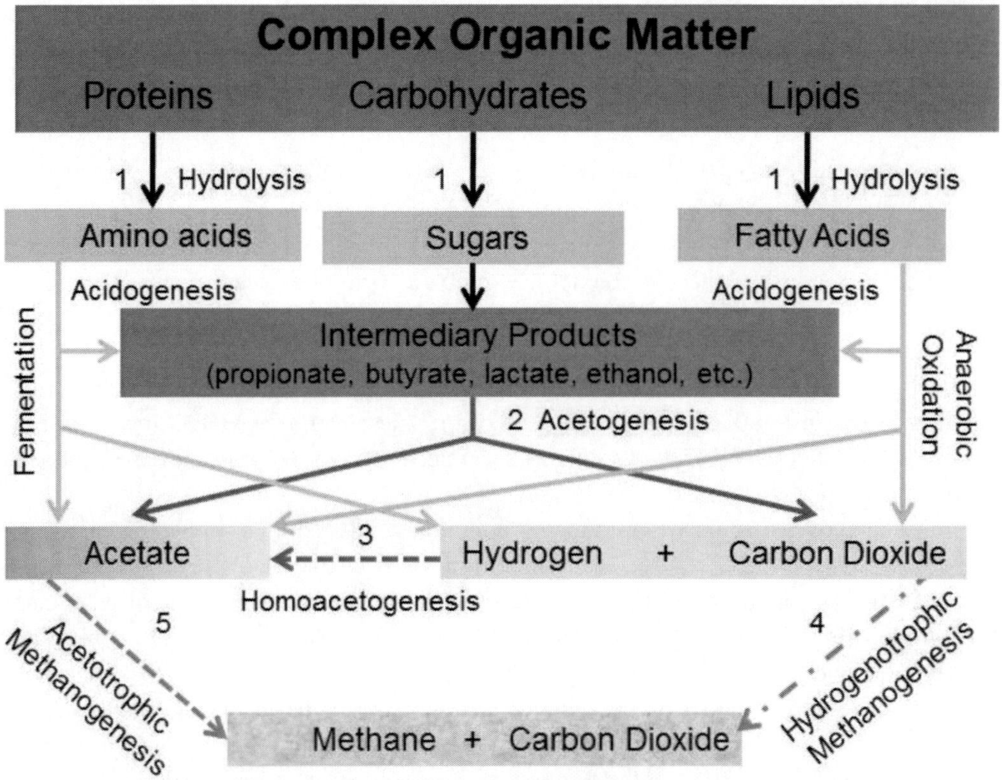

Figure 4. Metabolic pathway of anaerobic digestion (figure modified from Rittmann and McCarty, 2001).

Advantages of the Anaerobic Digestion Process. The advantages of anaerobic treatment can best be indicated by comparing this process with aerobic treatment. In aerobic treatment, as represented by the activated sludge and trickling filter processes, the waste is mixed with large quantities of microorganisms and air. Microorganisms use the organic waste for food, and use the oxygen in the air to burn a portion of this food to carbon dioxide and water for energy. Since these organisms obtain much energy from this oxidation process, their growth is rapid and a large portion of the organic waste is converted into new cells (biomass). The portion converted to cells is not actually stabilized, but it is simply engaged in this form. Although these cells can be removed from the waste stream, the biological sludge they produce still presents a significant disposal problem. In anaerobic treatment, the waste is also mixed with large quantities of microorganisms, but here, air is excluded (see the metabolic pathway of anaerobic digestion in Figure 4). Under these conditions, bacteria grow by converting the organic waste to biogas which is mainly composed of methane and carbon dioxide. Unlike aerobic oxidation, the anaerobic conversion to methane gas yields relatively little energy to the microorganisms. Thus, their rate of growth is slow and only a small portion of the waste is converted into new cells, the major portion of the degradable waste being converted to methane gas. Such conversion to methane gas represents waste stabilization since this gas is insoluble and escapes from the culmination of growth of a population of methane formers capable of fermenting one particular group of compounds. The process is not completely operational until all the groups of methane formers are finally established. This may take several weeks if the process is started without the benefit of "seed" sludge containing the required methane formers. While there are many different methane forming microorganisms, there are also many different acid forming bacteria. Waste stabilization requires a balance among all these organisms. The establishment and maintenance of this balance is normally indicated by one of the most important control tests, that for the concentration of volatile fatty acids (VFA) which represent the intermediate compounds of most importance in anaerobic treatment, and most of the methane formed from this process results from the degradation of these acids by the methanogenic microorganisms. When the system is in balance, the methanogenic bacteria use the VFA intermediates as rapidly as they appear. However, if the methane bacteria are not present in suitable numbers, or are being slowed down (inhibited) by unfavorable environmental conditions, they will not use the acids as rapidly as they are produced by the acid formers, and the VFA will suddenly increase in concentration. Thus, such an increase in acid concentration indicates that the methane formers are not in balance with the acid formers. An analysis for the individual acids present will indicate the particular methane bacteria not carrying out their portion of the treatment. Unfortunately, VFA does not indicate an unbalance in the acid forming organisms (Moletta, 2005).

Methane Formation. The methane producing microorganisms have proven to be very difficult to isolate and study. Consequently, relatively little is known of their basic biochemistry. The conversion of organic matter into methane proceeds through a long series of complex enzymatic and biochemical steps (see Figure 4). One source of methane is the direct cleavage of acetic acid into methane and carbon dioxide. This acid is one of the most important VFA's formed from the decomposition of complex organics and is the source of most methane produced by anaerobic treatment. Most of the remaining methane in anaerobic treatment is formed from the reduction of carbon dioxide.

Low sludge production, energy and nutrients requirements. As much as 80 to 90 percent of the degradable organic portion of a waste can be stabilized in anaerobic treatment by conversion to methane gas, even in highly loaded systems. This is in contrast to aerobic systems, where only about 50 percent of the waste is actually stabilized even at conventional loadings (McCarty, 1964; Gray, 2004; Kim and Gadd, 2008). Since only a small portion of the waste is converted to cells, the problem of disposal of excess sludge is greatly minimized. Also, the requirements for the nutrients, nitrogen and phosphorus, are proportionately reduced. This is especially important in the treatment of industrial wastes which lack these compounds. The sludge produced is quite stable and will not present a nuisance problem. Since anaerobic treatment does not require oxygen, treatment rates are not limited by oxygen transfer. The absence of a need for oxygen also reduces power requirements for treatment. In contrast, the methane gas produced by anaerobic treatment is a good source of fuel energy and is frequently used for heating buildings, tuning engines, or producing electricity. The anaerobic treatment process does have some disadvantages which may limit the use of this process for certain industrial wastes. The major disadvantage is that relatively high temperatures are required for optimum operation; temperatures in the range from 33 to 40°C are preferred. Diluted wastes may not produce sufficient methane for waste heating and this may represent a major limitation. Another disadvantage of anaerobic treatment is related to the slow rate of growth of the methane producing bacteria. Because of it, longer periods of time are required for starting the process. This slow rate of growth also limits the rate at which the process can adjust to changing waste loads, temperatures, or other environmental conditions (Mc Carty, 1964; Rittmannand McCarty, 2001).

The advantages of anaerobic treatment are quite significant, while the disadvantages are relatively few. The advantages normally far outweigh the disadvantages for more concentrated wastes, with BOD values greater than 10,000 mg/L. For less concentrated wastes, the disadvantages become more important, and may limit the use of this process. A noted exception is the successful anaerobic treatment of meat packing wastes with BOD concentrations as low as 1,000 mg/L. These wastes are fairly warm and the temperature requirement does not represent a limitation (McCarty, 1964; Rittmannand McCarty, 2001).

Process configuration. In anaerobic treatment, there are two basically different process designs. One is the "conventional process" most widely used for the treatment of concentrated wastes such as primary and secondary sludge at municipal treatment plants, and consists of a heated digestion tank containing waste and microorganisms responsible for anaerobic treatment. Raw waste is introduced either periodically or continuously and is preferably mixed with the digester contents. The mixed treated waste and microorganisms (biomass) are usually removed together for final disposal. Sometimes this mixture is introduced into a second tank where the suspended material is allowed to settle and concentrate for more efficient disposal. As the retention time in the conventional process is reduced, an increased percentage of biomass is removed from the tank each day with the effluent. The limiting retention time is reached when the microorganisms are being removed from the system faster than they can reproduce themselves, occurring after about three to five days at temperatures of operation from 38 to 40°C. For practical control and reliable treatment, a retention time much above this, or about ten to thirty days, is normally used. With diluted wastes, hydraulic retention times should be very short if the process is to be economical. These are possible in the anaerobic contact process, where the biomass lost is very low. In this case, a digester is used; however, it is followed by a settling tank which removes the active biological suspended solids from the effluent stream for recycle back to the digester. This system is similar in operation to the activated sludge process and permits the maintenance of a high biological population for rapid decomposition, while operating at a relatively low hydraulic detention time. Such a system has been found economical with wastes having BOD concentrations of about 1,000 mg/L and retention times of less than 6 to 12 hours (McCarty, 1964; Rittmannand McCarty, 2001). The gas produced in anaerobic treatment makes the suspended particles buoyant and difficult to settle. Therefore, a degasifier is frequently required between the digester and the settling tank in the anaerobic contact process to permit proper settling of the suspended solids. The important parameter governing the efficiency and operation of both the conventional process and the anaerobic contact process is the biological solids retention time (Moletta, 2005).

Microbiology and Biochemistry. Anaerobic treatment of complex organic materials is normally considered to be a two-stage process. In the first stage, there is no methane production and hence no waste stabilization. In this stage, the complex organics are changed in form by a group of facultative and anaerobic bacteria commonly termed the "acid formers". Complex materials such as fats, proteins, and carbohydrates are hydrolyzed, fermented, and biologically converted to simple organic materials. For the most part, the end products of this first stage conversion are organic fatty acids. Acid forming bacteria bring about these initial conversions to obtain the small amounts of energy released for growth, and a small portion of the organic waste is converted to cells (biomass). Although no waste stabilization occurs during the first stage of treatment, it is required to place the organic matter in a suitable form (bioavailable) for the second stage of treatment. It is in the second stage of methane fermentation that real waste stabilization occurs. During this stage, the organic acids are converted by a special group of microorganisms termed the "methane formers (archaea)" into the gaseous end products, carbon dioxide and methane. The methane formers are strictly anaerobic and even small quantities of oxygen are harmful to them. There are several different groups of methane formers. And each group is characterized by its ability to ferment a relatively limited number of organic compounds in acetotrophic and hydrogenotrophic (see Figure 4). Thus, in the complete methane fermentation of complex materials, several different methane formers are required. The methane formers which use materials such as formic acid and methanol grow very rapidly and can thrive at sludge retention times of less than two days. However, the most important methane formers, which live on acetic and propionic acids, grow quite slowly, and sludge retention times of four days or longer are required for their establishment. These microorganisms carry out the major portion of waste stabilization. Their slow growth and low rate of acid utilization normally represents the limiting step around which the anaerobic treatment process must be designed.

The many different methane forming organisms responsible for anaerobic treatment, their different sources of food, and their different rates of growth are responsible for some confusion as to when good waste treatment is well under way. For example, during the start-up of the anaerobic treatment processes, some methane formation is often noted during the early stages. However, this is produced only from certain materials that are fermented to methane readily. Significant methane production does not occur for several days or weeks, and when it does, it comes during the acid formation stage. A much larger portion (52 percent) is formed from the action of various methane producing bacteria which ferment propionic acid and other intermediates to acetic acid and methane. However, the largest percentage of methane will still result from acetic acid fermentation, which is the most prevalent volatile acid produced by fermentation of carbohydrates, proteins, and fats. Propionic acid, on the other hand, is formed mainly during fermentation of carbohydrates and proteins. The other volatile acids, although significant, are of minor importance. Thus, although many different organisms are required in anaerobic treatment, the two groups of methane formers, which handle acetic and propionic acids, are the most important in the methane fermentation. Unfortunately, they also appear to be among the slowest growing methane microorganisms and the most sensitive to environmental changes.

Waste Stabilization. Waste stabilization in anaerobic treatment is directly related to methane production. By using the formula (McCarty, 1964)

$$C_nN_aH_b + \left(n - \frac{a}{4} - \frac{b}{2}\right)H_2O \rightarrow \left(\frac{n}{2} - \frac{a}{8} + \frac{b}{4}\right)CO_2 + \left(\frac{n}{2} + \frac{a}{8} - \frac{b}{4}\right)CH_4$$

one is able to predict the quantity of methane from a knowledge of the waste chemical composition. From this formula, it can be shown that the ultimate oxygen demand of the waste being degraded is equal to the ultimate oxygen demand of the methane gas produced. This fact allows prediction of methane production in another way, that is, from an estimate of COD or BOD/L (ultimate BOD) stabilization. The ultimate oxygen demand of methane gas is as follows:

$$CH_4 + O_2 \rightarrow CO_2 + H_2O$$

This formula shows that one mol of methane is equivalent to two moles of oxygen. Converting to cubic meter of methane per kilogram of oxygen, the value for a relation between waste stabilization and methane production is obtained. Measured values for methane production per kilogram of COD or BOD, stabilization for a wide variety of wastes varying from pure laboratory substrates to complex waste sludge have shown the validity of this relationship and the close accuracy with which it can be used to predict methane production. The relationship between methane production and waste stabilization can also be used in another way in anaerobic waste treatment operation. Here, the methane production can readily be determined. Such a determination gives a direct and rapid measurement of actual waste stabilization and permits closely following the efficiency of waste treatment.

4.1. Key Performance Parameters for Optimal Anaerobic Digestion Operation

The anaerobic digestion process (AD), as other bioprocesses, depends on many process variables for its optimal operation. Temperature, pH, alkalinity, organic load, presence of toxic or inhibiting compounds are, among others, key variables that must be monitored and controlled in order to properly operate the bioprocesses. Furthermore, the operational stability of the AD strongly depends on the microorganisms whose metabolic behavior may be altered, positively or negatively, by the aforementioned process variables. Before start the review of past and current research on the AD treatment of tequila vinasses, let us point out the effect on such key variables on the performance of AD processes.

Temperature. Abrupt perturbations on the AD temperature may lead to process instability. There are three recommended temperature ranges for AD operation: psychrophilic (below 25°C), mesophilic (30 - 35°C) and thermophilic (50 – 60°C). It has been shown that higher overall AD rates and higher biogas production are attained at high temperatures but for economical reasons, it is recommended to operate the AD process in the mesophilic temperature range (Rittmannand McCarty, 2001).

Volatile fatty acids (VFA). As mentioned above, VFA are intermediate compounds whose excessive accumulation and limited buffering capacity may lead the AD process to a breakdown (Rozzi, 1991).

pH. The different microorganisms present in AD perform satisfactorily in the 6-8 pH range. Most methanogens perform better in the 7.0 to 7.2 ranges whereas the optimal pH operation of the hydrolytic and acidogenic bacteria is around 6. Nevertheless, since the growth of methanogens is lower than that of hydrolytic and acidogenic bacteria, the working pH of most AD processes is usually fixed around 7 (Rittmannand McCarty, 2001).

Alkalinity. Alkalinity or buffering capacity plays an important role in ensuring the operational stability of the AD process in the face of increasing VFA concentration preventing the AD breakdown. It is well known that a proper AD operation is guaranteed when operating at alkalinity (due to biocarbonates) readings higher than 2500 mg/L (Mc Carty, 1964). Other studies have shown that by maintaining the intermediate alkalinity – partial alkalinity (IA/PA) ratio between 0.1 and 0.35, the bioreactor acidification is prevented and as a consequence, the operational stability of the process is guaranteed (Zickefoose and Hayes, 1976; Ripley et al., 1986; Rozzi, 1991; Bernard et al., 2001).

Solids retention time (SRT). It is the biomass average residence time, which is also called "sludge age". At higher SRT, lower bioreactor volumes are needed with maximum removal organic matter rates (Rittmannand McCarty, 2001).

Hydraulic retention time (HRT). It is the wastewater average residence time and is given by the ratio between the reactor volume and the inlet flow rate (i.e., HRT= V/Qin) (Rittmannand McCarty, 2001).

Chemical oxygen demand (COD). It is the most common parameter used to describe the concentration of organic matter present in wastewaters and represents the oxygen concentration required to degrade (oxidize) the organic matter. Although the environmental laws establish the pollution level as certain biochemical oxygen demand (BOD) limits, its long sampling and characterization lab times make them unpractical for actual wastewater treatment operation, monitoring and control purposes. COD, in contrast, have shorter sampling times and may be related to DBO measurements.

Organic Loading Rate (OLR). It is given by the organic load fed to the bioreactor per volume and time, i.e., OLR = CODin/HRT. In the absence of inhibitors and under stable operating conditions, higher OLR yield higher biogas production but may lead to bioreactor acidification.

Suspended solids. These are classified as: a) total suspended solids (TSS) that are composed mainly by biomass, sand, etc. that may be eliminated by sedimentation, flocculation or filtration and, and b) volatile suspended solids (VSS) that is used to estimate the prevailing biomass concentration in the bioreactor.

Biogas composition and production. Biogas volume, flow rate and content are used to determine the actual performance of the AD process. It has been shown that under normal operating conditions, AD processes are able to produce biogas with methane contents between 60 to 80% depending on the temperature, pH and the nature of the wastewater to be treated. It has been also shown that higher methane production is attained at higher OLR if the operational stability is guaranteed.

Methane yield. It is given by the methane volume produced at normal conditions (Normal Liters (NL) which are computed at 1 atm and 0°C) per quantity of degraded organic matter (usually measured as COD). It has been shown that the theoretical methane yield is 0.348 NL of CH_4 per gram of COD removed (McCarty, 1964; Rittmann and McCarty, 2001).

Redox potential. It is recommended to operate the bioreactor between -300 to -330 mV to ensure the optimal functioning of the methanogenic archaea (Rittmann and McCarty, 2001).

5. Past and Current Research on the Anaerobic Digestion of Tequila Vinasses

In the past decades different research groups have addressed the pollution problem caused by tequila vinasses by implementing a number of basic and advanced approaches on bioreactor configurations, optimal operation and monitoring of tequila vinasses treatment processes to meet the stringent environmental norms while guaranteeing the so called operational stability of the bioprocess. In fact, the first reports on the anaerobic digestion of tequila vinasses began in the 1990's when packed bed and hybrid (UASB-anaerobic filter) bioreactors were used for such a purpose. Pinedo (1991) and Garcia-Dueñas (1991) reported 50-70% BOD reduction in the absence of methanogenic activity of the anaerobic sludge, whereas Alvarez et al., (1995) and Chavez (1997) attained up to an 80% BOD content removal in lab scale UASB's. Voellger (2000), on the other hand, treated 100% agave tequila vinasses in a lab-scale fluidized bed bioreactor which was initially operated batchwise by feeding vinasses with a COD content of 54 g/L and pH=3. Nutrients (namely, nitrogen and phosphorous salts), methanol and sodium acetate were added to facilitate the methanogenic bacteria growth and development. This author reported COD removal up to 90% for a load of 9.9 gCOD/L-d and a HRT of two days.

Méndez-Acosta et al., (2010), proposed an integrated system to comply with the Environmental Mexican Norms (NOM-SEMARNAT) and stabilize the anaerobic treatment of tequila vinasses with main design criteria: simple and easy operation, reduced operating time and associated costs (maintenance), integrated and compact design, minimal cost of set-up, start-up, monitoring and control. This system was composed of a 5.3 L fully instrumented and automated lab scale CSTR digester, on-line measuring devices of key variables (pH, temperature, flow rates, etc.), which are used throughout experiments with off-line readings of alkalinity, volatile fatty acids and biogas composition to guarantee the operational stability of the anaerobic digestion process (see Figure 5 for the process experimental set up).

The study aimed at the design and evaluation of an integrated system for the anaerobic treatment of tequila vinasses, a highly polluted wastewater with a temperature of around 90°C, pH<4 and COD concentrations ranging from 30 to 80 g/L. The system combined the well-known properties of anaerobic digestion (reduced waste biomass disposal costs, high organic loading rates, production of energy, elimination of off-gas pollution, etc.) with the proper on-line or off-line monitoring of key variables (pH, temperature, volatile fatty acids and alkalinity) to improve the treatment process and achieve better COD removal and methane production yields. This system was evaluated for 200 days and the experimental results showed that even under the influence of load disturbances, it was possible to reduce the COD concentration to 85% in the start-up phase and up to 85% during the normal operation phase while producing a biogas with 65% methane content. Moreover, it was found that the biogas production increased proportionally with the organic loading rate, reaching a maximum production of 537 NL of biogas per kg of COD removed. These results are in agreement with typical values reported for the anaerobic treatment of winery and distillery wastewaters, which are between 400 and 600 NL of biogas per kg COD removed with 60 to 70% methane content (Moletta, 2005). It was also shown that in order to maintain efficient treatment, the digester buffering capacity (given by the alkalinity ratio, α=intermediate alkalinity/ total alkalinity) must be closely monitored. However, hydraulic retention times (HRT) from 6-8 days were needed.

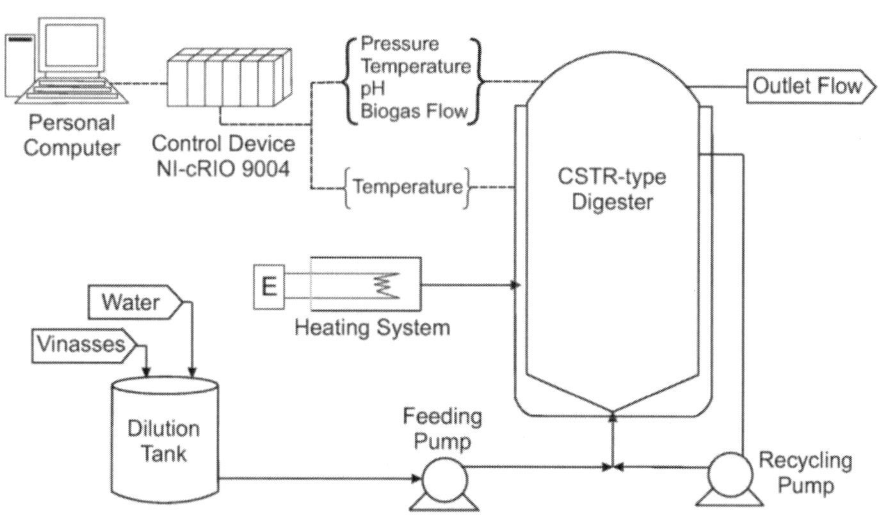

Figure 5. Process diagram of the CSTR-type digester used in the treatment of tequila vinasses (Méndez-Acosta et al., 2010).

Jáuregui-Jáuregui et al., (2012) also examined the performance of a lab-scale fixed-film digester (FBR) under the seasonal operating conditions prevailing in actual Tequila factories: start-up, normal operation and particularly, during the restart-up after a long starvation period. This author showed that a rather short start-up period (28 days) may be achieved despite the unbalanced vinasses COD/N/P ratio without previous acclimation of the inoculum. A high COD removal (90%) was attained at the end of this period even when the bioreactor was fed with raw vinasses with organic loading rates (OLR) up to 8 gCOD/L-d. Once the start-up procedure ended, the bioreactor operated normally for 77 days, maintaining a high COD removal (85 - 90%) by using shorter HRTs from 3-4 days. The biogas composition was monitored during the digester normal operation that yielded methane contents between 64 and 84% with traces of hydrogen (see Figure 6).

Thereafter, the bioreactor was shut-down for 6 months during which the inoculum starved and a restart-up procedure was conducted. It was found that this procedure was quite similar to the first one (i.e., around 28 days), but the COD removal was higher (up to 95%) and so were the methane yield (0.30 NL CH_4 per gram of COD removed) and methane content (75%). Therefore, the anaerobic fixed-film process was shown to be a robust and suitable solution for treatment of Tequila vinasses in small and medium size factories that operate seasonally with long stops and starvation periods by achieving high COD removal efficiencies and methane yields.

On the other hand, Rojo-Liera (2012) scaled-up the previous AD system to a 445L pilot fixed-bed digester, which was operated for 8 months. Figure 7 show an image of the pilot-scale anaerobic digester whose process diagram is similar to the CSTR and FBR lab-scale bioreactors (see Figure 5), as well as two pictures of the digester's support (Cloisonyle©PVC tubes) previous (above) and after (below) the digester start-up phase. As you can see in these last two pictures, after the start-up phase, a uniform biofilm is developed on the entire available surface of the support allowing a high solid retention time.

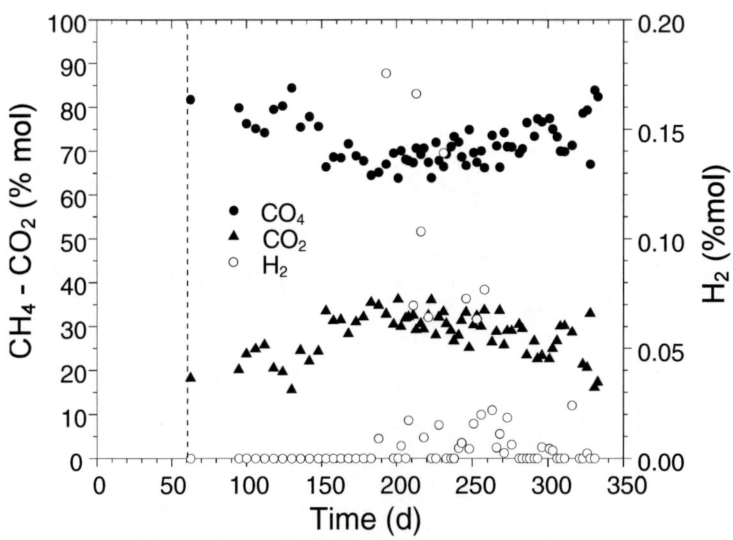

Figure 6. Biogas composition during the normal operation of the lab-scale FBR digester.

Figure 7. Pilot-scale anaerobic digestion process used for the treatment of tequila vinasses (Rojo-Liera, 2012).

The performance of the pilot-scale digester during different operating conditions are shown in Figure 8, where one can clearly see that the COD removal efficiency was higher than 80% regardless of the high organic loading rate (OLR) which reaches its maximum value at 12 gCOD/L-d (see Figures 8d and 8f). It can be also seen that the process was operated under mesophilic conditions around 35°C.

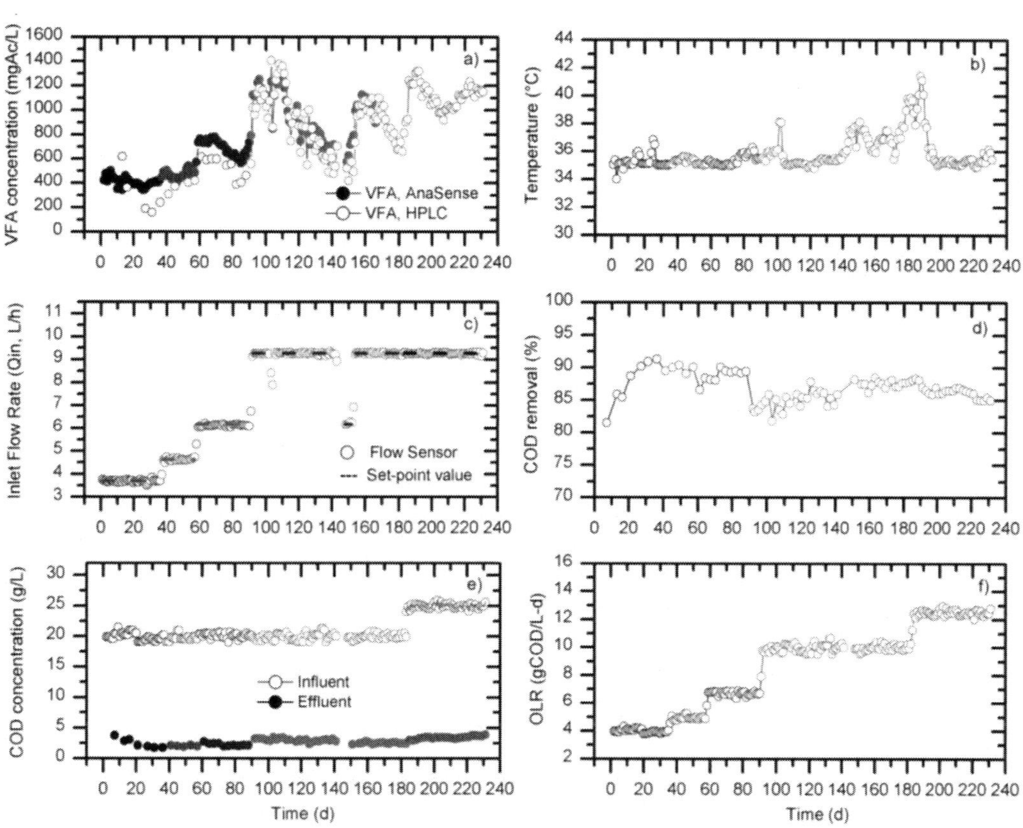

Figure 8. Results obtained in the pilot-scale digester used in the treatment of tequila vinasses under different operating conditions (Rojo-Liera, 2012).

Table 4. List of the overall efficiency results for the wide range of operating conditions evaluated in the pilot-scale digester (Rojo-Liera, 2012)

Operating condition	OLR (gCOD/ L-d)	HRT (d)	COD_{rem} (%)	VFA (mgAc/L)	Biogas flow (L/h)	CH_4 (%)	YCH_4 ($NL/gCOD_{rem}$)
1	3.94	5	90.3	393.9	38.6	67.3	0.291
2	4.96	4	89.9	464.9	51.6	67.9	0.327
3	6.68	3	89.3	625.9	60.6	67.1	0.285
4	10.01	2	87.6	746.8	87.1	68.8	0.285
5	12.49	2	85.1	1162.7	110.1	69.4	0.290

The system stability may be monitored by the dynamic behavior of key AD variables such as the VFA concentration, alkalinity factor (IA/PA) as well as the biogas production and content. From Figure 8a, it can be seen that the VFA concentration increases after the OLR is also increased. However, after a few days the VFA concentration decreases reaching steady-state values between 600-1000 mgAc/L. On the other hand, the methane content was stable and higher than 65% and the methane yield, reported as NL CH_4per gram of COD removed, was kept between 0.28 and 0.30, which is close to the theoretical value of 0.348 (McCarty, 1964) during all the experimental design.

In recent years Lizárraga-Palazuelos (2010) and Guevara-Santos (2011) used a two-stage lab-scale anaerobic digestion system to treat tequila vinasses (see Figure 9). It was conformed by a 4.5Lacidogenic bioreactor (where the organic matter is transformed into volatile fatty acids, hydrogen and carbon dioxide whose reaction is limited by hydrolysis) and a 8.7L methanogenic bioreactor (where VFA's are converted into methane and carbon dioxide whose reaction is controlled by the methanogens growth). One of the novelties in this work is related not only with the treatment of tequila vinasses but with the use of fixed-film anaerobic bioreactors, since most of the two-stage AD processes reported in the current literature are composed by CSTR or UASB bioreactors (for instance see El-Sheikh et al., 2011). Both acidogenic and methanogenic bioreactors were operated under mesophilic conditions around 35°C, while the pH for each bioreactor was regulated at 5.5 and 7.5, respectively. Results shown that the minimum global (acidogenic (1 day)-methanogenic (2 days)) HRT is about 3 days, which is similar to that obtained during the operation of the pilot-scale digester (3 days). The same results were obtained for the maximum OLR (9.8 gCOD/L-d), COD removal efficiency (76.3%), as well as for the methane content (>70%) and yield (0.28NL $CH_4/gCOD_{rem}$). However, the two-stage digester shows a better performance against load disturbances, which makes its application at full-scale promising.

On the other hand, Jaramillo-Gante (2012) evaluated the effects of pH and temperature on the anaerobic degradation of tequila vinasses by using an anaerobic sequential batch reactor (AnSBR), whose process lay-out is depicted in Figure 10. Results show that there is a substantial increase in the methane yield of about 0.3±0.03NL $CH_4/gCOD_{rem}$ when the process temperature is increased from 35°C to 38°C, while the digester pH is regulated at 7.0.

Other studies of treating tequila vinasses by anaerobic digestion include the production of another valuable biofuel: hydrogen. For instance, Espinoza-Escalante et al., (2009) studied the effect of three pretreatments (alkalinization, thermal treatment, and sonication) on Tequila's stillages hydrolysis process in the acidogenesis stage, through the following response variables: soluble chemical oxygen demand (COD's), total sugar and volatile fatty acids profile and the hydrogen production at the time. The stillages were subject to these pretreatments (according to a 23 factorial design); afterwards, they were transferred to a batch reactor at 35°C and inoculated with an anaerobic sludge. Multiple response optimization (MRO) analysis was done to find the global optimum for the response variables described above. This optimum was able to maximize simultaneously all these variables. It was found adequate to be useful hydrolyzing the organic matter present in tequila's stillages. In another study with batch experiments, Buitrón and Carvajal (2010) evaluated the production of H_2 at different pH's and initial temperatures with tequila vinasses, concluding that the tequila vinasses are a suitable substrate to produce hydrogen without a previous pre-treatment. Moreover, Espinoza-Escalante et al., (2009) studied the effect of three parameters of operation (hydraulic retention time, pH and temperature) in semi-continuous experiments to make the production of methane as much as H_2, and found the best results for the production of H_2 atpH = 5.5, hydraulic retention time = 5 days and temperature = 55 C. The previously mentioned studies showed evidence that there is a great potential for the production of H_2 from Tequila vinasses as well as for the optimization of the fermentation conditions for the scaling up of this process.

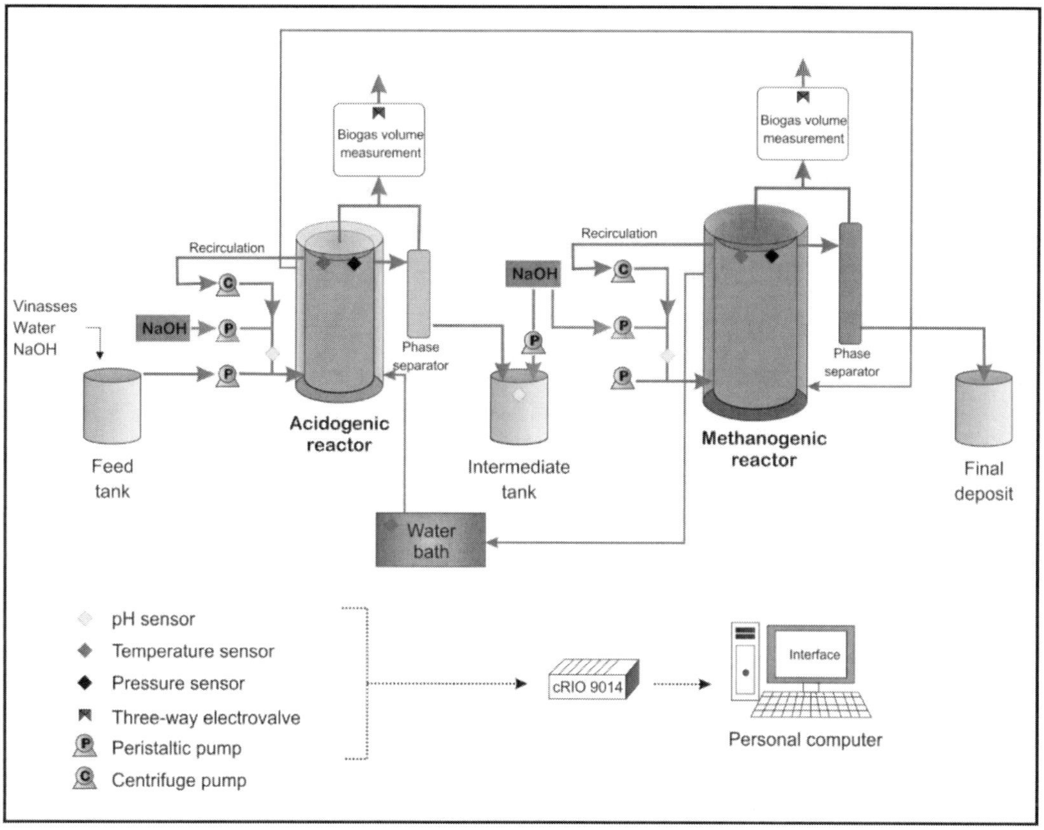

Figure 9. Two-stage anaerobic digestion process used in the treatment of tequila vinasses (Guevara-Santos, 2011).

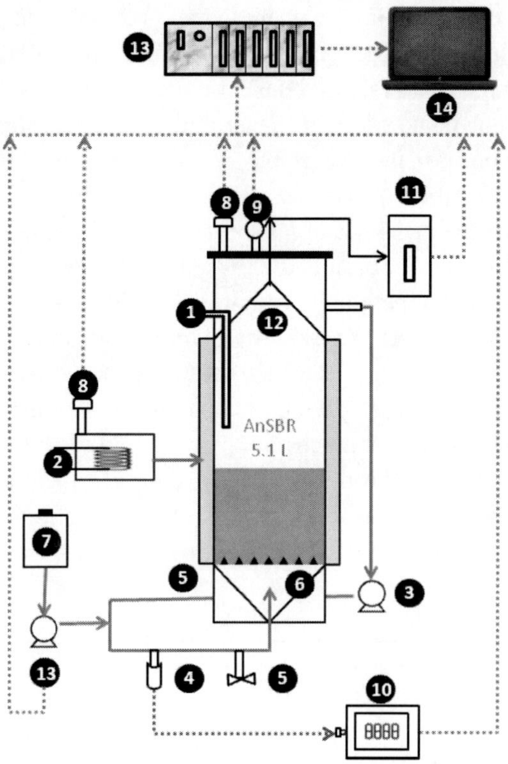

Figure 10. Experimental set-up of the An SBR digester: (1) feeding/drain pipe; (2) immersion circulator; (3) recycling pump; (4) pH/ORP sensor; (5) sampling probe; (6)diffuser/decanter; (7) NaOH solution; (8) temperature sensor; (9) pressure sensor; (10)pH/ORP transmitter; (11) biogas counter; (12) phase separator; (13) cRIO 9004controller; and (14) PC. Liquid streams (gray line), gas stream (black line), and data transmission (dotted line) (Jaramillo-Gante, 2012).

6. Concluding Remarks

Anaerobic digestion of tequila vinasses is a promising method of sustainable methane production. Results of current research efforts show that it is possible to scale-up a number of bioreactor configurations to treat 440 m^3 /year of tequila vinasses of a small tequila distillery. Thus, based on the obtained methane yield of the reviewed studies (0.29 NL $CH_4/gCOD_{removed}$), it is possible to generate 27 m^3/day of methane which are equivalent to 162 kW/day that may be used to continuously operate a centrifuge pump for 590 hours of collected biogas or used in kettles or boilers to provide heat in the plant while reducing the effect of fossil biofuel combustion and operation costs.

References

Alvarez, E., Linerio, J., Espinoza, A., Briones, R., Ilangovan, K. & Noyola, A. (1995). Anaerobic treatment of tequila vinasses in an UASB (in Spanish). In *Proceedings Biotechnolgy and Bioengineering Congress*. Ixtapa, Mexico.

Bernard, O., Polit, M., Hadj-Sadok, Z., Pengov, M., Dochain, D., Estaben, M. & Labat, P. (2001). Advanced monitoring and control of anaerobic wastewater treatment plants:software sensors and controllers for an anaerobic digester. *Wat Sci Technol.*, *43*, 175–182.

Buitrón, G. & Carvajal, C. (2010). Biohydrogen production from tequila vinasses in an anaerobic sequencing batch reactor: Effect of initial substrate concentration, temperature and hydraulic retention time. *Bioresource Technol., 101*, 9071-9077.

Cedeño-Cruz, M. & Álvarez-Jacobs, J. (2004). Production of tequila from agave: historical influences and contemporary processes. In: Gschaedler-Mathis, AC. (ed) Ciencia y tecnología del tequila; Avances y perspectivas. Centro de Investigación y Asistencia en Tecnología y Diseño del Estado de Jalisco (CIATEJ). Guadalajara, Mexico, 225–241.

Camara Nacional de la Industria del Tequila (CNIT. National Chamber of the Tequila Industry). (2012). Tequila, appellation d'origin controlee (in Spanish), www.tequileros.org, Agosto.

Chavez, L. M. (1997). *Effect of the HRT and recycling in an UASB during the treatment of tequila vinasses.* B.S. Thesis, Chemical Engineering, University of Guadalajara, Guadalajara, Jalisco, Mexico.

Consejo Regulador del Tequila (CRT). Tequila, appellation d'origine controlee, Statistics. (in Spanish), www.crt.org.mx, August, (2013).

El-Sheikh, M. A., Saleh, H. I., Flora, J. R. & AbdEl-Ghany, M. R. (2011). Biological tannery wastewater treatment using two stage UASB reactors. *Desalination.*, *276*, 253-259.

Espinoza-Escalante F. M., Pelayo-Ortiz C., Navarro-Corona J., González-García Y., Bories A. & Gutiérrez-Pulido H. (2009). Anaerobic digestion of the vinasses from the fermentation of *Agave tequilana Weber* to tequila: the effect of pH, temperature and hydraulic retention time on the production of hydrogen and methane. *Biomass Bioenerg. 33*, 14–20.

Estrada-Gasca C. A. & Islas-Samperio J. (2010). Alternative energy: Proposals for research and technological development for Mexico (in Spanish), 1st. Edition, Mexican Academy of Sciences, Mexico.

Fricke K. (2008). *The tequila industry and the treatment of its residues in Mexico*. Technical Report, Technische Universität Braunschweig.

García-Dueñas D. A. G. (1991). *Comparative study of the organic load reduction of tequila vinasses in two different anaerobic reactors* (in Spanish). M.Sc. Thesis, University of Yucatan, Merida, Yucatan, Mexico.

García-Maldonado J. G. (2007). Reporte de Análisis (in Spanish), Instituto Tecnológico Agropecuario Celaya, Guanajuato, Mexico.

Guevara-Santos, N. E. (2011). Performance evaluation of a two stage anaerobic digester for the treatment of tequila vinasses: Operation phase (in Spanish). M.S. Thesis, Bioprocess Engineering. University of Guadalajara, Guadalajara, Jalisco, Mexico.

Gray, N. F. (2004). *Biology of wastewater treatment*. Imperial College Press, 2nd. Edition.

Idarraga, G., Ramos, J., Zuñiga, V., Sahin, T. & Young, R. A. (1999). Pulp and paper from blue agave waste from tequila production. *J. Agric Food Chem.*, *47*, 4450-4455.

Iñiguez-Covarrubias G, Lange, S. E. & Rowell, R. M. (2001a.Utilization of byproducts from the tequila industry. Part 1: Agave bagasse as a raw material for animal feeding and fiberboard production. *Bioresource Technol., 77*, 25-32.

Iñiguez-Covarrubias, G, Díaz-Teres, R., Sanjuan-Dueñas, R., Anzaldo-Hernández, J. & Rowell, R. M. (2001b). Utilization of by products from the tequila industry. Part 2: Potential value of Agave tequilana Weber azul leaves. *Bioresource Technol., 77*, 101-108.

Iñiguez, G., Acosta, N., Martínez, L., Parra, J. & González, O. (2005). Utilización de subproductos de la industria tequilera. Parte 7: Compostaje de bagazo de agave y vinazas tequileras (in Spanish). *Revista Internacional de Contaminación Ambiental.*, *21*, 37–50.

Jaramillo-Gante, N. E. (2012). *Kinetic study of the tequila vinasses degradation in sequential batch reactors* (in Spanish). M.S. Thesis. Bioprocess Engineering. University of Guadalajara, Guadalajara, Jalisco, Mexico.

Jáuregui-Jáuregui, J. A., Méndez-Acosta, H. O., Snell-Castro, R., Alcaraz-González, V., & González-Álvarez, V. (2010). *Anaerobic treatment of tequila vinasses in an up-flow fixed-bed reactor: start-up, operation and restart-up.* In Proceedings 12th World Congress on Anaerobic Digestion, Guadalajara, Jalisco, Mexico.

Kim B. H. & Gadd G. M. (2008). *Bacterial physiology and metabolism.* Cambridge University Press, 1st. Edition.

Linerio-Gil, J. E. & Guzmán-Carrillo, A. (2004). Tratamiento de efluentes y aprovechamiento de residuos (in Spanish). In: Gschaedler-Mathis, AC. (ed) Ciencia y tecnología del tequila; Avances y perspectivas. Centro de Investigación y Asistencia en Tecnología y Diseño del Estado de Jalisco (CIATEJ). Guadalajara, Mexico, 258–284.

Lizárraga-Palazuelos, J. E. (2010). *Performance evaluation of a two stage anaerobic digester for the treatment of tequila vinasses: Start-up phase.* M.S. Thesis, Chemical Engineering, University of Guadalajara, Guadalajara, Jalisco, Mexico.

López-López, A., Davila-Vazquez, G., León-Becerril, E., Villegas-García, E. & Gallardo-Valdez, J. (2010). Tequila vinasses: generation and full scale treatment processes. *Rev Environ Sci Biotechnol., 9*, 109–116.

McCarty, P. L. 1964. *Anaerobic waste treatment fundamentals.* Part: I-IV. Public Works.95, 91–126.

Méndez-Acosta H. O., Snell-Castro, R., Alcaraz-Gonzalez, V., Gonzalez-Alvarez, V. & Pelayo-Ortiz, C. (2010). Anaerobic treatment of tequila vinasses in a CSTR-type digester. *Biodegradation.*, *21*, 357–363.

Moletta, R. (2005). Winery and distillery wastewater treatment by anaerobic digestion. *Water Sci Technol.*, *51*, 137–144.

Pinal, L., Cedeño, M., Gutiérrez, H. & Alvarez-Jacobs, J. (1997. Fermentation parameters influencing higher alcohol production in the tequila process. *BiotechnolLett.*, *19*, 45-47.

Pinedo L. E. (1991). *Anaerobic treatment of tequila vinasses in a packed bed reactor.* B.S. Thesis. Biology. University of Guadalajara, Guadalajara, Jalisco, Mexico.

Prado-Ramírez, R., González-Álvarez, V., Pelayo-Ortiz, C., Casillas, N., Estarrón, M., & Gómez-Hernández, H. E. (2005). The role of batch distillation on the quality of tequila. *Int. J. Food Sci. Technol.*, *40*, 701–708.

Ripley, L. E., Boyle, W. C. & Converse, J. C. (1986). Improved alkalimetric monitoring for anaerobic digestion of high-strength wastes. *J. Water Pollut. Control Federation. 58*, 406–411.

Rittmann, B. E. & McCarty, P. L. (2001). *Environmental biotechnology: Principles and applications.* McGraw-Hill.New York.

Rojo-Liera, N. M. (2012). Performance evaluation of a pilot-scale anaerobic digester for the treatment of tequila vinasses (in Spanish). M. S. Thesis, Bioprocess Engineering, University of Guadalajara, Guadalajara, Jalisco Mexico.

Rozzi A. (1991). Alkalinity considerations with respect to anaerobic digester. In Proceedings 5th Forum Applied Biotechnol. *Med FacLand bouww Rijksuniv Gent.*, *56*, 1499–1514.

Valenzuela-Zapata, A. G. (2003). Tequila's agave: cultivation and industry in Mexico (in Spanish), 1st. Edition, Mundi-Prensa.

Voellger, Z. M. (2000). Determination of the kinetics of an anaerobic digestion of tequila vinasses carried out in a fluidized bed bioreactor. M.S. Thesis, *Chemical Engineering*. University of Guadalajara, Guadalajara, Jalisco, Mexico.

Zickefoose, C. & Hayes, R. (1976). *Anaerobic Sludge Digestion: Operations Manual*. Office of Water Program Operations, US Environmental Protection Agency.

In: Energy and Environment Nowadays
Editors: Luis G. Torres and Erick R. Bandala
ISBN: 978-1-63117-398-1

Chapter 8

PV Systems Design and Applications

Pedro Bañuelos Sánchez
Universidad de las Américas Puebla, Puebla, Mexico

1. Introduction

Photovoltaic (PV) refers to the direct generation of electricity by solar irradiation. Over the period 2000 to 2011, PV systems were the fastest-growing renewable power technology worldwide. Cumulative installed capacity of PV reached roughly 65 GW at the end of 2011, up from only 1.5 GW in 2000. Germany and Italy accounted for over half global cumulative capacity, followed by Japan, Spain, the United States and China. According to International Energy Agency (IEA) analysis, under extreme assumptions, solar energy could provide up to one-third of the world's final energy demand after 2060. These assumptions include policy makers successfully reducing greenhouse gas emissions beyond the current international targets, and electricity-driven technologies fostering significant energy efficiency improvements and displacing fossil fuels in many uses in buildings, industry and transportation (IEA, 2013). As an example of photovoltaic system, figure 1 shows a 4 kW PV system. It has been installed near Raley Hall on the Appalachian State University campus. The system will convert solar energy into approximately 5250 kilowatt hours of electricity each year, about enough energy to power an energy-efficient home (ASU, 2013).

2. Fundamental Concept

Solar cells, also called photovoltaic (PV) cells, convert sunlight directly into electricity. PV gets its name from the process of converting light (photons) to electricity (voltage), which is called the *PV effect*. The PV effect was discovered in 1954, when scientists at Bell Telephone discovered that silicon (an element found in sand) created an electric charge when exposed to sunlight. Soon solar cells were being used to power space satellites and smaller items like calculators and watches. Nowadays, thousands of people power their homes and businesses with individual solar PV systems. Utility companies are also using PV technology for large power stations.

ASU, 2013.

Figure 1. Photovoltaic system example.

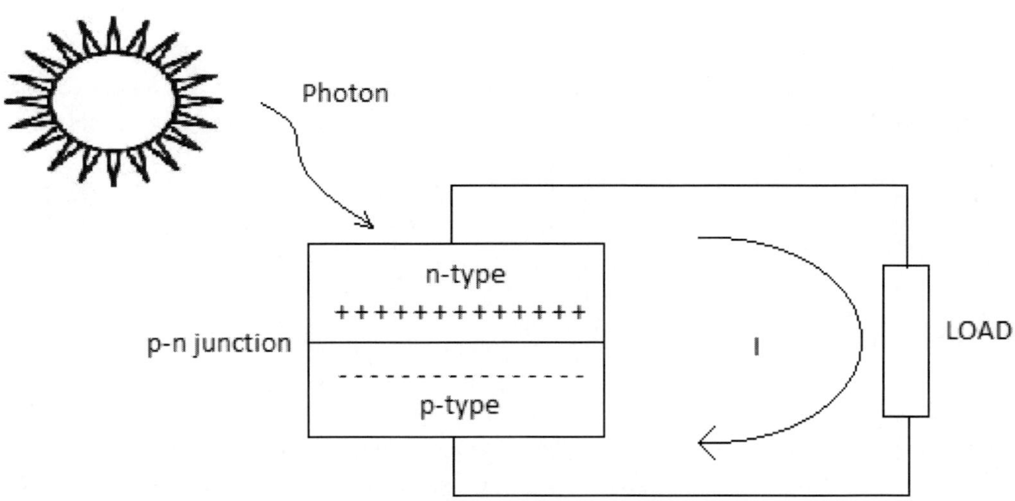

Figure 2. Simple photovoltaic cell schematic.

A photovoltaic cell can be seen as the simple arrangement shown in figure 2. When an incident photon, with enough energy to dislodge a valence electron, hits the cell, the electron will jump to the conduction band and a current flow is initiated. The electron must posse at least the band gap energy to go from valence band to conduction band.

The energy of a photon is given by

$$E \quad = hv \tag{1}$$

where h=6.625x10^{-34}Js is the Planck constant, and v is the frequency (Hz). The frequency, wavelength (λ), and speed of light (c=3x10^{8} m/s) are related by

$$v = \frac{c}{\lambda} \tag{2}$$

Not all of the light is useful for a photovoltaic cell. Just light with a specific wavelength has enough energy to dislodge a valence electron. Considering the process that happens in silicon, from eqs. (9.1) and (9.2), we can have:

$$\lambda = \frac{hc}{E_g} \tag{3}$$

where E_g is the band gap energy (eV) in silicon. Then, considering 1 eV=1.6x10^{-19} J, and E_g=1.111 eV for silicon:

$$\lambda = \frac{(6.6250x10^{-34})(3x10^{8})}{(1.111)(1.6x10^{-19})} = 1.1183x10^{-6}\, m = 1.12\, \mu m \tag{4}$$

As shown in eq. (9.4), light with a wavelength of 1.12 μm has enough energy to dislodge a valence electron in silicon. If wavelength is greater than 1.12 μm, the energy is insufficient to dislodge a valence electron in silicon. Hence, if wavelength is less than 1.12 μm, photons possess more energy than is required to dislodge a valence electron in silicon. To use a larger spectrum of light energy, new materials have been explored as PV cell materials.

3. New Technologies on Photovoltaics

Solar cells based on silicon (Si) semiconductors account for nearly 90% of 2011 sales of photovoltaic (PV) products. Annual production of Si-based PV in 2011 reached more than 15 gigawatts—an order of magnitude higher than other PV technologies. Silicon in PV takes many forms, including the industry-dominant single-crystalline (c-Si) and multi-crystalline wafers sawn from ingots; melt-grown ribbons; thin hydrogenated amorphous silicon (a-Si:H); and microcrystalline Si layers grown from gaseous precursors. Many improved technologies are on the horizon, including thin c-Si by epitaxial growth, crystallized polysilicon layers, and nanoscale Si (NREL, 2013).

National Renewable Energy Laboratory (NREL), for example, has world-leading research capabilities and expertise in silicon materials and devices, especially photovoltaic cells. NREL focus on basic and applied research for high conversion efficiency at low cost.

NREL has significant and long-term capabilities in both cadmium telluride (CdTe) and copper indium gallium diselenide (CIGS) thin-film PV research and device development. These PV systems are considered as Second Generation PV systems and have an early market deployment.

There are a wide range of PV cell technologies on the market today. Different types of materials have been tested and proved their efficiency in PV conversion. PV cell technologies are usually classified into three generations, depending on the basic material used on the level of commercial maturity (IRENA, 2012).

First generation PV systems.- These PV systems are fully commercial nowadays. They use the wafer-based crystalline silicon (c-Si) technology, either single crystalline (sc-Si) or multi-crystalline (mc-Si).

Second generation PV systems.- These PV systems are in development. They are based on thin-film PV technologies and generally include three main families: 1) amorphous (a-Si) and micromorph silicon (a-Si/μc-Si); 2) Cadmium-Telluride (CdTe); and 3) Copper Indium Selenide (CIS) and Copper Indium Gallium Diselenide (CIGS).

Third generation PV systems.- Theses systems include technologies such as concentrating PV (CPV) and organic PV cells that are still under demonstration or have not yet been widely commercialized.

3.1. Second Generation PV Systems

CdTe Research

Cadmium telluride (CdTe) is a semiconductor with the band gap of 1.47 -1.48 eV, optimal for solar cells. CdTe is a direct-gap semiconductor, so that the thickness of only a few microns is sufficient for almost complete absorption of solar radiation (97 – 98 %) with photon energy $h\nu > E_g$. As the temperature increases, the efficiency of CdTe solar cell is reduced less than with silicon devices, which is important, given the work of solar modules in high-power irradiation. Compared to other thin-film materials, technology of CdTe solar modules is simpler and more suitable for large-scale production. Figure 3 shows the spectral distribution of quantum efficiency for CdTe semiconductor. Practical efficiencies are not longer than 0.7 (Kosyachenko, 2011).

Cadmium telluride based (CdTe-based) thin-film solar cell modules currently represent one of the fastest-growing segments of commercial module production. This is due partly to the simplicity of the two-component absorber layer (i.e., CdTe contains only cadmium and tellurium) and the ability of bulk cadmium telluride source material (in the form of high-purity powders) to be reconstructed into the CdTe thin films needed to produce PV modules. During most of the 20 years of research undertaken by the CdTe Group at NREL, much effort has been directed at producing CdTe structures that allow more light to penetrate the top layers of the device (the transparent conducting contacts and cadmium sulfide [CdS] layers) to achieve high efficiency. This understanding has been transferred to commercial processes for use in producing higher-performance modules.

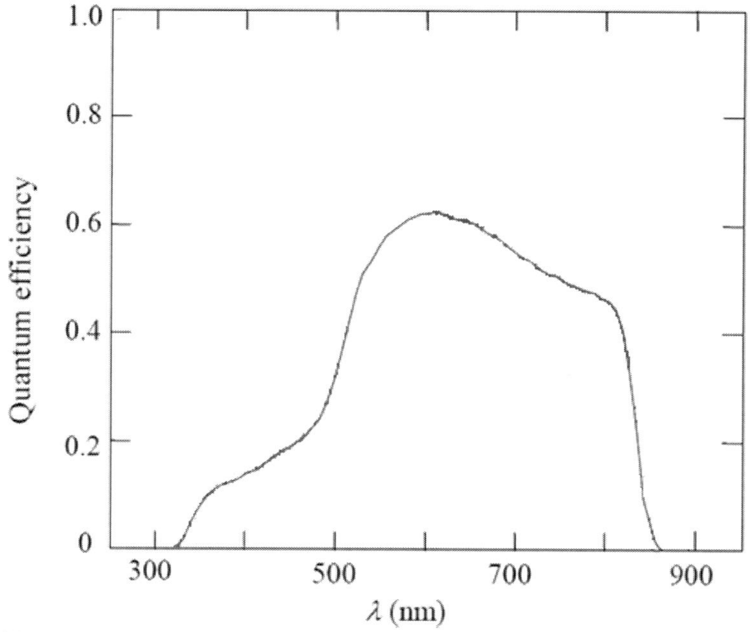

Kosyachenko, 2011.

Figure 3. Spectral distribution of quantum efficiency of CdTe solar cells.

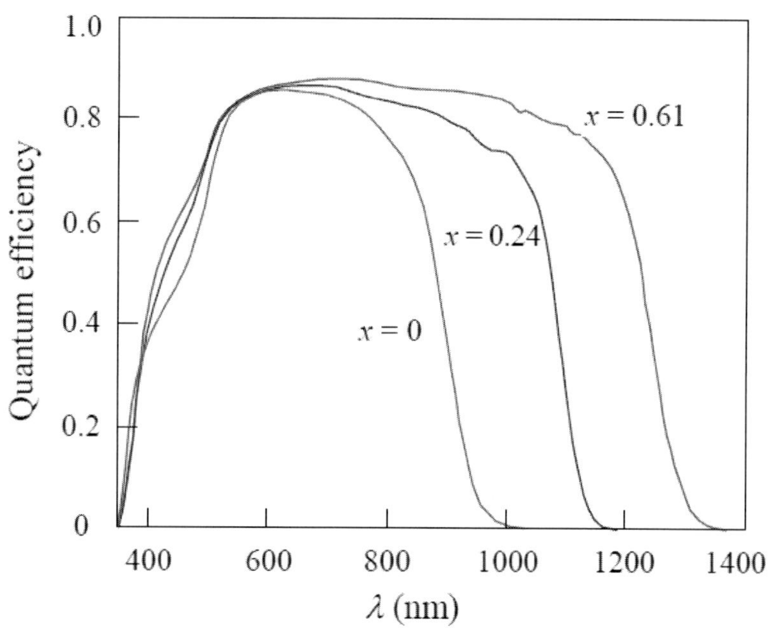

Kosyachenko, 2011.

Figure 4. Spectral distribution of quantum efficiency of CIGS solar cells.

CIGS Research

Copper Indium Diselenide ($CuInSe_2$, CIS) and Copper Gallium Diselenide ($CuGaSe_2$, CGS) form $CuIn_xGa_{1-x}Se_2$ alloy (CIGS) in any ratio of components. Varying the ratio of $CuInSe_2$ and $CuGaSe_2$, there is a slight change in the material parameters except the band gap. In the case of CIS semiconductor, its band gap is in the range of 1.0 to 1.04eV. And for CIGS, band gap goes to 1.68 eV. The dependence of the band gap of $CuIn_xGa_{1-x}Se_2$ on x is described by eq. (9.5) (Kosyachenko,2011):

$$E_g(x) = 1.010 + 0.626x - 0.167x(1-x) \quad (5)$$

CIGS alloy has been considered as promising material for high performance thin-film solar cells and fabrication of monolithically interconnected modules intended for cost effective power generation. Parameters of CuInxGa1-xSe2 can be fitted as optimal for the photoelectric conversion.

Copper indium gallium diselenide based (CIGS-based) thin-film solar cell modules currently represent the highest-efficiency alternative for large-scale, commercial thin-film solar cells. Several companies have confirmed module efficiencies exceeding 13%. Most of these companies are using ideas and intellectual properties that were developed by the NREL CIGS Group during the past 20 years of research. Central to this understanding was the group's development of the "three-stage process." This process enables the formation of a CIGS thin-film layer that is of the proper composition and structure to allow the charges generated by sunlight (i.e., electrons and holes) to exist long enough in the CIGS layer of the device so that they can be separated and collected at the front and back contacts. This separation and collection is critical for demonstrating high conversion efficiency. One of the main characteristics of a solar cell is the spectral distribution of quantum efficiency (spectral response), which is ultimately determined the short-circuit current density of CdTe. As we can see in figure 4, depending on the value of x, wavelength (λ) changes using a larger spectrum of light energy and improving the efficiency.

3.2. Third Generation PV Systems (IRENA, 2012)

Concentrating Photovoltaic Technology

Concentrating PV (CPV) systems use optical devices, such as lenses or mirrors, to concentrate direct solar radiation onto very small, highly efficient multi-junction solar cells made of a semiconductor material. The sunlight concentration factor ranges from 2 to 100 suns (low to medium concentration) up to 1 000 suns (high concentration). To be effective, the lenses need to be permanently oriented towards the sun. Asingle or double axis tracking system has to be used for low and high concentrations, respectively. Cooling systems (active or passive) are needed for some concentrating PV designs, while other novel approaches can get round this need.

Commercial CPV modules with silicon-based cells offer efficiency in the range of 20% to 25%. CPV based on multi-junction solar cells using III-V semiconductors have achieved laboratory efficiency of more than 40% (IEA, 2010). Commercial multi-junction devices manufactured by Sharp, Emcore, Spectrolab and Azur have efficiencies of around 35% - significantly higher than conventional single-junction c-SI solar cells. Continued Research and Development holds the promise of increasing CPV efficiencies up to 45% or even 50% (Cotal, 2009).

Organic Solar Cells

Organic solar cells are composed of organic or polymer materials. They are inexpensive, but not very efficient. They are emerging as a niche technology, but their future development is not clear. One can observe a strongly increasing Research and Development effort in the domain of solar cells based on organic layers. This progress is essentially based on the introduction of nanostructured material systems to enhance the photovoltaic performance of these devices. The growing interest is fuelled by the potentially very low cost of organic solar cells thanks to the low cost of the involved substrates, the low cost of the active materials of the solar cell, the low energy input for the actual solar cell/module process and last, but not least, the asset of flexibility.Organic PV module efficiencies are now in the range 4% to 5% for commercial systems and 6% to 8% in the laboratory.In addition to the low efficiency, a major challenge for organic solar cells is their instability over time.

4. Types of PV Systems

Energy generated by PV systems can be used to provide DC and/or AC power service. In some cases, PV systems are interconnected with the utility grid, providing more energy to the utility. In others cases, PV systems are independent of the utility grid. This is the case of rural communities where utility grid doesn't exist. Recently, PV systems are connected with other alternative energy, like wind energy or fuel cell energy. This is the way to generate electricity reducing the global warming.

PV systems can be mainly classified as grid-connected and stand-alone systems (FESC, 2013). Grid connected PV systems are connected in parallel with the electric utility grid. A PV array produces DC power. However, main industrial, commercial, and home applications use AC power. An inverter is then necessary. The inverter converts the DC power produced by the PV array into AC power consistent with the voltage and power quality requirements of the utility grid, and stops supplying power to the grid when the utility grid is not energized. This last features is a safety one. In all grid-connected PV system, it ensures that the PV system will not continue to operate and feed back into the utility grid when the grid is down for service or repair. Figure 5 shows a grid connected PV system sketch. As we can see, a grid connected PV system can supply energy either to an independent AC load, or to the utility grid, or both of them at the same time. In the case indicated in Figure 5, when there is no sunlight, AC load is fed by the utility grid through the distribution panel.

Stand-alone PV systems operate independently to the electric utility grid. They are designed and sized to supply a DC and/or AC electrical loads. Figure 6 shows a direct-coupled PV system. In this case, there is no electrical energy storage (batteries), meaning, it operates only during sunlight hours. Ventilation fans and water pumps are examples where direct-coupled PV systems are used. This kind of PV systems have to produce all the power they are capable of. A type of electronic DC-DC converter, called Maximum Power Point Tracking (MPPT), is used between the array and the load to help better utilize the available array maximum power output. MPPT is not a mechanical tracking system that physically moves the array to make it point directly at the sun. MPPT is a fully electronic system that varies the electrical operating point of the array so that it is able to deliver maximum available power.

Figure 7 shows a stand-alone PV system with a battery. This system can feed the load even there is no sunlight. PV array charges up the battery through the charge controller. Battery output voltage is a DC voltage, then, a DC load can be directly fed by the battery. But if the load is an AC load, an inverter is necessary. Again, inverter converts DC voltage into AC voltage feeding the AC load, as in the case of grid connected PV systems. Total amount of power supplied by the PV system depends on the battery's maximum output power rating. Charge controller limits the rate at which electric current is added to or drawn from battery. It prevents overcharging and may prevent against overvoltage, which can reduce battery performance or lifespan. It may also prevent completely draining the battery to protect battery life.

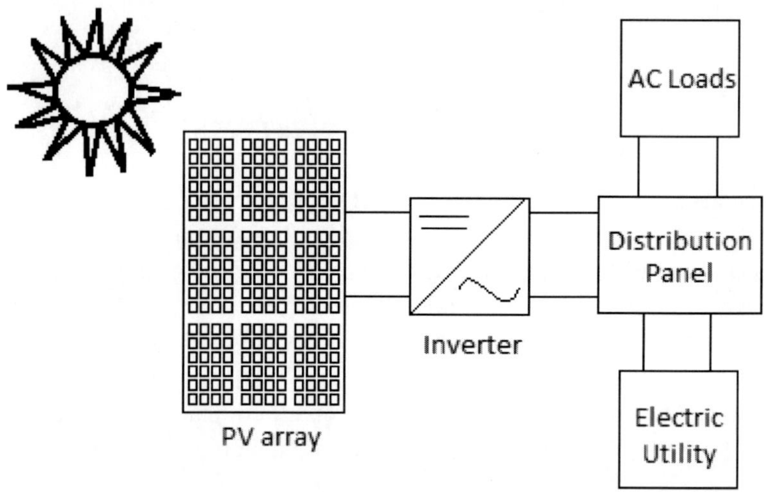

Figure 5. Grid connected photovoltaic system.

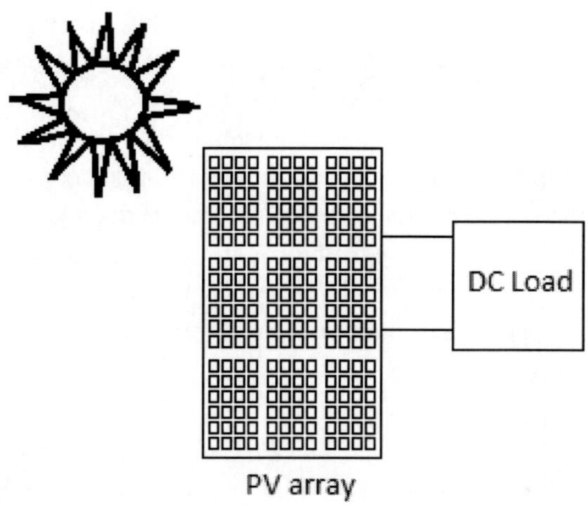

Figure 6. Direct-coupled PV system.

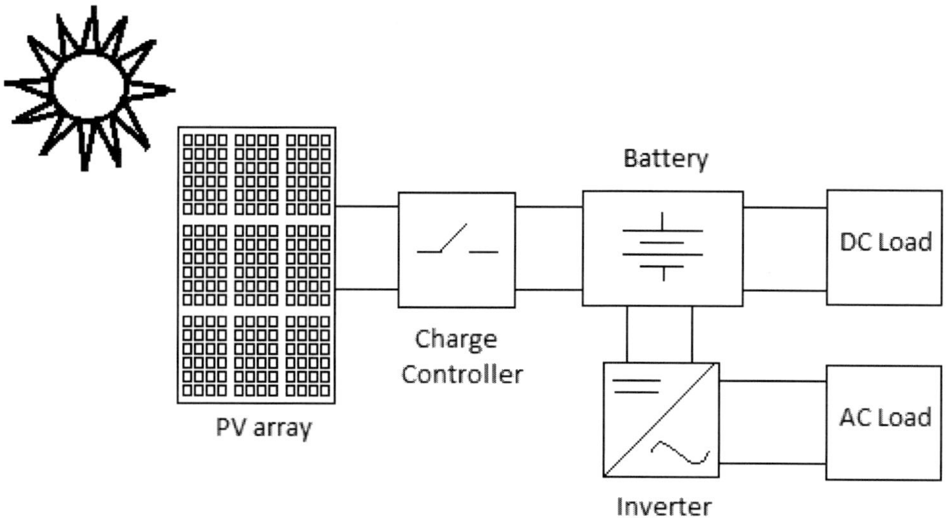

Figure 7. Stand-alone PV system feeding AC and DC loads.

5. Designing a PV System

Battery storage, in a photovoltaic system, permits operation during the night hours or when the solar irradiation is insufficient to meet the required electrical demand. Battery is a key component to ensure the independence of a PV system. Lead acid battery is commonly used because of its features such as wide operation temperature range, low self-discharge, long service life and maintenance free. The installation cost of the battery is high compared to the PV installation because of its limited service time. Battery life time is reduced if there is low PV energy availability for longer period or improper charging-discharging. So, the battery charging needs control for achieving high State of Charge and longer battery life. Hence proper controller for battery charging is an inevitable need for this hour. The main function of the battery charger in stand-alone PV system is to fully charge the battery without permitting overcharging while preventing reverse current flow at night and deep discharge under load conditions.

Batteries used in PV system applications have to meet specific characteristics and requirements. Typical battery characteristics are the voltage, the charge capacity, the cycle capacity, and the depth of discharge (DOD). The voltage across the battery terminals is specified in volts (V), and the charge capacity of a battery is rated in units of amp-hours (Ah). The product of voltage and charge capacity yields the energy stored in the battery. For example, a 12 V battery with a charge capacity of 56 Ah stores 672VAh=0.672 kWh of energy. Cycle capability refers to the number of charge/discharge cycles expected from the battery. The depth of discharge specifies how much of the stored energy at full capacity may be extracted without damage to the battery. If we say a battery is 100% fully charged, it means the DOD of this battery is 0%. If we say the battery have delivered 30 % of its energy, we say the DOD of this battery is 30%. And if a battery is 100% empty, the DOD of this battery is 100%. Batteries for photovoltaic applications require high cycle capacity as well as high depth of discharge capability. For lithium batteries, it is not suggested fully discharge them to 100% DOD, it would shorten the cycle life of batteries.

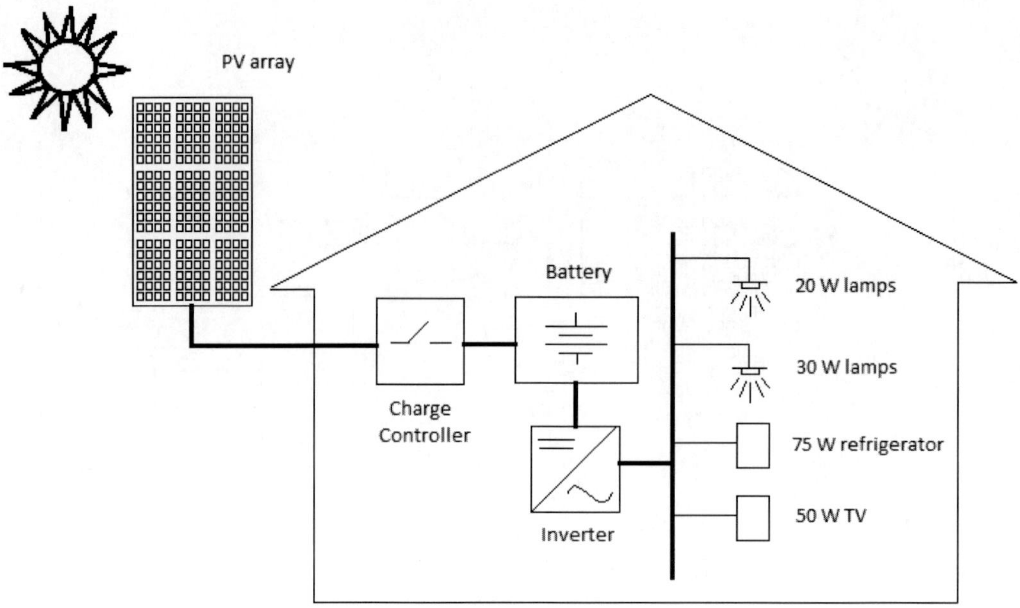

Figure 8. PV system with electrical loads.

Example 1

A house has fluorescent lamps, a refrigerator, and a television set, see figure 8. Design a solar PV system for supplying electrical energy to all these electrical loads. Indicate battery size, number of PV arrays, inverter size, and charge controller size. The house has two types of fluorescent lamps that work during different time. Characteristic's loads are as follows:

- Fourteen 20 W fluorescent lamps with electronic ballast used 2 hours per day
- Five 30 W fluorescent lamps with electronic ballast used 3 hours per day
- One 75 W refrigerator that runs 24 hours per day with compressor runs 12 hours and off 12 hours
- One 50 W television used for 6 hours per day

1. Power Consumption Demands

The first step in designing a solar PV system is to find out the total power and energy consumption of all loads that need to be supplied by the solar PV system. In this case:

Electrical load	**Energy**
Fluorescent lamps:14 x 20W x 2h	560 Wh/day
Fluorescent lamps: 5 x 30W x 3h	450 Wh/day
Refrigerator: 1 x 75W x 12h	900 Wh/day
Television: 1 x 50W x 6h	300 Wh/day
Total appliance use	2210 Wh/day

PV system has to supply 2210 Wh per day. Battery size and number of PV arrays depend on this total energy per day.

2. Battery Sizing

The battery type recommended for using in solar PV system is the deep cycle battery. This type of battery is specially designed for to be discharged to low energy level and rapid recharged or charged and discharged day after day for years. The battery should be large enough to store sufficient energy to operate the appliance at night and cloudy days. For this example, consider 10 days of autonomy, meaning that even when the days are completely cloudy and then PV arrays don't work, the solar PV system can supply energy to the loads. Then, considering to use a 12 V battery, we have:

- Total appliance use = 2210 Wh/day
- Nominal Battery voltage = 12 V
- Days of autonomy = 10 days
- Battery capacity = (2210 Wh/day) (10 days) / (12 V) = 1841.66 Ah
  So the battery should be rated 12 V, 2000 Ah for 10 days autonomy.

3. PV Array Sizing

Different size of PV arrays will produce different amount of power. To find out the sizing of PV arrays, the total energy (E) produced by a single PV array is needed. The total energy(E) produced depends on size of the PV array and climate of site location. A commonly used equation to calculate E is as follows (CENSOLAR, 2001):

$$E=(5 - L/15)x(1+L/100)xP \tag{6}$$

where, L is the latitude of the place where PV system will be installed; and P is the total power of a single PV array. Considering P = 280 W, L = 19° as the latitude of Mexico City, and following our example, total energy can be calculated as:

$$E=(5 - 19/15)x(1 + 19/100) \text{ x } 280 = 1243.94 \text{ Wh} \approx 1244 \text{ Wh}$$

Then,

$$Number\ of\ PV\ arrays = \frac{Total\ appliance\ use}{Total\ energy\ produced} = \frac{2210}{1244} = 1.77 \cong 2\ panels$$

So, this system should be powered by at least 2 modules of 280 W PV array.

4. Inverter Sizing

The input rating of the inverter should never be lower than the total watt of loads. The inverter must have the same nominal voltage as the selected battery. For stand-alone systems, the inverter must be large enough to handle the total amount of watts you will be using at one time. The inverter size should be 25-30% bigger than total watts of loads. But, if the load is a motor or a compressor, then, the inverter size should be at least 3 times the capacity of those appliances and must be added to the inverter capacity to handle surge current during starting.

On the other hand, for grid-connected systems, the input rating of the inverter should be same as PV array rating to allow safe and efficient operation. Then, for this example:

- Total watts of all appliances = (14 x 20) + (5 x 30) + (1 x 75) + (1 x 50) = 555 W
- For safety, the inverter should be considered 25 – 30 % bigger size.
- The inverter size should be about 555 W x 1.3 = 720 W or greater.

5. Charge Controller Sizing

The charge controller is rated against amperage and voltage capacities. Charge controller must be selected to match the voltage of PV array and the batteries. Make sure that the selected charge controller has enough capacity to handle the current from PV array. For the series charge controller type, the sizing of the controller depends on the total PV input current which is delivered to the controller and also depends on PV panel configuration (series or parallel configuration). A common practice for sizing the charge controller is to take the short circuit current (Isc) of the PV array, and then multiply it by 1.3. For our example, consider the following PV characteristics, found in reference (LG Solar, 2013):

- Pm = 280 Wp (Nominal power)
- Vm = 31.5 Vdc (Voltage at nominal power)
- Im = 8.97 A (Current at nominal power)
- Voc = 38.9 V (Open-circuit voltage)
- Isc = 9.56 A (Short-circuit current)
- Charge controller rating = (2 panels x 9.56) x 1.3 = 24.85 A
  So the charge controller should be rated 25 A at 12 V or greater.

6. Needs for Carbon Credits

Human beings have a lot of needs in which energy is important. Transportation, industrial, commercial, and residential needs are growing up each day everywhere around the world. The main sources of energy have been petroleum, natural gas, coal, and nuclear power. Petroleum, natural gas, and coal are fossil fuels, meaning they are non-renewable resources and can run out in some years. Other relatively recent source of energy is the so called renewable energy, where wind, solar, biomass, geothermal, and ocean energy are used. This kind of energy is replenished naturally, meaning it's infinite. However, even the developed countries are using, in a large percentage, non-renewable energies to solve their energy needs. In 2011, in the United States of America (USA), 79.78% of its energy came from fossil fuels, and only 9.14% from renewable energies (EIA, 2011), see figure 9. Generation of electrical energy contributes to increase global warming, because of we generate electricity by burning fossil fuels. But not only USA increases global warming. In Mexico for example, 86.81% of electricity is generated by burning fossil fuels, as indicated in table 1.Unfortunately, non-renewable energy has impact on air pollution, water pollution, and climate change.

Renewable energies have much less impact to the environment than non-renewable energies. They could be the answer to future energy needs. Today, solar energy represents a small portion of the total energy, even though this kind of resource is enormous. The average intensity of light outside the atmosphere is about 1353 W/m^2. In order to produce 1 GW of power, an area of about 5 km^2 would be needed, assuming a conversion of 20%. The earth receives more energy from the sun in 1 hr than the global population uses in an entire year (Gonen, 2012).

Solar photovoltaic (PV) industry is growing very fast. More than 100 countries have installed PV systems to increase their electrical energy capacity during 2010. Solar energy is becoming an important alternative for the future energy needs. An estimated 17 GW of PV capacity was added worldwide in 2010 (compared with just under 7.3 GW in 2009), bringing the global total to about 40 GW, more than 7 times the capacity in operation five years earlier. Total existing capacity of all PV grew 72% relative to 2009, with the average annual growth rate over the 2005 to 2010 period exceeding 49%. The PV market was driven by falling costs, new applications, strong investor interest, and continued strong policy support, but also by accelerated tariff digressions in some countries (REN, 2011).

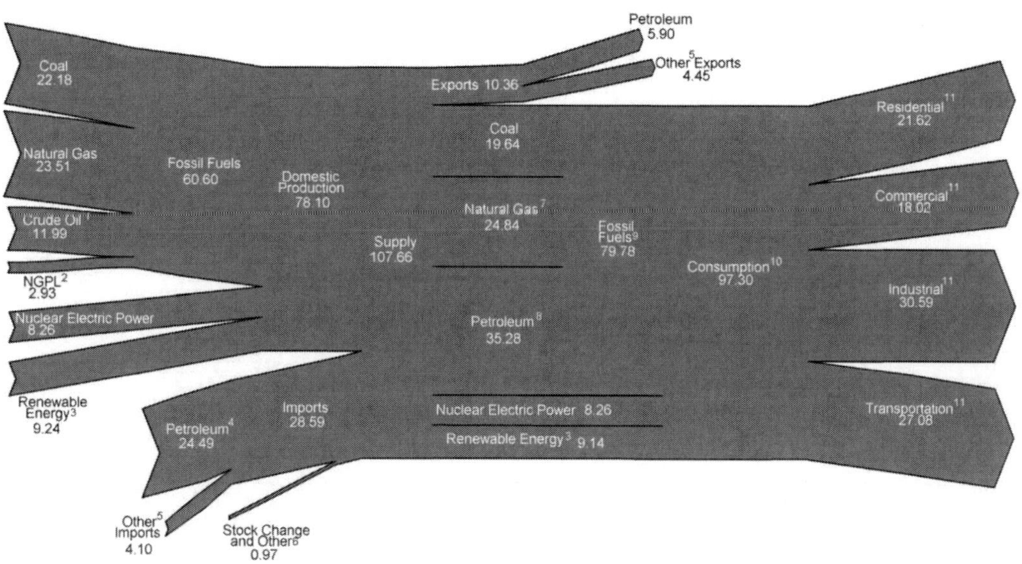

EIA, 2011.

Figure 9. United States of America energy flow diagram for 2011.

Table 1. Sources to generate electrical energy in Mexico (CFE, 2013)

Type of generation	Percentage
Geothermal	2.52%
Coal	6.77%
Nuclear	5.01%
Wind	0.10%
PV	0.005%
Petroleum (independent power producers)	34.84%
Hydroelectric	5.55%
Petroleum	45.20%

Table 2. Total Carbon Dioxide Emissions from the Consumption of Energy (Million Metric Tons) (IES, 2013)

	2007	2008	2009	2010	2011
North America	7065.083	6877.108	6407.535	6617.038	6506.96
Bermuda	0.7509	0.75025	0.71151	0.69942	0.77688
Canada	593.0896	578.2483	549.6838	546.652	552.5565
Greenland	0.59981	0.6417	0.64787	0.64841	0.61108
Mexico	444.2703	452.7937	421.1238	432.21	462.2929
Saint Pierre and Miquelon	0.08914	0.09159	0.08899	0.08899	0.09143
United States	6026.284	5844.582	5435.279	5636.739	5490.631

Table 3. Per Capita Carbon Dioxide Emissions from the Consumption of Energy (Metric Tons of Carbon Dioxide per Person) (IES, 2011)

	2007	2008	2009	2010	2011
North America	15.94832	15.37151	14.18846	14.5209	14.16161
Bermuda	11.22069	11.13345	10.48889	10.24562	11.31172
Canada	18.00736	17.41046	16.41474	16.19242	16.23705
Greenland	10.42658	11.14759	11.24774	11.24983	10.59609
Mexico	4.08709	4.11798	3.78668	3.84293	4.06503
Saint Pierre and Miquelon	14.61311	15.14135	14.84191	14.97427	15.52824
United States	20.00551	19.21966	17.71768	18.2224	17.62122

Energy supply systems are facing significant changes in many countries around the world. As an example, German power system has experimented important changes. German renewable energy resources are now contributing 25% of the power needed to meet electricity demand, compared with 5% only 20 years ago. PV systems have been privileged. In September 2012, about 1.2 million PV systems were installed, with a total installed peak capacity of more than 31 GWp. During some hours of 2012, PV systems contributed about 40% of the peak power demand. The installed capacity of PV systems is now in Germany higher than that of any other renewable energy source, as well as that of any conventional power source (van Appen, 2013).

As described above, a standalone photovoltaic power system consists of solar array, controller with maximum power point tracker, batteries, inverter and loads. Since the solar array is a sole energy source, the power of the system will change significantly with the variation of solar radiation, temperature, load conditions and battery state of charge (SOC)(Khabit, 2010). Thus it is very crucial to optimize sizes of solar array and battery to meet load demand. Actually, PV cells have inherent poor sunlight energy conversion efficiency, around 20% (NREL Poly, 2013). However, PV energy systems represent one of the most promising renewable energies. These systems can significantly reduce the emission of carbon dioxide into the atmosphere. Some countries have been agree to reduce all greenhouse gas emissions by some international treaties that sets binding obligations on industrialized countries, such as the Kyoto Protocol to the United Nations Framework

Convention on Climate Change (UNFCCC). The UNFCCC is an environmental treaty with the goal of preventing dangerous anthropogenic interference of the climate system. Some others mechanisms have been implemented.

A carbon credit is a generic term for any tradable certificate or permit representing the right to emit one ton of carbon dioxide or the mass of another greenhouse gas with a carbon dioxide equivalent to one ton of carbon dioxide. Carbon credits and carbon markets are a component of national and international attempts to mitigate the growth in concentrations of greenhouse gases (GHGs). The goal is to allow market mechanisms to drive industrial and commercial processes in the direction of low emissions or less carbon intensive approaches than those used when there is no cost to emitting carbon dioxide and other GHGs into the atmosphere. GHG mitigation projects generate credits, this approach can be used to finance carbon reduction schemes between trading partners and around the world.

Table 2 shows the total carbon dioxide emissions from the consumption of energy in North America. As we can see, United States, Canada and Mexico are producing a huge amount of carbon dioxide, to supply basic energy needs. But, unfortunately, they are burning fossil fuels to produce this energy. If these countries stimulate PV systems, like in the case of Germany, carbon dioxide emissions, and then global warming, will reduce too. But not only industrialized countries emit greenhouse gases. Table 3 shows the per capita carbon dioxide emissions from the consumption of energy. Bermuda and Saint Pierre and Miquelon emit more than Mexico.

Example 2

Let's take the same electrical system as indicated above in the example 1. The house has fourteen 20 W fluorescent lamps with electronic ballast used 2 hours per day, five 30 W fluorescent lamps with electronic ballast used 3 hours per day, one 75 W refrigerator that runs 24 hours per day with compressor runs 12 hours and off 12 hours, and one 50 W television set used for 6 hours per day. Consider that the thermal energy content of coal is 6150 kWh/ton. Although coal fired power generators are very efficient, they are still limited by the laws of thermodynamics. Only about 40 percent of the thermal energy in coal is converted to electricity. So the electricity generated per ton of coal is 0.4 x 6150 kWh or ***2460 kWh/ton***. Find out how many tons per year of coal are burned for this house.

Then:

- Total energy consumption per year: (2.21 kWh/day) x (365 days) = 806.65 kWh
- Thermal energy content of coal: 6150 kWh/ton
- Electricity generated per ton: 0.4 x 6150 = 2460 kWh/ton
- Total burned tons: (806.65 kWh) / (2460 kWh/ton) = 0.3279 tons
- We need to burn 328 kg of coal per year to power our example house.

But let's look at what else is produced to power that house. A typical 500 megawatt coal power plant produces 3.5 billion kWh per year. That is enough energy for 4.34 million of our houses to operate year round. To produce this amount of electrical energy, the plant burns 1.43 million tons of coal. It also produces:

Pollutant	Total for a powerplant	One house year's worth
Sulfur Dioxide – Main cause of acid rain	10,000 Tons	4.6 pounds
Nitrogen Oxides – Causes smog and acid rain	10,200 Tons	4.7 pounds
Carbon Dioxide – Causes greenhouse gas	3,700,000 Tons	1697 pounds

1697 pounds = 0.8484 tons = 848.4 kg of carbon dioxide emission for each house! That's a pretty big amount, considering we are millions of people.

Summary

PV arrays are made by semiconductors, like silicon, gallium, and telluride. Recent researches are looking for higher efficiencies. The task is not easy but international laboratories, as NREL, are working on it. The use of PV systems can contribute to diminish global warming due to a lot of tons of carbon dioxide can be avoided.

References

ASU, (2013). Appalachian State University. http://www.appstate.edu

CENSOLAR, (2001). Centro de Estudios de la Energía Solar. "La Energía Solar Aplicaciones Prácticas". PROGENSA (Promotora General de Estudios, S.A.). Spain.

CFE, (2013). Comisión Federal de Electricidad, CFE. "Generación por fuente". http://www.cfe.gob.mx/ConoceCFE/1_AcercadeCFE/Estadisticas/Paginas/Generacion.aspx

Cotal, H. et al., (2009), III–V multijunction solar cells for concentrating photovoltaics, *Energy Environment Science*, 2009, 2, pps 174–192, Sylmar, CA.

EIA, (2011). "Energy Flow, 2011". Energy Information Administration(EIA) of the United States Department of Energy(USDOE). http://www.eia.doe.gov

FESC, (2013). Florida Solar Energy Center, FSEC.http://www.fsec.ucf.edu/en/consumer/solar_electricity/basics/types_of_pv.htm

Gonen T., (2012). "Electrical Machines with MATLAB". CRC Press.

IEA, (2010). International Energy Agency (IEA) (2010), Technology Roadmap: Solar Photovoltaic Energy, IEA/OECD, Paris.

IEA, (2013). International Energy Agency (IEA). "Solar (PV and CSP)". http://www.iea.org/topics/solarpvandcsp/

IES, (2011). International Energy Statistics. U. S. Energy Information Administration. http://www.eia.gov/cfapps/ipdbproject/iedindex3.cfm?tid=90&pid=45&aid=8&cid=&syid=2007&eyid=2011&unit=MTCDPP

IES, (2013). International Energy Statistics. U. S. Energy Information Administration. http://www.eia.gov/cfapps/ipdbproject/IEDIndex3.cfm?tid=90&pid=44&aid=8

IRENA, (2012)."Renewable Energy Cost Analysis – Solar Photovoltaics". International Renewable Energy Agency, IRENA. Report June 2012.

Khabit, T., (2010). "A review of designing, installing and evaluating standalone photovoltaic power systems". *Journal of applied sciences* 10(13). pp. 1212-1228.

Kosyachenko, L. A., (2011). "Solar Cells – Thin – Film Technologies". Publishier InTech. Open access book.

LG Solar, (2013). Solar Panel. Mod. Mono X NeoN LG280N1C. http://www.wholesalesolar.com/products.folder/module-folder/LG/LG280N1C.html

NREL Poly, (2013). National Renewable Energy Laboratory (NREL), Polycristalline Thin-film Photovoltaic System. http://www.nrel.gov/pv/thinfilm.html.

NREL, (2013). National Renewable Energy Laboratory.

REN, (2011). "Renewables 2011. Global Status Report". REN 2011. http://www.ren21.net/portals/97/documents/GSR/GSR2012_Master18.pdf.

van Appen J., et al., (2013). "Time in the Sun. The Challenge of High PV Penetration in the German Electric Grid". *IEEE Power & Energy magazine*. Vol. 11. Number 2. March/April.

In: Energy and Environment Nowadays
Editors: Luis G. Torres and Erick R. Bandala
ISBN: 978-1-63117-398-1

Chapter 9

ANALYSIS OF AN AFFORDABLE RURAL ECOLOGIC HOME FOR MARGINAL COMMUNITIES IN THE SIERRA NEGRA ZONE, STATE OF PUEBLA, MÉXICO

***Benjamín Sánchez Andrade*[1], *Marie Hoepfl*[2], *Juan Gabriel Garcia Maldonado*[3] *and Benito Corona Vásquez*[4]**

[1] Department of Civil Engineer and Sustainable Development, Instituto Tecnológico y de Estudios Superiores de Monterrey campus Puebla, Puebla, Puebla, México
[2] Department of Technology and Environmental Design, Appalachian State University
[3] Mexican Institute for Water Technology, Jiutepec, Morelos, México
[4] Department of Civil Engineer and Environmental, Fundación Universidad de las Américas Puebla, Cholula, Puebla, México

INTRODUCTION

The design process of a building is an integral part of the larger and more complex building procurement process through which an owner defines facility needs, architectural possibilities, and services (Stein et al., 2006). Because the expected lifetime of a building is usually long, decisions about a building's design have long-term consequences. For instances, an appropriate building design can reduce the energy consumption by incorporating passive solar techniques, or by creating a good thermal boundary between interior and exterior and taking advantage of any natural possibility to assist space heating, cooling or lighting. Moreover, a good design must be oriented to give the inhabitants a safe, healthy and comfortable living space during any season of the year. Of course, building designs can change and must be adapted according to the place and the environmental characteristics. This is very important since it could determine the success or failure of a particular project.

The main objective of this study was to develop a rural home design for marginal communities in the region of Sierra Negra, State of Puebla, México. The idea is focused on providing an energy-efficient home design, at the lowest possible cost. Additional considerations included the use of local materials and labor.

In addition to improving the living conditions of the groups mentioned above, it is expected that the results of this project could provide useful information to improve the way in which different public aid programs in Puebla are applied. These programs are focused on supporting the construction, expansion, and improvement of rural homes, with the goal of reducing asymmetries in life conditions between rural and urban populations. The financial resources of these programs come from federal and state subsidies, and they are applied by public agencies like the Department of Social Development of México (SEDESOL), which have the mandate of giving economic support to people who need financial aid. Also, these programs incorporate different approaches supplying worthy roofs and floors, temporary employment, work and training, among others (Schumacher, 2006). The problem is that most of the time the economic resources are applied without an energy efficient approach, so inevitably the budget is spent in an improbable way. This is important because the same resource assigned to the rural home construction could achieve a better level of service to the habitants, just by proposing a suitable design that takes advantage of natural conditions.

The Sierra Negra inhabitants live in marginal conditions. According to Sotelo Díaz (2004), 49% of population does not have access to clean drinking water, 63% does not have a suitable drainage system, which is only available in few parts, generally in the municipality centers, and 56% of inhabitants lack access to electrical energy. Concerning home building, 76% of the population live in houses with earthen floors, adobe or wooden walls in bad conditions without any security and are highly uncomfortable. It is also common to find ceilings of pegboard, tile, sheet or tejamanil roofs, and sometimes the use of sheer maguey leaves. The principal sources of house heating are wood stoves most of times even located inside the building, and it is quite common to find wood cookers. In winter season snow water is wetting the firewood, often the unique resource of the habitants to generate heat in their homes.

In the state of Puebla, the regional economic unbalance is one of the fundamental problems. As we can analyze, there are many needs and necessities that require immediate attention in these communities. The problematic of Sierra Negra is focused in marginized living conditions that are generated by poverty, lack of productive alternatives and a higher percentage of illiteracy.

Bioclimatic Dwelling Design in Mexico

With the commitment of increasing the quality of life of people living below poverty levels, the Mexican Institute for Water Technology (IMTA) developed a model and constructed a prototype of an ecologic housing named "La Casa Ecológica" ("The Ecologic House"). The objective of this model was to evaluate the suitability of appropriate technology concerning the water supply and its sanitation; and to further develop technologies that could help to improve the quality of life in rural homes. This housing model is located inside the IMTA facilities in Jiutepec, Morelos, México (See Figures 1 and 2).

Figure 1. Exterior of The Ecologic House by IMTA.

Constructed wetland household and home garden.

Figure 2. Back of The Ecologic House by IMTA.

It represents an ideal ecological housing model for rural, indigenous and marginal communities. The material used in the construction of this dwelling is adobe, which is of common use in rural zones. The house is made with adobe bricks with double width. The walls have reed sticks as vertical and horizontal reinforcement. The adobe provides benefits like thermal and acoustic insulation, due to its low caloric conductivity rate and good acoustic absorption. Also, this material is commonly elaborated by the habitants and no additional energy is required to transform it into a construction material. For these reasons is low cost and easy to furnish. The ceiling inside of the house is made of wood while outside of the house is made with fiber-cement sheets. Polystyrene sheets were placed between these two materials as insulation.

As a first step of this project, the thermal behavior of the prototype of the ecologic home built by IMTA was tested with Revit Architecture©. The results of the analysis were expected to illustrate differences in energy efficiency and the cost savings that can be achieved principally through variations in the application of passive solar techniques and an appropriate bioclimatic design.

ENERGY AND THE BUILDING SHELL

Heat flows through the building by two mechanisms: transmission and air leakage. Heat transmission is driven by the temperature difference between indoors and outdoors. A building's thermal resistance determines how much heat transmits through the shell. The resistance to heat transmission means that the shell resist conduction, convection, and radiation heat flow. This heat transmission is directly linked with the conductivity of the building materials on walls, roofs windows and floors. Thermal conductivity allows comparing the level of heat conductivity of common building materials. For this study, the thermal conductivity information for the different building materials used in The Ecologic House was taken from the official Mexican standards NOM-008-ENER-2001 ("Energy efficiency in buildings, thermal envelope for non-residential buildings").

Building-shell components, such as wall and roof assemblies, contain a variety of materials in their cross-sections. For that reason, the average thermal resistances for each component are an area-weighted average and it is necessary to calculate the overall coefficient of heat flow, well known as "U-value". The U-value is the rate of heat flow through an assembly (window, wall, among others.) bounded by air on both sides, includes the effects of all materials, air films, and air spaces (Stein et al., 2006). The Imperial unit heat flow is Btu/h-ft^2-°F, and for the International System of Units (SI) is W/m^2*°C. The procedure to calculate the overall coefficient of heat flow is described in the Mexican Official Standard NOM- 008-ENER-2001. Table 1 shows a summary of the characteristics of building materials of the Ecologic House. Tables 2 and 3 illustrate the assembly U-value and R-value with the area-weighted average, for roof and external walls of The Ecologic House, respectively.

According to Krigger & Dorsi (2009), blower- testing of doors is the most practical way to predict energy savings from air-sealing methods. The concept is simple: the blower door pressurizes the home by blowing air in or depressurizes it by sucking air out. The combined area of the building's holes and the pressure difference between indoors and outdoors determine how much air volume the blower door blows. This airflow is measured in cubic feet per minute (CFM). The standard measure for a home's air leakage is the airflow through a blower door at 50 pascals of house pressure. Tables 4 and 5 show the experimental measurements obtained from The Ecologic House blower-door test (See Figures 3 and 4). Table 6 summarizes all the calculated factors for the Ecologic House.

Table 1. Building materials specifications of The Ecologic House

Material	Conductivity (W/m.K)	Thermal resistance R- value (m^2.K/W)	Heat Value U-Value (W/m^2.K)	Thermal Resistance R-Value (ft²-hr °F/Btu)	Conductivity (Btu/ft-hr °F)	Thickness (m)	Density (Kg/m^3)
Adobe*	0.930	0.419	2.38	2.381	0.537	0.390	-
Mortar gypsum*	0.372	0.067	14.88	0.382	0.215	0.025	800
Cement-sand mortar*	0.630	0.008	126.00	0.045	0.364	0.005	2000
Expanded polystyrene**	0.039	0.128	7.80	0.728	0.023	0.005	-
Wood*	0.162	0.154	6.48	0.876	0.094	0.025	663
Door*	0.162	0.278	3.60	1.577	0.094	0.045	663
Windows***	0.025	0.162	6.18	0.919	0.014	0.004	-
Concrete*	1.740	0.086	11.60	0.490	1.005	0.150	2300

*Diario Oficial de la Federación, 2001.
**Secretaría de Estado de Vivienda y Actuaciones Urbanas, 2008.
***2011 Autodesk©, 2011.

Table 2. Assembly U-value and R-value for the roof of The Ecologic House

	R-Values (ft²-hr °F/Btu)		
Roof Component	**Frame**	**Insulation**	**Gaps**
Afi	0.6800	0.6800	0.6800
Wood ceiling	0.8763	0.8763	0.8763
Framing/Insulation/Gaps	0.8763	0.7280	1.0000
Fiber-cement sheet	0.0451	0.0451	0.0451
Afo	0.1700	0.1700	0.1700
Sum R	2.6477	2.4994	2.7714
Sum U	0.3777	0.4001	0.3608
Area %	0.1170	0.8510	0.0320
Area Wtd U	0.0442	0.3405	0.0115
Assembly U	0.3962	BTU/F-ft²-h	
Assembly R	2.5239	°F-ft²-h/BTU	

Table 3. Assembly U-value and R-value for external walls of The Ecologic House

Wall Component	R-Values (ft²-hr °F/Btu)	
Afi	0.6800	
Finish	0.0451	
Adobe	2.3813	
Afo	0.1700	
Assembly R	3.2764	°F-ft^2-h/BTU
Assembly U	0.3052	BTU/F-ft^2-h

Table 4. Air flow rate calculation

A (m^2) =	0.022
V (m/s) =	26.36
Q (m^3/s)=	0.58
Q (CFM)=	1234.26

Table 5. Pressure difference between the home and outdoors

Micromanometer			
50 pa = 0.51 cm H_2O			
h_1 (cm)	5.759	5.661	5.675
H_2 (cm)	5.817	5.712	5.737
H cm H_2O	-0.058	-0.051	-0.062
Pressure (pa)	-5.686	-5.000	-6.078
Average (pa)	-5.588		

Table 6. Air leakage factors for the Ecologic House

$CFM_{5.6}$	1234.26
CFM_{50}	5128.71
CFM_{nat}	271.36
CFM_{ASHRAE}	43.24
Degree of Pro	18.90
Footprint (f^2)	573.52
Volume (ft^3)	5517.90
Number of person	5.00
ACH_{50}	55.77
ACH_{nat}	2.95
ACH_{ASHRAE}	0.47
Effective Leakage (in^2)	492.56

SOURCE: Own creation.

Figure 3. Blower door testing – “The Ecologic House”.

Figure 4. Blower door testing – “The Ecologic House.”

LOCATION SELECTION

In order to run an energy analysis it was necessary to establish the location for the project and provide local weather information. Nineteen communities are settled in the Sierra Negra zone. The criteria to choose the location of the simulations was through determining the community with most heating needs. The need of heating in this zone is important due very low temperatures during the winter season and the lack of economic resources, which obligates the inhabitants to put improvised fireplaces inside their buildings to generate heat (Quiroz, 2010). Unfortunately these practices have caused deaths because intoxication with carbon monoxide. In other cases, it has been reported casualties by hypothermia. A heating degree-day (HDD) is a unit of measurement used by heating engineers to describe how long the outdoor air temperature is below 65°F (18.3°C), which is the minimum limit of the comfort range, during each day, month, or year (Krigger & Dorsi, 2004). To define a comfort range between 19°C to 23°C to rural dwelling design does not make a lot of sense, because the economical resources of the families in these dwellings are not enough to afford it, in a strict way. For this reason the range could be extended to 16°C and 26°C between 8:00 to 22:00 hours depending if the dwelling is inhabited or not (CIBSE, 2006). Also, this proposed schedule takes into account the period of time that people are awake, because in the sleeping hours people are tending to use blankets and covers, increasing hereby their tolerance to low temperatures.

With the weather information from 2009 and 2010, provided by the Mexico's National Commission of Water (CONAGUA), it was possible to determine the heating degree days for all the communities which have installed a weather station in the zone, in order to pick the one with the worst conditions. As displayed below, the worst scenario was for the community of Zoquitlán (HDD 2082 °C = 3747 °F). The Table 7 summarizes the calculations to determine the heating degree-days with the weather information of 2009 of the weather stations. With the software mentioned it was possible to specify the location of the project and select the appropriate weather station using the Internet Mapping Service. Weather stations include "actual year" virtual weather stations and typical year weather stations (TMY2 and other formats) based on 30-year averages of weather data, typically taken from airport locations (Autodesk©, 2011).

Heat flowing out of and into buildings is a major energy consumer. The need for heating, called heating load, is how many BTU's per hour (BTU/h) in Imperial units that need to be added or removed to provide comfort. In the International System, the unit used for this purposes is the Joule (J). As it was mentioned above, the two major components of the heating load are heat transmission and air leakage. Internally generated heat and solar heat are subtracted from heat load if they are significant. Heat loss is the number of BTU's flowing through the building shell monthly or annually. Heat loss is a measurement of energy expended to heat the building. Cooling load is the number of BTU/h the cooling system needs to remove during the hottest summer weather; it is used to determine the power of the cooling system. The approach of this project is to focus the attention in how to reduce the heating load of the building, because the low temperatures are the principal factor that makes people to use inappropriate practices of heating. Also, low temperatures represent a higher death risk. On the other hand, this kind of rural dwelling is directed to people with low economical income, so the implementation of a cooling system is not affordable.

Table 7. Heating Degree Days for Sierra Negra Communities

	Municipality	Weather Station	HDD (°C)
1	Palmar del Bravo	Palmar del Bravo	396.50
2	Caltepec	Acatepec	376.00
3	Tecali de Herrera	Ahuatepec	113.00
4	Alcomunga	Alcomunga	1264.50
5	Coxcatlan	Calipan	14.00
6	Caltepec	Caltepec	121.75
7	Chapulco	Chapulco	190.00
8	Vicente Guerrero	San Bernardino Lagunas	453.70
9	Tecamachalco	Tecamachalco	497.15
10	Tehuacan	Tehuacan	189.50
11	Telpatlan	Telpatlan	999.00
12	Tlacotepec de P. Diaz	Tlacotepec de P. Diaz	0.00
13	Xochitlan Todosantos	Xochitlan Todosantos	25.00
14	Zapotitlan Salinas	Zapotitlan Salinas	8.00
15	Zoquitlan	Zoquitlan	2082.00
16	Molcaxac	Molcaxac	77.00

SOURCE: Prepared with information from CONAGUA (2011).

VALIDATION OF THE MODEL

The Figure 5 is the first model tested in the software Revit Architecture© that is the original prototype of The Ecologic House by IMTA. Before started the optimization simulation process, it was necessary to validate the original model through the comparison of the thermal performance results generated by Autodesk© and the results obtained analytically. It is important to say that once the study location was established, this software does not allow to modify the climatological information that it uses to do the modeling.

The first stage of the validation process was to compare some weather information that the software uses and the weather information reported by CONAGUA. Revit Architecture© uses the climatologically information of a virtual weather station selected to do the modeling; and for the analytical procedure the weather information provided by CONAGUA was used. Because Autodesk© has rights reserved of the weather information it manages, the only information that its servers shows related to climate data, is the chart of monthly design data, see Figure 6. The Figure 6 displays the comparison between these two sources of information. Both graphics display the same tendency on the monthly average daily min/max temperature. However there are little offsets, both positive and negative on a monthly basis. This has to be considerate in the data analysis of the validation simulation.

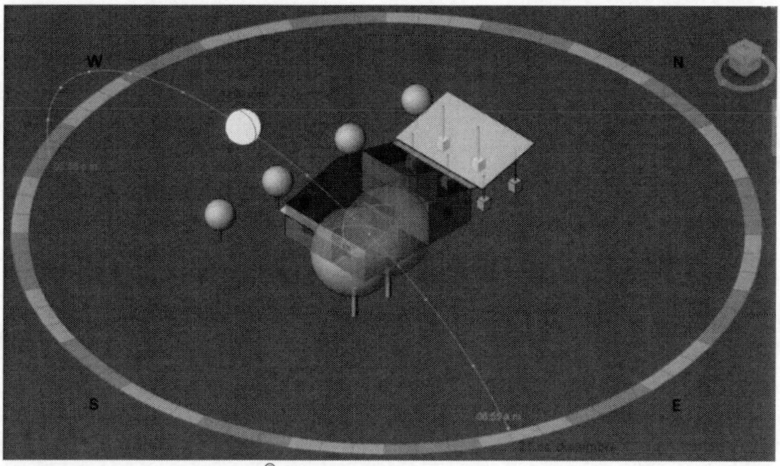

SOURCE: Autodesk Revit Architecture©, 2012.

Figure 5. Original virtual model - The Ecologic House.

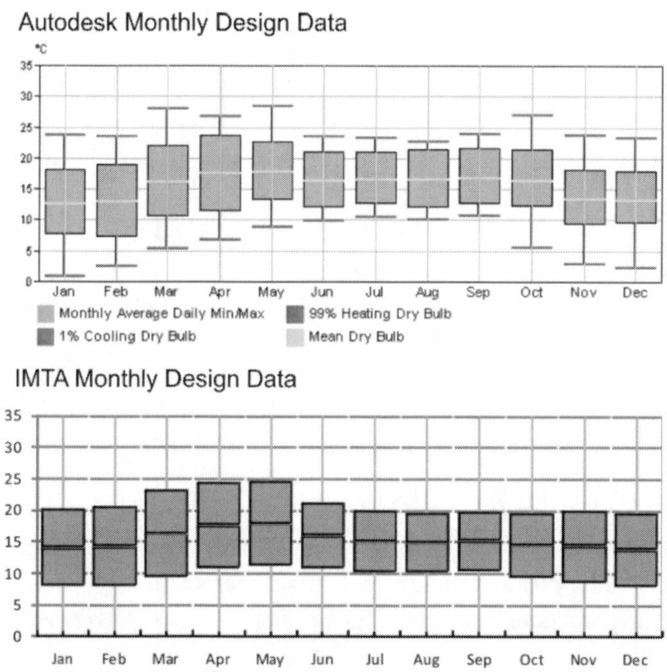

SOURCE: Autodesk Revit Architecture©, 2012.

Figure 6. Monthly average daily min/max temperatura design data.

Regarding the thermal characteristics of building materials, Table 8 displays the differences between the thermal characteristics of the building materials available on the software and the actual ones in local building stores. In order to be consistent, it was necessary to use the software data of the building materials, to do the calculations by the analytical way to determine the monthly heating load, and to compare them with the results obtained by Revit Architecture©. Figure 7 summarizes and compares the results of the monthly heating load, calculated the analytical way, and the results obtained by software modeling. In both graphics the results show the same tendency that match the two procedures applied to determine the monthly heating load. Also the top three elements that cause the major percentage of heat loss are roofs, walls, and windows, in this order. Regardless that both graphics are showing the same tendency, there are significant differences on the percentage distribution and the accumulated heat loss per month.

Table 8. Building materials thermal characteristics

Mass Surface	Revit R-value (ft^2-hr °F/Btu)	Calculated R-value (ft^2-hr °F/Btu)
Mass exterior wall	3.27	2.81
Mass roof	2.52	1.99
Mass glazing	0.9188	0.9188
Mass floor	4.20	0.49

SOURCE: Prepared with information from Autodesk Revit Architecture©, 2012.

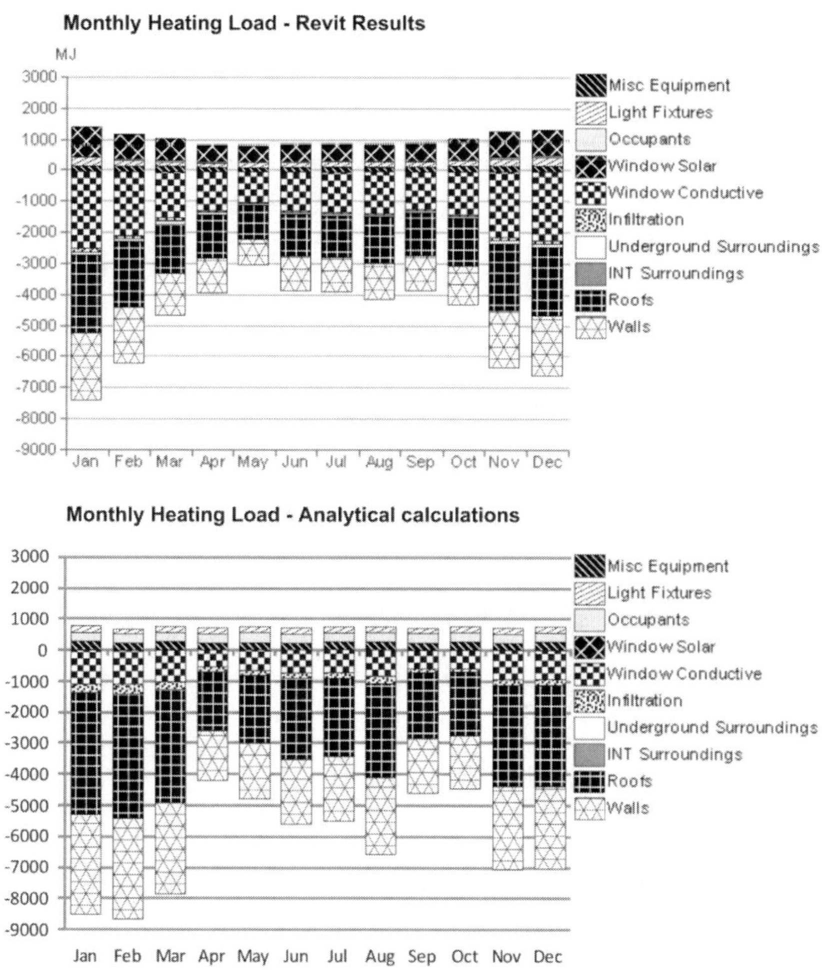

SOURCE: Prepared with information from Autodesk Revit Architecture©, 2012.

Figure 7. Monthly Heating Load for the Ecologic House.

It is necessary to say that, because of the lack of official climatological information for the chosen place, in some cases it was not possible to determine the heating loads that Revit Architecture© analyzes: Like the window-solar heating gain, underground surroundings and INT surroundings. In the same way, in order to validate the comparison in a better way, it was necessary to use the same air-infiltration of 21.79 CFM_{nat} proposed by the software, in both cases, the software results and the analytical calculations. This is one of the principal limitations of the software, it is not possible to use a specifically calculated CFM_{nat} for buildings. Because the difference between the assumed CFM_{nat}, and the real one is substantial, the calculations results are displayed in Figure 7 with the same CFM_{nat} value of 21.79, just with illustrative ends. The charts in Figure 7 show the cumulative heating loads for the analyzed model in each month. Heat loss through roofs, walls, windows, and infiltration represent the monthly demand for heat. However, miscellaneous equipment, occupancy, light fixtures, and window solar gains reduce the demand for heat. In order to reduce the heating load on the project, it is possible to use the graphs in Figure 7 to identify the critical components and focus the attention and efforts to improve their performance.

All the simulations were done under the same conditions, in order to be consistent with the input data. However, it was important to make calculations to determine the monthly heating load with the real information of locally available building materials, their thermal characteristics and real CFM_{nat}. Figure 8 presents the results of these computations. As it was mentioned above, Revit Architecture© uses specific values for air infiltration depending the building type selected. For this case the air infiltration value is 21.79 CFM_{nat}. The real air infiltration value for the Ecologic House is 271.36 CFM_{nat}, according to the calculations in the blower door testing section. This characteristic is one of the major differences between the real model and the simplified model created in the software, and that was the principal cause of the difference between the final results. The other elements of the model show the same tendency but in different percentages.

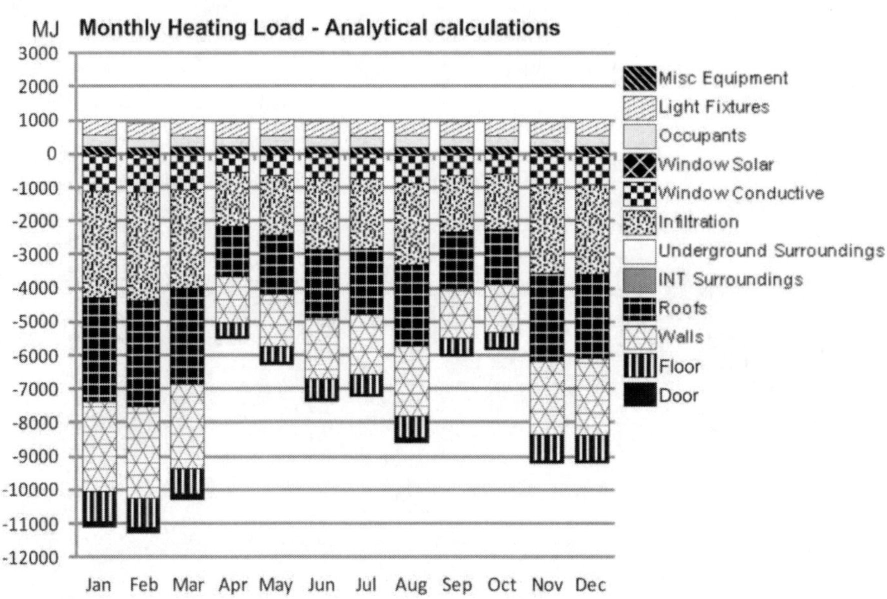

Figure 8. Monthly Heating Load for the Ecologic House (Real thermal values).

SOFTWARE SIMULATION FOR PROTOTYPE OPTIMIZATION

Modeling systems and computer software tools are commonly used in the construction industry in order to assist in the evaluation of a building's thermal performance and its associated environmental impacts. Since the early 1960s, their use has gradually increased in particular during design, construction, and operation stages of the building life cycle. For the purpose of this study, the evaluation of the thermal performance and its associated environmental impacts were modeled with the software Revit Architecture© as the basic framework to determine the relationship between sustainable, energy efficient features and the thermal performance of the building. The modeling software needs input data to do the thermal analysis.

The first parameter to modify in the simulation process was the orientation of the building. According to Morrissey et. al. (2001), the correct orientation of a building could impact significantly its thermal performance. The first step was to match the prolonged axis of the housing to match the east-west orientation. The original model has the extended axis of the building to the north-south direction, completely in the opposite direction of the commonly recommended orientation. This change was made by rotating the original model clock wise by 30°, 60, and 90°, in order to analyze the response of the building due to the gradual orientation change. Also this change represents a better bioclimatic design, because the bedrooms are located in the sun-heated zone, which is the south face of the building.

The first simulations, from first to fourth, with a gradually modified orientation did not show any improvement on the thermal performance of the dwelling. This failure was somehow expected because regardless that the building now has the appropriate orientation, the orientation of the roof slope and the overhanging of the original design does not match with this criterion. For this reason the logical next step was to modify the orientation of the roof slope and its respective awning. In order to analyze the thermal performance between the first and the second design, it was necessary to run other four new simulations in the same conditions than the firsts four. Table 9 presents a summary of the number of simulation and the changes accumulated during the optimization process, showing the energy use intensity results for each case.

After these next four simulations, the building showed a better thermal performance along the year. Comparing the results of the original case and the modified one in the simulation #8, the energy use intensity decreases from 819 to 584 MJ, in total. The thermal performance of walls and roofs improved in both, the cooling and heating load. On the other hand, with the new design the major heat losses and gains are through window conductance and solar gain, respectively. These results make sense because in the last simulation a major glazing percentage is facing the south. At this point, there are opportunity areas to improve the thermal performance, on controlling the glazing percentage per face of the building, the glazing type, and the overhanging length. To deal with problems of glazing, it was necessary to reduce its percentage to the minimum. It is true that the excess on glazing helps to gain heat along the winter season, but the results shows that there are severe problems with overheating of the building during the day, along the year in these conditions. In the simulation #9, the glazing on the east side was eliminated. In the simulation #10, the glazing on the west side was eliminated.

Table 9. Number of simulation versus conditions modeled

	Energy Use Intensity Results (MJ/sm/yr)																						
Number of simulation	Original Case	Orientation 30º clockwise spin	Orientation 60º clockwise spin	Orientation 90º clockwise spin	Original case with reoriented roof	Reoriented roof, 30º clockwise spin	Reoriented roof, 60º clockwise spin	Reoriented roof, 90º clockwise spin	Without east glazing	Without west glazing	External sheets reoriented	Proposed overhang 16 cm length	Proposed overhang 26 cm length	Proposed overhang 36 cm length	Proposed overhang 56 cm length	Proposed overhang 76 cm length	Maximum percentage of glazing area, north side	Maximum percentage of glazing area, south side	Minimum percentage of glazing area, south side	Bathroom window	Internal ceiling attached	Relocation of the vegetation	New vegetation as windbreak
1	761																						
2	X	826																					
3	X		898																				
4	X			906																			
5					819																		
6					X	806																	
7					X		723																
8					X			584															
9					X			X	581														
10					X			X		413													
11					X			X		X	413												
12					X			X		X	X	413											
13					X			X		X	X		413										
14					X			X		X	X			413									
15					X			X		X	X				413								

Table 9. Number of simulation versus conditions modeled

	Energy Use Intensity Results (MJ/sm/yr)																						
Number of simulation	Original Case	Orientation 30° clockwise spin	Orientation 60° clockwise spin	Orientation 90° clockwise spin	Original case with reoriented roof	Reoriented roof, 30° clockwise spin	Reoriented roof, 60° clockwise spin	Reoriented roof, 90° clockwise spin	Without east glazing	Without west glazing	External sheets reoriented	Proposed overhang 16 cm length	Proposed overhang 26 cm length	Proposed overhang 36 cm length	Proposed overhang 56 cm length	Proposed overhang 76 cm length	Maximum percentage of glazing area, north side	Maximum percentage of glazing area, south side	Minimum percentage of glazing area, south side	Bathroom window	Internal ceiling attached	Relocation of the vegetation	New vegetation as windbreak
16					X			X		X	X					413							
17					X			X		X	X	X					413						
18					X			X		X	X	X					X	420					
19					X			X		X	X	X					X		413				
20					X			X		X	X	X					X		X	413			
21					X			X		X	X	X					X		X	X	477		
22					X			X		X	X	X					X		X			413	
23					X			X		X	X	X					X		X	X		X	413

After the simulation #10, the thermal performance of the house improved moving the energy use intensity from 584 to 413 MJ. The elimination of the glazing on the east and west decreased the heating and cooling load, improving the performance on window permittivity and window solar gains. At this stage the monthly heating load chart resulted more balanced between heat losses and gains, not the case with the monthly cooling load. The simulation #11 consisted in the reorientation of the external sheets. The external sheets are not attached to the dwelling; they are used just as shadows devices for the inhabitants. This change was made only to be coherent to the overall design. This means, that regardless that the reorientation does not help to improve the thermal performance of the building, they represent more so a future renewable energy potential, for example through the installation of photovoltaic solar panel system, solar thermal system, among other possibilities.

Regardless that the performance of solar gains and losses through the glazing has improved; an appropriate overhang design could take advantage of solar gains during the winter and reject them during the summer. For this reason the next step was to design an optimum overhanging for the Ecologic House. In this modeling software it is possible to develop solar studies. By showing the impact of natural light and shadows on the project, solar studies yield valuable information that can help support effective passive solar design. Also it is possible to visualize how shadows from terrain and surrounding buildings affect the site, and where natural light penetrates a building during specific times of the day and year. An appropriate shadow device is the one that allows the entrance of the sunlight through glazing during the winter season and block it during the summer season. The overhang has to allow the full entrance of the sunlight at the time of the winter solstice when the position of the sun is low, and long enough to block it completely during the time of the summer solstice when the position of the sun is high.

Because Revit Architecture© is a simulator, it is possible to manipulate the settings to specify the location of the building in the globe, and the specific times of the year to visualize the solstice sunlight, and the shadow patterns that it creates. With the help of this tool, it was possible to determine that the original overhang length of 25 cm does not satisfy the criterion of an appropriate overhang length. In contrast, a new proposed length of 16 cm showed an appropriate interaction between the position of the sun and the shadow patterns created along the year.

The thermal simulations from #12 to #16 were concerning to the overhang length variation, to analyze the impact of this characteristic in the energy load. The simulations #12, #13, #14, #15 and #16 were performed increasing the overhang length gradually from 16, 26, 36, 56 and 76 cm respectively. According to the results, the overhang length that fits better to the proposed design is the simulation #12 (16 cm length). This overhang length takes advantage of the direct solar gain during the winter season and rejected it during the summer season. The above, helps to balance the heating load along the year. Also this design promotes natural lighting inside the home. In the same way, simulations #12 to #16 showed that there are potential on reducing the heating load caused by the windows conduciveness. However, the only way to improve this characteristic is through changing the window type. There are several options on the market that could reduce the energy consume of the house, like double and triple panel glazing and storm windows. However these items are above the cost effectiveness for this low-cost rural dwelling design.

The next stage of the modeling was to continue modifying the percentage of glazing by face of the building. It is important to say that the overhang length rules the windows height.

To modify the percentage of glazing, the windows width was investigated. Simulation #17 performed the thermal behavior of the dwelling with the maximum percentage of glazing on the north face of the building, 5% of the footprint of the building. The location of this glazing did not affect the current thermal performance of the dwelling. What is more, the position of these glazing is an important source of natural lighting diffusion, without solar gains. For this reason we consider the proposed change acceptable.

Simulations #18 and #19 performed the thermal behavior of the home with the maximum and minimum percentage of glazing on the south face of the building, 12% and 5% of the footprint of the building respectively. The maximum percentage, impacts the thermal performance increasing the heat gains through window conductance, and the minimum shows the opposite. This makes sense because the glazing is the house element with the lowest thermal resistance values. So the increase or decrease of its percentage will impact directly the thermal performance of the building. The minimum percentage of glazing on the south face of the dwelling is the one that balances in a better way the thermal performance and the service that it provides, like natural lighting and ventilation. In simulation #20, the window of the bathroom was reinstalled because of the need of ventilation in this room. The results did not show a significant increment on the heating or cooling load because it is a small area. For this reason this modification was acceptable.

The next suggested change for the Ecologic House was the installation of an internal ceiling that acts like a thermal buffer between the roof and the different zones of the home, to reduce the heating and cooling load of the roof. The results of this simulation #21, displayed that there was no decrease on heating or cooling loads because of the roof, as it was expected. In contrast, the internal ceiling caused a problem of overheating on internal and external walls and the general energy consumption increased. Also, the thermal behavior of the other elements of the dwelling was modified. The installation of an internal ceiling did not give any benefit to the thermal performance of the dwelling. On the other hand, this represents an extra cost for the building. For that reason this modification was dismissed.

The last modifications for the Ecologic House model were regarding the surroundings. In simulation #22, it was suggested the relocation of the existing vegetation in a favorable place. This modification had the objective of improving the thermal performance of the dwelling during the summer and winter season. The existing trees had to be located in the south face of the house near to the glazing. The distance between the trees and the house had to be enough to take advantage of the shadows created by them. The leaves blocking the entrance of sunlight through the glazing, depending on the tree height and wide. It was critical for this modification to suggest deciduous trees at this face of the building. In this way, the foliage that the trees reach during the summer, block the sunlight avoiding overheating problems in this season. On the contrary, during the winter when this kind of vegetation loses their foliage, the sunlight enters freely, increasing heat gains. It is important to clarify that Revit Architecture© does not simulate the dynamic of the vegetation's foliage mentioned above. The representation of vegetation in this software only functions as permanent shadow devices. For that reason the results did not show the real impact of this retrofit on the thermal performance.

In the last simulation #23, Figure 9, it was proposed to set new vegetation as windbreak. According to the information gathered from Revit Architecture©, the predominant winds for this location come from the northeast. Figure 10 shows the wind information to perform in the modeling. The reason to set some new vegetation on this side of the home is to block the cold wind on winter season principally. The appropriate vegetation for this objective has to be evergreen trees or conifers at the north east of the building. The evergreen trees do not lose their foliage during the winter season and they act as a natural and esthetic windbreak.

Energy Use Intensity

Electricity EUI:	77 kWh / sm / yr
Fuel EUI:	137 MJ / sm / yr
Total EUI:	413 MJ / sm / yr

Monthly Heating Load

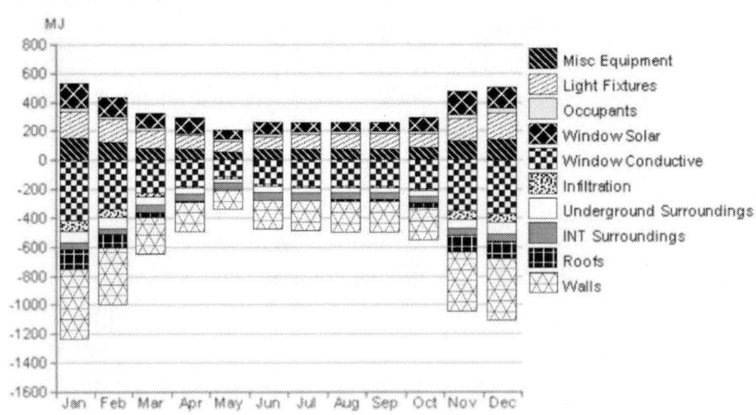

Monthly Cooling Load

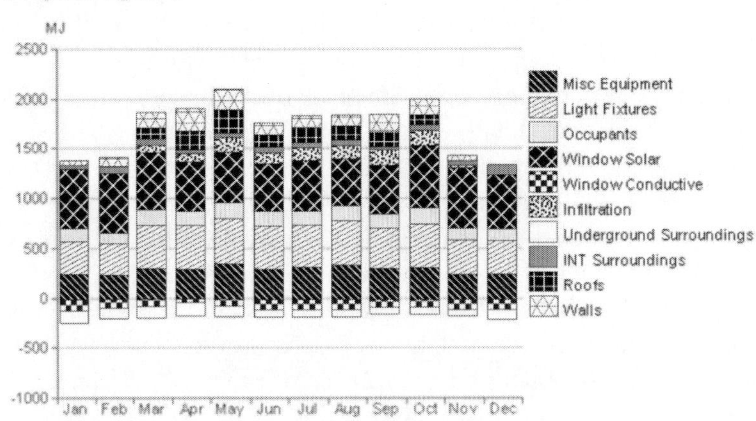

SOURCE: Autodesk Revit Architecture©, 2012.

Figure 9. Simulation 23 – New vegetation as windbreak.

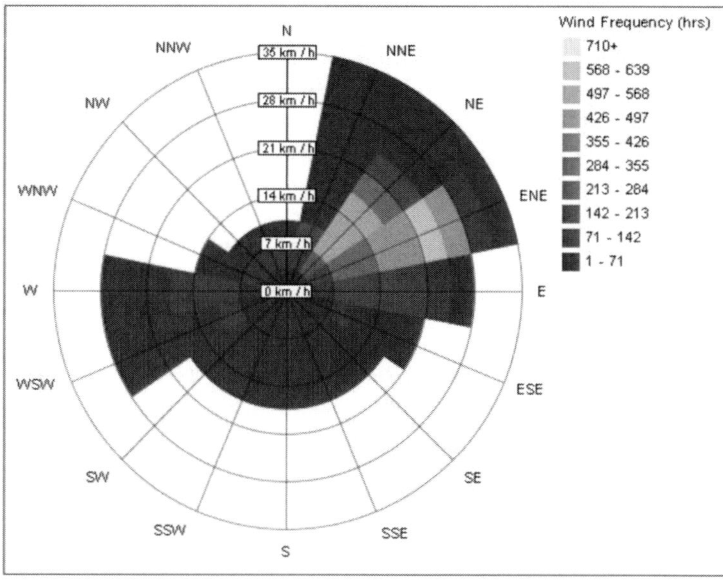

SOURCE: *idem.*

Figure 10. Annual wind rose (frequency distribution).

CONCLUSION

Results from software simulations suggest that it is possible to improve the thermal performance of a building along the year through the appropriate application of bioclimatic techniques.

The final comparison of the monthly heating loads calculated by the software (see Table 20) shows the improvement on the thermal performance of the dwelling throughout the year. On a monthly basis, the load balance is the result of the difference between the heat gains and losses. In both cases, the original and the final design, there are more heat losses than gains. However, the proportion difference is more than significant. Month by month the percentage difference between each case is about 156% - 175%, showing an improvement on the thermal performance of the final design. The grand total percentage difference between the heat gains and losses is 164%, this number summarizes the final improvement on the thermal performance of the design, concerning heating load.

As illustrated in the graphical and numeric analyses, the improvement on the thermal performance is a balance between heat gains and losses. This means, an appropriate dwelling design would have to take into account both, the techniques to avoid heating losses through the susceptible elements, as well as the techniques applied to gain heat in the same proportion of the losses, through sensible elements. In essence, these are the reasons why "small" changes, well addressed, can be translated into significant benefits on the thermal performance of a building.

It is necessary to say that this study was focused on the heating loads more than on the cooling loads due to the socio-economic conditions of the population and the health problem addressed since the beginning. In other words, the need of heating in this kind of rural dwellings has been responsible of human casualties because of the use of inappropriate practices to keep warm inside a house. For that reason, the cooling loads analysis was out of the scope of this study. Nevertheless, the comparison between the original case and the final design shows a decrease on the monthly cooling loads as a collateral effect of applying an appropriate bioclimatic dwelling design.

The thermal simulation of the construction showed that it is possible to improve the thermal comfort of the inhabitants without the application of artificial means of heating or cooling. What is more, the appropriate bioclimatic techniques do not imply extra costs. Usually, their application only implies a change in the shape and disposition of the different elements that conform the building. The appropriate bioclimatic architecture applied to homes, takes advantage of the sun, vegetation, wind, between others to decrease the environmental impact through reducing the energy consume.

Because this bioclimatic re-design suggests significant changes just in the form, geometry and shapes of the building, it does not represent an increment of the initial capital investment. Furthermore, the subsidies that the government gives to attend the lack of housing on this kind of marginized communities could be applied with an energy-efficiency approach. This in turn, should help the fight against the emission of greenhouse gases to the atmosphere and increase the social benefits for the mentioned communities.

The results of this study suggest that there are possibilities of improvement on the thermal performance of the dwelling through the use of other construction materials or retrofits. However, as it was pointed out before, the use of alternate materials for construction could imply an increase in building cost, which is basically against the implications of the approach of a low-cost solution. For that reason, this topic is considered as a future research objective.

In the same way, there are alternative construction materials that could be attached to the new design in order to make it even more efficient in the resources use, at a low cost. However, the limitations of the software input manipulation and the lack of information about the thermal characteristics of these new alternative materials makes it difficult to incorporate them into the final design. Nevertheless, this topic is equally worthy of future research to delve in the knowledge of the use of this kind of materials.

Finally, it is worthy to say that this study could be replicated in other underdeveloped communities that are spread along the Mexican territory in order to attend the specific needs on rural home construction. There are several marginal communities in México with the same characteristics and housing problems as those observed in the Sierra Negra zone. The methodology applied in this research could be replicated in each community of interest, in order to address the problems of rural housing need. The results of this research suggest that there is a significant potential on the application of bioclimatic building design to meet energy efficiency on these buildings. Therefore, the incorporation of these construction techniques could be applied not only on rural building programs but in the construction building codes for any city in México. Since there are no mandatory energy efficiency building codes in México, the traditional way of construction does not take into account the objectives of efficient resource use. Resulting in energy waste at almost all regular homes in the country. The results of this study suggest that the application of bioclimatic homes design could help to decrease the monthly heating and cooling loads of any building. For that reason, the application of this kind of studies to urban dwelling design in México should be worthy of future research.

ACKNOWLEDGEMENT

We would like to give special thanks to Juan Gabriel Garcia Maldonado and Miguel Ángel Córdova, specialist and subcoordinator of the Mexican Institute for Water Technology respectively, for their attention, help and unconditional support during the development of this research.

REFERENCES

[1] Comisión Nacional del Agua (CONAGUA). *Observaciones climatológicas hechas a las 8 horas de los años 2009 y 2010, de las estaciones climatológicas*. Jefatura de Redes de Medición y sistemas. Información solicitada por el Sr. Benjamín Sánchez Andrade. [2011, July] Puebla, Puebla.

[2] Hirsch J. & Associates. (2011). *Autodesk Revit Architecture© 2012* (Student Version) [Computer software]. USA: 2012 Autodesk, Inc.

[3] Krigger, J & Dorsi, Chris. (2009). *Residential Energy: Cost Savings and Comfort for Existing Buildings* (5th ed.). Helena, Montana: Saturn Resource Management.

[4] México. Norma Oficial Mexicana. NOM-008-ENER-2001. Eficiencia energetic en edificaciones, envolvente de edificios no residenciales. *Diario Oficial de la Federación*. 25 de Abril de 2001. pp. 98-100.

[5] Morrissey, J.; Moore, T. & Horne, R.E. (2010). Affordable passive solar design in a temperate climate: An experiment in residential building orientation. *Renewable Energy. 36 (2011)*, 568 - 577.

[6] Quiroz, P. (2010, january 13). *Sufre la sierra negra el invierno más crudo de los últimos 5 años*. TV Azteca Puebla. [On line], Spanish. From: http://www.tvaztecapuebla.com.mx/ unete.php [2011, february 03].

[7] Schumacher González, M. (2006). *Vivienda rural para campesinos, Barrio la Soledad, Estado de México*. Tesis Licenciatura. Arquitectura. Departamento de Arquitectura y Diseño, Escuela de Ciencias Sociales, Artes y Humanidades, Universidad de las Américas Puebla, Puebla, México. May. p. 51

[8] Secretaría de Estado de Vivienda y Actuaciones Urbanas. (2008). Catálogo de elementos Constructivos del Código Técnico de la Edificación. Instituto Eduardo Torroja de ciencias de la construcción con la colaboración de CEPCO y AICIA. España. Mayo. p. 21

[9] The Chartered Institution of Building Services Engineers (CIBSE) (2006). *Environmental Design CIBSE Guide A* (7th ed.). London, England: CIBSE.

[10] Stein, B.; Reynolds, J. S.; Grondzik, W.T. & Kwok, A. G. (2006). *Mechanical and electrical equipment for buildings* (10th ed.). Hoboken, New Jersey: Wiley.

[11] 2011 Autodesk, Inc. (2011). *Autodesk WikiHelp*. Retrieve October 4, 2011, From http://wikihelp.autodesk.com/Revit/enu/2012/Help/Revit_User's_Guide/2427-Analyze_2427/2428-Conceptu2428/2437-Results_2437

In: Energy and Environment Nowadays
Editors: Luis G. Torres and Erick R. Bandala
ISBN: 978-1-63117-398-1

Chapter 10

A BIOREFINERY BASED ON SEEDS AND VEGETABLE RESIDUES WITH AN INDUSTRIAL ECOLOGY VISION

Luis G. Torres*[*] *and Gemma Cervantes
Depto. de Bioprocesos, UPIBI-Instituto Politécnico Nacional, Mexico

1. INTRODUCTION

Nowadays, there is an increasing interest in the production of coagulant-flocculant agents derived from natural sources. These products are employed in the treatment of municipal or industrial wastewaters. They are preferred instead of the use of Fe/Al salts+synthetic polymers for the following reasons:

1) They are more friendly to the environment and do not introduce metals or hazardous products. Polyacrylamides are partially degraded to their monomer, i.e., acrylamide, which is neurotoxic. On the other hand, industrial Al and Fe compounds contain high concentrations of other metals aside from Fe and Al, which, in turn, have been related to Alzheimer disease.
2) Less sludges are produced with these materials and they are more biodegradable than those produced with synthetic polymers.
3) Their production is not based on the petro-chemical industry (i.e., synthetic polymers).

The biorefinery concept has gained great interest that has been widely developed in the last years. This concept offers a systemic vision of the use of plants and other natural resources, while industrial ecology (IE) offers a systemic vision of human and industrial systems (Cervantes 2007). Through the methodology called industrial symbiosis (IS) (Boons 2011) it is possible to achieve a way to close the material cycles. Applying IE and IS to the biorefinery concept can improve biorefinery efficiency, diminish the use of natural resources, and decrease the economic costs of the biorefinery.

[*] Corresponding author

Although some authors have come close to this approach by using the *feed+fuel* biorefinery concept (Mathews 2011) or by applying IS to industries that may be considered in the biorefineries field, such as the forest industry (Soka 2011), these two concepts have not been linked in the literature. This chapter offers, as a novelty, the IE vision applied to a biorefinery.

2. ANTECEDENTS

Annona muricata and *Annona chirimola* are subtropical fruits that are consumed as fresh fruits in Latin America, known as *guanabana* and *chirimoya*. They are usually sold, after peeling, in markets as fresh fruit pulp. Very rarely are they used to prepare desserts or beverages consisting in water or milk (even soymilk) dilutions (Figure 1). *Prosopis laevigata* and *Delonix regia* are shrubs or trees grown in subtropical altitudes. In Mexico, they are known as *mezquite* and *flamboyan* trees (Figure 1). These leguminous trees produce pods that contain the seeds. These pods are rarely used as food. *P. laevigata* pods have been used to feed goats. The wood of *Prosopis* trees is very appreciated because of its hardness.

Figure 1. Trees, pods, and fruits of a) Prosopis laevigata, b) Delonix regia, c) Annona muricata, and d) Annona cherimola.

Flamboyan trees are very famous for their red-orange flowers that practically cover their crown during the flowering season. Once the pods become mature, the seeds contain small amounts of oil, as well as variable amounts of fiber, protein, ash, and carbohydrates. Basically, the carbohydrates are known as galactomananns, very similar to those found in guar and locust bean gums. These galactomannans have been used as coagulant-flocculant aids.

On the other hand, *Opuntia ficus-indica* is a shrub that grows even in desertic zones of Mexico. There are many species of *Opuntia*, but the most interesting are those grown for their juicy fruits (green-purple color, known as prickly pear) to be sold in public markets. Other ficus-indica-like species are grown to utilize the cladodes as food. They are cleaned, thorns are removed, and the cladodes are cut in strips, boiled, and consumed as part of fresh salads, soups, or meat dishes (Figure 2).

Figure 2. Opuntia ficus-indica shrub.

In recent years, exploitation of *O. ficus* mucilage has been reported. The mucilage contained inside the pods can be used as a texturizing material, to protect fruits against mosquito bites and for other applications. *O. ficus* mucilage use as a coagulant-flocculant aid has been proposed. Our research group (Torres et al. 2013) reported recently different ways to extract the mucilage, as well as the coagulant-flocculant characteristics of the final products.

Opuntia ficus-indica fiber has been employed as nutritional food based on its capability to maintain low levels of glucose in people with diabetes, aside from its reported capability of lowering triglicerides and colesterol levels in blood. Some small factories have dried and milled the *Opuntia* shrub, elaborating capsules that are sold as food supplements.

2. METHODOLOGY

3.1. Industrial Symbiosis Diagram of the Biorefinery's Set Up

This work is based on an industrial ecology approach, using the industrial symbiosis methodology. First of all, a general scheme was drawn for the production of coagualant-floculant aids, and oils as raw material for biodiesel, heat, and energy, A reuse or use as raw material was searched for every residue or by-product. The symbolism of colors was based on that of the Research Group on Industrial Ecology (GIEI) in UPIBI: raw materials in green, energy flows in red, water flows in blue, residue flows used as raw material in brown, products in violet. The main blocks correspond to *Annona cherimola* and *muricata* fruits production (white), *Prosopis laevigata* and *Delonix regia* culture (grey), and *Opuntia ficus-indica* production (black).

3.2. Material and Energy Flows

Using real data, previously reported by our working group or other researchers, the material and energy flows were calculated on the basis of 1 ton of fresh fruit, pods, or shrubs. As in many cases there are more than one alterntative, one of them was selected for comparison purposes.

4. RESULTS AND DISCUSSION

4.1. The Proposed Biorefinery's IS Diagram. Synergies in Material and Energy Flows

The following scheme (Figure 3) was proposed for a biorefineries' vision based on an IE approach. Main raw materials are *Annona cherimola* and *muricata (white), Prosopis laevigata* and *Delonix regia* (grey), and *Opuntia ficus-indica* production (black)

Main products are:

1) coagulant-floculant aids from *A. muricata* and *A.chirimola*,
2) coagulant-floculant aids *from P. laevigata and D. regia,*
3) coagulant-flocculant *from Opuntia ficus-indica,*
4) oil for biodiesel production,
5) desserts from the fruits of *A muricata* and *A.chirimola,*
6) nutritional food from *Opuntia ficus-indica,*
7) compost
8) ashes
9) heat and energy (including biogas).

Some synergies were proposed in this biorefinery scheme for water, energy, and materials including residues. In the case of water, water from centrifugation and drying may be used to cultivate the vegetables. For the energy vector, heat from combustion is used for distillation of wasted solvents. For solid materials, seeds from *guanabana* and *chirimoya* fruits and from mezquite and *flamboyan* pods are pressed obtaining oil for biodiesel production and a cake for the coagulants. Wasted solvents from solvent washing of this paste are recovered and reused through distillation. Pods and integuments from mezquite and *flamboyan* are burnt and the residual ashes used as fertilizer for the plants. Residual biomass from the Opuntia (nopal) shrub is used in three different processes: combustion, composting, and anaerobic digestion to obtain biogas. The produced compost is used as fertilizer. Applying this IE vision to the biorefinery, every material, water, or energy flow is used as a raw material trying to close the cycle.

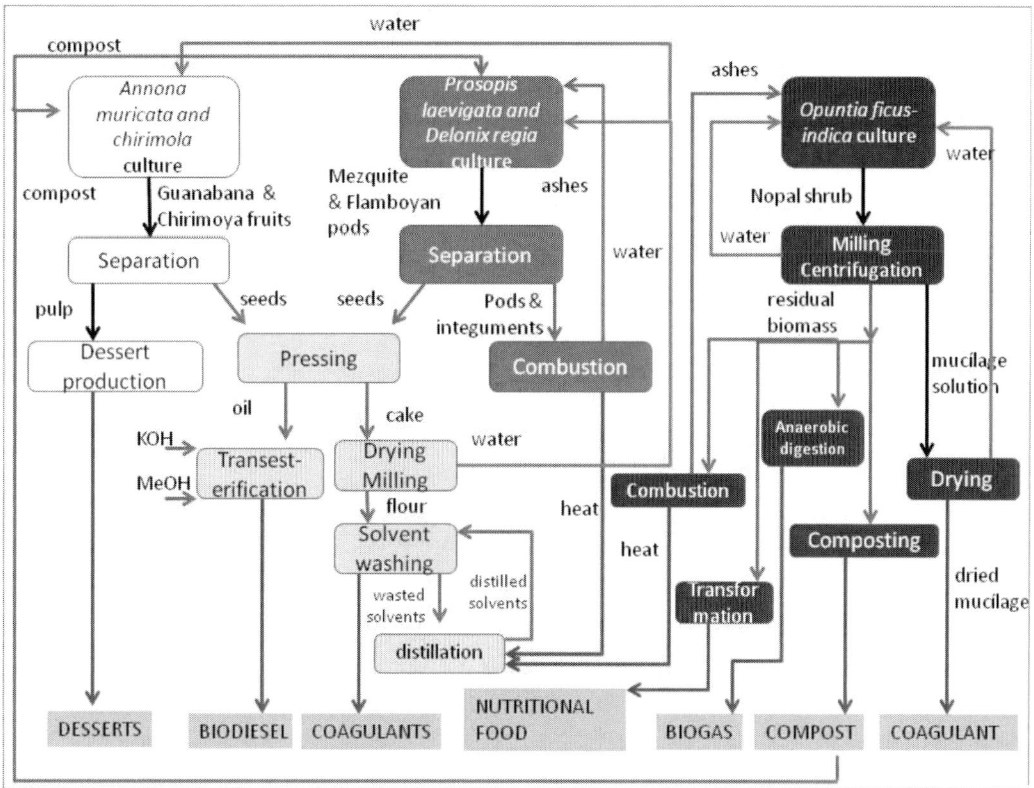

Figure 3. Synergies' diagram in a biorefinery for coagulant-flocculant-aids, oil as raw material for biodiesel production and energy.

4.2. Material Balances

Table 1 shows the weight proportions for fruits/pods and seeds. As noticed, in the case of the tropical fruits from *Annona muricata* and *cherimola*, between 58 and 75% of the mature fruit is fresh pulp, which will be employed for the production of desserts, such as ice creams, fruit beverages, jellys, and other food specialities. Between 4 and 6% of the fruits correspond to seeds (including teguments), which will be employed in coagulant-flocculant production. Regarding the pods of *Prosopis laevigata,* 77.8% of the dry pod correspond to seeds (including teguments) and the rest (22.2%) is a vegetal residue. On the contrary, for *Delonix regia,* only 18% of the pod corresponds to seeds (including teguments) and 82% is a vegetal residue.

Table 1. Weight proportions between fruits/pods and the seeds.

	Pulp (%)	Skin (%)	Seeds (%)*	Residues of the pod (%)	Reference
Annona muricata	75.6	12.7	4.8	-	FAO (2006)
Annona chirimola	58.6	33.2	6.1	-	Rivas (2010)
Prosopis laevigata	-	-	77.8	22.2	Rios et al., (2010)
Delonix regia	-	-	18	82	Corzo-Rios (2012)

*including tegument

Table 2 shows the proximal analysis for the seedsand of the *Opuntia* shrub, regarding humidity, protein, fiber, fat, ash, and carbohydrate contents. It is clear that the seed with the highest oil level is that arising from *Annona muricata* (41%), followed by *Annona cherimola* (19.5%), and *Opuntia* shrub (7.6%). *Prosopis laevigata* seed showed a low oil content, of about 2.8%, and *Delonix regia* seed showed the lowest oil content (0.5%). This means that, starting with 1 ton of the previously mentioned raw materials, it is possible to get between 5.6 and 411.1 kg of oil.

Table 2. Proximal analysis for the seeds and the shrub (*Opuntia ficus-indica*).

	Humidity	Protein	Fiber	Fat	Ash	Carbo-hydrates	Reference
Annana muricata	5.13	18.74	16.29	41.12	2.55	16.99*	Torres et al., (2012)
Annana cherimola	4.43	19.53	18.71	35.75	2.65	23.36*	Torres et al., (2012)
Delonix regia	3.96	2.25	1.87	0.56	0.20	95.12*	Torres et al., (2012)
Prosopis laevigata	NR	11.67	NR	2.87	NR	53.97*	Reveles et al., (2010)
Opuntia ficus-indica	88.7	3.81	1.38	7.62	13.1	74.13	Rossel and Ortiz (2011)

*as NFE = Nitrogen-free extract.

If carbohydrate contents are considered, seeds and shrubs can contain between 16.6 and 95.1% of carbohydrates. The case of seeds and shrubs will be discussed separately. Regarding seeds, *Annona muricata* yields 169 kg of flour per ton of seed, whereas *Delonix regia* yields 950 kg of flour per ton of seed. Torres et al., (2012) reported that the mucilage can be recovered at a proportion of 1 kg of dry product per ton of shrub. Note that the major component of *Opuntia ficus* shrub is water (88.7%), which will be recovered and oriented as water service at other points of the general scheme.

4.3. Energy Balances

Regarding energy recovery from the production steps, there are at least three energy recovery schemes:

a) Energy recovered by the anaerobic digestion of vegetal residues, after hydrolysis or similar processes.
b) Energy recovered from the combustion of dry materials.
c) Energy recovered from the gasification of residual materials.

In this chapter, the two first recovered energies (a and b) will be discussed.

In the case of energy recovery from direct burning of dry material, Table 3 shows a list of products and energy contents (related to the kind of residue and humidity content), as well as the characteristics of ash production and emissions of some elements due to the burning process. There are no values for the specific vegetal residues (fruits teguments/skin and seed teguments/pods), but they can be inferred from the mentioned table. It s remarkable that vegetal residues, i.e., wheat straw, switchgrass, and corn-cob, show final humidities between 7 and 8%. These products contain 18 MJ/kg. Corn stover and hard wood pellets contain 19 MJ/kg, but low-sulphur coal contains 25 MJ/kg. For the vegetal residues included in this work, it will be appropriate to establish 18 MJ/kg, only when humidity content is not higher than 8%. For environmental issues, it must be noted that, besides COx (not reported herein), other products such as NOx, SOx, and Cl will be produced during combustion. The combustion systems employed may have been provided with gas caption devices.

Table 3. Some energy data for the calculations. Adapted from Clarke et al., (2011).

Material	Humidity (%)	Energy MJ/kg	Ash content (%)	Nitrogen (%)	Sulfur (%)	Chlorine (μg/g)
Wheat straw	7	18	7.7	0.8	0.1	525
Corn stover	20	19	5.1	0.5	0.1	1,380
Switchgrass	8	18	5.7	0.9	0.1	1,980
Corn cob	7	18	1.5	0.4	0.1	2,907
Hard wood pellets	7	19	0.4	0.2	0	472
Low sulphur sub-bit coal	NR	25	6	0.9	0.4	35

Ash content is a very interesting feature, since ashes contain diverse alkaline earth and heavy metals that can be used as macro- and micro-nutrients in tree and shrub cultivations. For example, Carpinteyro-Urban (2011) reported the presence of Na, K, Ca, and Mg, as well as some heavy metals, i.e., Al, Cu, Fe, Ni, Pb, and Zn, in the cotyledons of the locust bean, guar, *and Prosopis laevigata,* as well as in the *Opuntia ficus* mucilage. In the case of the skin, teguments and pods, these residual matter may also contain those elements, at different levels. If these metals are present in the vegetal residues, even at lower concentrations, they will be an excellent source of micronutrients, like Ca, K, Mg, and Zn. Other heavy metals are also useful for the cultivation of trees and shrubs.

If energy production is based on the anaerobic digestion of vegetal residues for methane production, the biogas yield for different waste materials has been calculated by some researchers. Table 4 shows some of these values reported by Rossel and Ortiz (2012). The agrochemical residues shown are sugar cane and agave bagasses, maize silage, and specifically prickly cactus (*Opuntia ficus*). As observed, the biogas yield for these materials ranges from 155 (sugar cane) to 557 L/kg (maize silage). Since not all the gas corresponds to methane, the second row shows the methane yields. In this case, the lowest value corresponds again to sugar cane (84 L/kg) and the highest corresponds to maize silage (306 L/kg). In absence of specific values, methane yield from teguments and pods will be assumed to correspond to the lower values of the table (84 L/kg). For *Opuntia ficus* biomass, the specific value of 203 L/kg will be employed.

Table 4. Metal contents in selected vegetal galactomannans (all in mg/kg). From Carpinteyro-Urban (2011).

Product	Al	Ca	Cu	Fe	K	Mg	Na	Ni	Pb	Zn
Locust bean gum	<11.2	611	3.2	126	2071	377	10077	4.5	29.8	25
Guar gum	<11.2	1500	2.7	141	1806	338	5633	4.6	25.8	27
Prosopis laevigata gum	14.8	3977	11.4	157	613	487	3739	4.0	25.4	45
Opuntia ficus mucilage	1114.0	56760	37.2	330	292618	48416	9675	37	39	156

Table 5. Calculation of biogas and methane production by using Buswell simplified equation from some organic materials. Adapted from Rossel and Ortiz (2012).

Material	Biogas yield (L/kg dwb)	Methane yield (L/Kg dwb)	Methane ratio (%)
Sugar cane bagasse	155.83	84.08	53.95
Agave bagasse	352.10	190.84	52.40
Cactus prickly	356.46	203.62	57.12
Maize silage	557.15	306.86	55.08

Once the oil is obtained from the seeds, it is important to characterize it in terms of the features that are related with biodiesel characteristics. Table 5 shows some general characteristics of *Annona muricata* and *cherimola* seed oils in comparison with other well known oils, i.e., *Cleome viscosa* and soybean. Unfortunately, there are no reports regarding the quality of *Prosopis laevigata* or *Delonix regia* seed oils. It is remarkable that *Annona muricata* seed oil has a value of specific gravity similar to that found in *C. viscosa* and soybean oils. Regarding the refractive index, values for all the seed oils reported herein are very similar (about 1.47). The acid value for *Annona muricata* and *chirimola* are lower than that reported for *C. viscosa*, but higher than that reported for soybean. The saponification values for the annonas' seed oils are much lower than those reported for *C. viscosa* and soybean oils. The unsaponifiable matter values for *A. muricata* and *chirimola* are about 60.4 and 1.3 mg KOH/g, respectively, but these values have not been reported for the other two vegetal oils. Regarding the iodine index, *A. muricata and chrimola* showed values of 1.1 and 24 mg/kg, respectively.

Table 6 shows the products obtained in the biorefinery scheme, for the five cultivated species. The basis for calculation was 1 ton of fruit or pods or shrub (in the case of Cactaceae). This information will be very useful for selecting the best options from the economical point of view. Before any considerations regarding the cost of the produced materials, let us review some of the yields. Starting with 1000 kg of *Annona muricata* (guanabana fruit) the yield will be 756 kg of fruit pulp, almost 20 kg of oil, and nearly 27 kg of cuagulant-flocculant aid. Besides, about 63.5 kg of compost will be produced to be used for the cultivation of the same trees.

Table 6. Characteristics of some seed oils.

Parameter	*Annona* muricata*	*Annoma** chirimola*	*Cleome viscosa****	Soybean*
Color	Pure yelow	NR	NR	Pale yellow
Odor	Repugnant	NR	NR	Odorless
Specific gravity	0.922	0.740	0.920	0.922
Refractive index	1.464	1.469	1.470	1.465
Acid value (KOH/g)	23.56	11.04	49.90	0.60
Saponification value (KOH/g)	100.84	52.11	212.30	246.25
Unsaponifiable matter (%)	60.46	1.28	NR	NR
Peroxide value (mg/kg)	1.1	24.04	NR	NR

From *Onimawo (2002), **Amoo et al., (2008), and ***Kumari et al., (2012).

On the other hand, 1000 kg of the chirimoya fruit (*Annona chirimola*) will give 586 kg of pulp, 27.8 kg of oil, and around 30 kg of coagulant-flocculant. Furthermore, if some residues are burned, 15.3 kg of ashes (to be employed as nutrients for trees cultivation) and a certain amount of heat will be obtained too.

If 100 kg of mezquite pods are considered, 772 kg of coagulant-flocculant aid will be produced as well as 5 kg of oil; 7.8×10^4 L of biogas will be obtained if residues are transformed by anaerobic digestion. When starting with flamboyan pods, 171 kg of coagulant-flocculant will be obtained plus 1 kg of oil. If the residues are anaerobically digested, as much as 2.88×10^5 L of biogas can be produced. Finally, for the case of 1000 kg of cactacea shrub, 104 kg of dry nutritional supplement will be obtained plus 8.8 kg of coagulant-flocculant aid. This process will generate 990 L of water that can be recycled to the process, and used for *Opuntia* cultivation (See Table 7).

5. Conclusions

This biorefinery based on four different seeds and one cactus shrub, including the residual fruit skins, pods, and integuments, is a suitable proposal to produce: 1) four different coagulant-floccualnt aids, 2) oil for biodiesel production, 3) fruit pulp for different sweet foods, 4) nutritional food, 5) biogas, 6) compost, and 7) ashes.

Table 7.Resume of the mass balance for the biorefinery. Amount of products generated by Tonn of fruit, pods or shrubs (in kilograms, except for biogas, in L)

	Pulp 1	Pulp 2	Oil 1	Oil 2	Oil 3	Oil 4	CF aid 1	CF aid 2	CF aid 3	CF aid 4	CF aid 5	Biogas	Water	Ash	Comp	NF
Annona Muricata	756.0		19.7				26.8								63.5	
Annona Chirimola		586.0		27.8				30.3						15.3		
Prosopis Laevigata					5.16				772.8			$7.8x10^4$				
Delonix regia						1.0				171.7		$2.88x10^5$				
Opuntia Ficus-indca											8.8		991.0			104.2

CF: coagulant-flocculant, Comp: compost, NF: nutritional food

During the process, some water will be produced and recycled for the trees and shrubs cultivation. Regarding energy flows, heat and energy will be produced. Part of this energy will be recycled to the system. Although some authors have come close to this approach by using the feed+fuel biorefinery concept or by applying IS to industries that may be considered in the biorefineries field, these two concepts have not been linked in the literature. This chapter offers, as a novelity, the IE vision applied to a biorefinery.

Acknowledgements

This work was partially supported by ICyT-DF Grant PICSO 10-8. The participation of Luis J. Corzo (UPIBI-IPN) and the donation of samples of purified *Delonix regia* seed gum are thanked. The kind donation of *Prosopis* seeds and powder by Jorge Yañez (UPIBI-IPN) is also acknowledged. The participation of some students is also thanked (Sandra Carpinteyro, Alma Carrillo, Giselle Cadena, Marco Antonio Martinez, all from UPIBI-IPN). Finally, the participation of Carlos Orozco (UPIBI-IPN) in the spray drying of *O. ficus* mucilage is acknowledged.

References

Ahmad M., Khan M.A., Zafar M., Sultana S. (2012) Biodiesel from non edible oil seeds: A renewable source of bioenergy. Online resources: *www.intechopen*.com. Consulted on November 7th, 2012.

Boons F., Spekkink W., Mouzakitis Y. (2011) The dynamics of industrial symbiosis: a proposal for a conceptual framework based upon a comprehensive literature review, *Journal of Cleaner Production* 19, 905-911

Carpinteyro-Urban S. (2011) Tratamiento de aguas residuales empleando polímeros naturales y biodegradabilidad de los lodos generados (2011) MSc Thesis. *Graduate studies in Bioprocesses*. UPIBI-IPN. Mexico.

Cervantes, G. (2007) *Ecologia Industrial*. Barcelona: Fundació Pi i Sunyer

Clarke S., Preto F. (2011) Biomass burn characteristics. Ontario Ministry of Agriculture, *Food and Rural Affairs*. Fact sheet. Order No. 11-033. AGDEX 737/120. Canada.

Corzo-Rios L. (2012) *Personal communication*.

FAO (2006) Fichas técnicas. Productos frescos y procesados. Consultado en: *http:/www.fao.org/inpoh_archive/content/documents/Vlibrary/AE620s/Pfrescos/GUANABANA.html*. Consultado en nov de 2012.

Haque M.A., Islam M.P., Hussain M.D., Khan F. (2009) Physical, mechanical properties and oil content of selected indigenous seeds available for biodiesel production in Bangladesh. Agricultural Engineering International. *The CIGR E-Journal*. Manuscript 1419. Vol XI.

Kumari R., Jarim V.K., Kumar S. (2012) Biodiesel production from seed oil of Cleome viscosa L. *Indian Journal of Experimental Biology*.50: 502-510.

Mathews, J., Tan H., Michael J. Moore, M., Bell, G. (2011) A conceptual lignocellulosic "feed+fuel" biorefinery and its application to the linked biofuel and cattle raising industries in Brazil. *Energy Policy* 39 (9), 4932–4938

Reveles F.O., Rosales R., Nava C.A., Delgado E., Cuellar E.I., Carrete F.O., Rios J.C (2010) Identification of plant species with potential use in liquid biofuels production in Durango, Mexico. (In Spanish.) *Revista Mexicana de Ciencias Agricolas*. 1(Jan-Mar): 45-54.

Ríos J.C., Merlín E., Soto E., Rosales R. (2010) Seedlings growth and mesquite productivity in the State of Durango, Mexico. (In Spanish.) Memorias del VII Simposio Internacional sobre la Flora Silvestre en Zonas Áridas. Hermosillo, Sonora, Mexico. March 17-19, 2010. Internet resources: *www.dictus.uson.mx/florazonasaridas*. Consulted Nov 2012.

Rossel D., Ortiz H. (2012) *Calculating biogas production by conversion of organic lignocellulosic biomass*. Memorias del 1er. Congreso Iberoamercano Sobre Biorefinerías. Los Cabos, Baja California, Mexico. October 24-26, 2012.

Sokka, L., Pakarinen, S., Melanen, M. Industrial symbiosis contributing to more sustainable energy use –an example from the forest industry in Kymenlaakso, Finland (2011) *Journal of Cleaner Production* 19. 285–293.

Torres L.G., Cadena G., Carpinteyro-Urbán S., Corzo L.J. (2012) New galactomannans and mucilages with coagulant-flocculant activity, for an environment-friendly coagulation-flocculation of wastewaters. *Current Advances in Environmental Science*. Submitted.

About the Authors

Chapter 1
Nathalie Cabirol
Facultad de Ciencias
Departamento de Ecología y Recursos Naturales
Universidad Nacional Autónoma de México (UNAM)
Ciudad Universitaria, C.P. 04510, México, D.F., México

Marcelo Rojas-Oropeza
Facultad de Ciencias
Departamento de Ecología y Recursos Naturales
Universidad Nacional Autónoma de México (UNAM)
Ciudad Universitaria, C.P. 04510, México, D.F., México

Bernd Weber
Facultad de Ingeniería
Universidad Autónoma del Estado de México (UAEM)
C.P. 05130, Toluca, México

Chapter 2
Marcia Morales
Departamento de Procesos y Tecnología
Universidad Autónoma Metropolitana
UAM-Cuajimalpa, Artificios No. 40, Col. Miguel Hidalgo, C.P. 01120, México DF,
Telephone: + (52) 5526363801
Fax: + (52) 5526363834
Email: mmorales@correo.cua.uam.mx

Alma Toledo-Cervantes
Doctorado en Ciencias Biológicas y de la Salud
Universidad Autónoma Metropolitana UAM- Iztapalapa, Mexico

Chapter 3

Horacio Bach
Department of Medicine, Division of Infectious Diseases
University of British Columbia, Vancouver, Canada

Luis R. Hernández
Departamento de Ciencias Químico Biológicas,
Universidad de las Américas Puebla,
Santa Catarina Mártir s/n, 72810, Cholula, Puebla, México

José D. Lozada-Ramírez
Departamento de Ciencias Químico Biológicas
Universidad de las Américas Puebla
Santa Catarina Mártir s/n, 72810, Cholula, Puebla, México

Eugenio Sánchez-Arreola
Departamento de Ciencias Químico Biológicas
Universidad de las Américas Puebla
Santa Catarina Mártir s/n, 72810, Cholula, Puebla, México

Chapter 4

Gemma Cervantes
UPUBI- Instituto Politécnico Nacional
Calle acueducto s/n. Col Barrio La Laguna-Ticoman
Del. Gustavo A. Madero, Mexico 07340 DF Mexico
E mail: gemma.cervantes@gmail.com

Mariana Ortega
UPUBI- Instituto Politécnico Nacional
Calle acueducto s/n. Col Barrio La Laguna-Ticoman
Del. Gustavo A. Madero, Mexico 07340 DF Mexico

Chapter 5

Erika Bustos Bustos
Centro de Investigación y Desarrollo Tecnológico en Electroquímica SC
Parque Tecnológico Querétaro SN, 76703 Pedro Escobedo
Querétaro de Arteaga, Mexico

Luis A Godinez
Centro de Investigación y Desarrollo Tecnológico en Electroquímica SC
Parque Tecnológico Querétaro SN, 76703 Pedro Escobedo
Querétaro de Arteaga, Mexico

Juan Manriquez Rocha
Centro de Investigación y Desarrollo Tecnológico en Electroquímica SC
Parque Tecnológico Querétaro SN, 76703 Pedro Escobedo
Querétaro de Arteaga, Mexico

Victor A. Ramìrez Coutiño
Centro de Investigación y Desarrollo Tecnológico en Electroquímica SC
Parque Tecnológico Querétaro SN, 76703 Pedro Escobedo
Querétaro de Arteaga, Mexico

Adrián Rodríguez García
Centro de Investigación y Desarrollo Tecnológico en Electroquímica SC
Parque Tecnológico Querétaro SN, 76703 Pedro Escobedo
Querétaro de Arteaga, Mexico

Francisco J. Rodríguez Valadez
Centro de Investigación y Desarrollo Tecnológico en Electroquímica SC
Parque Tecnológico Querétaro SN, 76703 Pedro Escobedo
Querétaro de Arteaga, Mexico
Email: frodriguez@cideteq.mx

Chapter 6

Guillermo E. Baquerizo Araya
Bioengineering Department
Unidad Profesional Interdisciplinaria de Biotecnología
Instituto Politécnico Nacional. Av.
Acueducto s/n Col. Barrio la Laguna Ticomán, Mexico City 07340, Mexico
E-mail: gbaquerizo@ipn.mx

Chapter 7

V. González-Álvarez
University of Guadalajara
CUCEI, Department of Chemical Engineering
Calz. Gral. Marcelino Garcia Barragan 1421
Guadalajara, Jal. 44430 Mexico

H.O. Méndez-Acosta
University of Guadalajara
CUCEI, Department of Chemical Engineering
Calz. Gral. Marcelino Garcia Barragan 1421
Guadalajara, Jal. 44430 Mexico

Chapter 8
Pedro Bañuelos Sánchez
Universidad de las Américas Puebla
Puebla, Mexico. Sta. Catarina Mártir
Cholula 82710 Mexico. Tel: +52 222 2292652
E-mail: pedrobañuelos@udlap.mx

Chapter 9
Benito Corona Vásquez
Department of Civil Engineer and Environmental
Fundación Universidad de las Américas Puebla
Cholula, Puebla, México, C.P. 72810
E-mail: benito.corona@udlap.mx

Juan Gabriel Garcia Maldonado
Subcoordinator of Environmental Hydraulics
Mexican Institute for Water Technology
Jiutepec, Morelos, México, C.P. 62550
E-mail: Gabriel_garcia@tlaloc.imta.mx

Marie Hoepfl
Department of Technology and Environmental Design
Appalachian State University
Boone, North Carolina, U.S.A., 28608
E-mail: hoepflmc@appstate.edu

Benjamín Sánchez Andrade
Department of Civil Engineer and Sustainable Development
Institito Tecnológico y de Estudios Superiores de Monterrey campus Puebla
Puebla, Puebla, México, C.P. 72453
E-mail: bsanche@itesm.mx

Chapter 10
Luis G Torres
Depto. de Bioprocesos
UPIBI-Instituto Politécnico Nacional
Calle acueducto s/n. Col Barrio La Laguna-Ticoman
Del. Gustavo A. Madero. Mexico 07340 DF Mexico
E-mail: ltorresbustillos@gmail.com

Gemma Cervantes
Depto. de Bioprocesos
UPIBI-Instituto Politécnico Nacional
Calle acueducto s/n. Col Barrio La Laguna-Ticoman
Del. Gustavo A. Madero. Mexico 07340 DF Mexico

Editors

Erick R. Bandala
Grupo de investigación en Energía y Ambiente
Universidad de las Américas Puebla. Sta. Catarina Mártir
Cholula 82710 Mexico
Tel: +52 222 2292652
E-mail: erick.bandala@udlap.mx

Luis G Torres
Depto. de Bioprocesos
UPIBI-Instituto Politécnico Nacional
Calle acueducto s/n. Col Barrio La Laguna-Ticoman
Del. Gustavo A. Madero. Mexico 07340 DF Mexico
E-mail: ltorresbustillos@gmail.com

INDEX

C

F

I

J

K

L

M

N

O

P

Q

R

S

T